PLENUM PRESS HANDBOOKS OF HIGH-TEMPERATURE MATERIALS

No. 3 – THERMAL RADIATIVE PROPERTIES

PLENUM PRESS HANDBOOKS OF HIGH-TEMPERATURE MATERIALS

No. 1 — MATERIALS INDEX
Peter T. B. Shaffer

No. 2 — PROPERTIES INDEX
G. V. Samsonov

No. 3 — THERMAL RADIATIVE
PROPERTIES
W. D. Wood, H. W. Deem,
and C. F. Lucks

PLENUM PRESS HANDBOOKS OF
HIGH-TEMPERATURE MATERIALS

No. 3
THERMAL RADIATIVE
PROPERTIES

by
W. D. Wood, H. W. Deem,
and C. F. Lucks

Springer Science+Business Media, LLC
1964

ISBN 978-1-4899-5456-5   ISBN 978-1-4899-5454-1 (eBook)
DOI 10.1007/978-1-4899-5454-1

Softcover reprint of the hardcover 1st edition 1964

Library of Congress Catalog Card Number: 64-17206

CONTENTS

*Detailed tables of contents for these sections will be found on pages 33, 51, 95, 159, 195, and 417, respectively.

SUMMARY

This report is a compilation of data on thermal radiative properties. It also includes a brief discussion of the basic fundamentals of thermal radiation and of the methods of measuring these properties. Much of the information has previously been distributed in DMIC memoranda; however, it is consolidated in this report for the benefit of those with a broad interest in radiant heat transfer.

Thermal radiative data are included for the following materials: titanium and its alloys; stainless steels; iron-, nickel-, and cobalt-base superalloys; the refractory metals (chromium, columbium, molybdenum, tantalum, and tungsten) and their alloys; coated materials for elevated-temperature service; and ceramics and graphite.

INTRODUCTION

There has been, for many years, a broad interest in data on the thermal properties of materials. Many of these data have been reported in an excellent manner by
Armour Research Foundation in WADC TR 58-476, Volumes I through IV. The Armour
work covered data available through 1957.

The rapid development of space and missile technology has created an increased
and specific need for data on radiant heat transfer and thermal-radiation properties.
The Defense Metals Information Center, therefore, has assembled the available information, with special emphasis on data reported since 1957.

For the most part, only those materials within the assigned scope of DMIC are
covered. However, some data on allied materials, such as ceramics and graphite, were
identified during the normal search of the literature and have been included to complete
the record.

Method of Presentation

The authors have attempted to evaluate the sources of the data according to the
methods and techniques described by the various investigators. In many cases, full
details were not given, and a complete evaluation was impossible. With these considerations in mind, the authors have shown curves which, in their estimation, indicate the
most probable values for the various conditions and materials.

The data have been separated according to material and to the type of measurement, whether spectral or total. In those sections of this report dealing with metals,
emittance data are given as the complement of the reflectance, which assumes a sample
opaque to the radiation concerned in each case. Many ceramic materials and coatings,
however, transmit considerable amounts of incident radiation and must be relatively
thick to be effectively opaque. The data for ceramics are, therefore, given only in the
units measured by the investigators, since for most cases the emittance must be considered as the complement of the sum of the reflectance and the transmittance.

The emittance values for coatings reported by one investigator may vary considerably from those for the same coating material reported by another. The main reasons
for such variances, assuming reasonable care in obtaining data, are differences in the
surface condition and in the thickness of the coatings. Differences in surface condition
result from differences in the methods of coating the substrate material and from the
numerous methods of bonding and curing the coatings.

An innate property of most ceramic materials is their varying degree of transpar-
ency to radiation at different temperatures and different wavelengths. Since most refractory coatings must be applied in relatively thin sections (a few mils thick) because of
strength and thermal-shock considerations, the emittance of the underlayer or substrate
material may, and usually does, have considerable effect on the measured emittance of
the coated body.

The data for refractory coatings are given separately for each investigator since,
for the reasons given above, a direct comparison of values reported for the same coating
material by different investigators is apt to be misleading.

All data have been plotted with each reference shown by a different symbol. The "reference information" accompanying each graph gives the names of the investigators and the number of the reference from which the data were obtained. These references are identified at the end of each section. Notations of composition and surface conditions of the sample tested, and a brief notation as to methods and conditions of measurement, are given when available.

FUNDAMENTALS AND DEFINITIONS

Thermal Radiation

The process of emission of radiant energy by a body, which depends on its temperature, is called thermal radiation*. Each body, by virtue of its temperature, is constantly emitting electromagnetic radiation from its surface into the space about it and is absorbing radiations originating elsewhere and incident upon it. Electromagnetic radiation is composed of all wavelengths, including extremely short-wave secondary cosmic rays and the longest radio waves. Theoretically, all bodies emit radiation over the entire electromagnetic spectrum. The amount of energy emitted generally varies with wavelength in a manner similar to that shown in Figure 1. Three curves are shown giving the spectral distribution of blackbody radiation (defined later) at different temperatures.

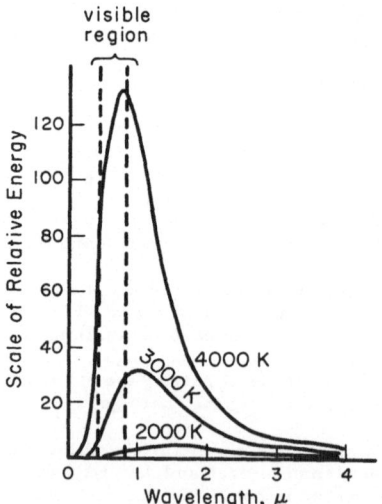

FIGURE 1.   SPECTRAL DISTRIBUTION OF RADIATION FROM A BLACKBODY AT 2000, 3000, AND 4000 K

*Referred to hereafter simply as radiation.

It can be seen from Figure 1 that the maximum energy emitted by a body increases as the temperature increases and that the wavelength at which the maximum energy is radiated becomes shorter as the temperature is increased. For comparison, the region normally associated with the visible wavelengths is shown within the vertical dotted lines.

Classifications

Radiation is usually classified as to the distribution of energy under consideration. The two main classifications ordinarily used are total radiation and spectral radiation.

Total Radiation. Total radiation is radiation over the entire spectrum of emitted wavelengths.

Spectral Radiation. Spectral radiation refers to radiation emitted within a specified wavelength interval. This interval is ordinarily so small as to be considered as radiation at a specific wavelength. Radiation of this type is called monochromatic.

Solar Radiation. Another classification of radiation which should be mentioned is that of solar radiation. Obviously, solar radiation refers to radiation from the sun. Radiation reaching the earth's surface from the sun is reduced considerably by the earth's atmosphere. The atmosphere is nearly transparent to some wavelengths of solar radiation and is nearly opaque to other wavelengths. The insolation (irradiation of a body by the sun) of a rocket or satellite is, therefore, quite different at sea level and outside the atmosphere.

The sun does not radiate as a blackbody. However, the average observed solar-energy curve from about 0.45 μ to about 20 μ closely approximates the spectral-energy distribution of a blackbody at 6000 K. It is therefore not unusual to speak of the "blackbody temperature" of the sun as 6000 K.

Figure 2 shows the general spectral-distribution curves for a blackbody at 6000 K and for the sun outside the earth's atmosphere.

Many factors affect the amount of solar energy received by the earth's surface and this amount varies considerably from day to day. On a cloudless day, in the middle lattitudes, about 80 per cent of the incident solar energy reaches the ground. With average cloudiness, however, only about 50 per cent is received by the earth's surface.

Figure 3 shows a relative spectral-distribution curve for the solar energy received by the earth at sea level.

Directional Radiation. Radiation emitted by a body is also classified by its direction of propagation. The amount of energy radiated varies with the angle of emission from the surface of the body. The most common classifications are normal, hemispherical, and directional radiation. Normal radiation refers to radiation emitted in a direction normal to the surface of the body. Hemispherical radiation refers to radiation emitted in all possible directions from a flat surface. Directional radiation, in general, refers to radiant emission at a specified angle with the surface. Normal radiation is obviously a special case of directional radiation.

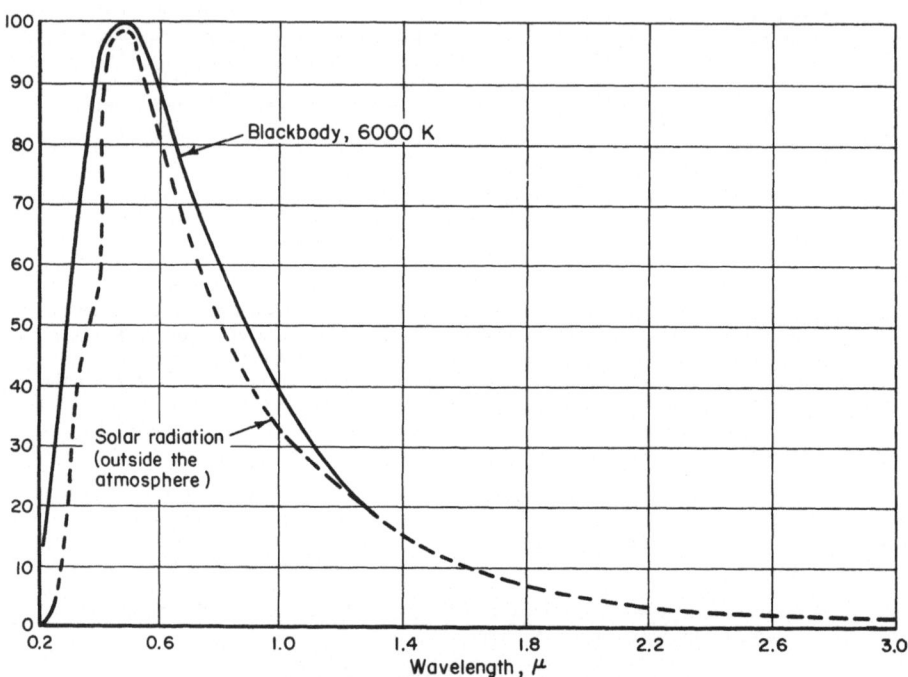

FIGURE 2. SPECTRAL DISTRIBUTION OF A BLACKBODY AT
6000 K AND THE SUN OUTSIDE THE EARTH'S
ATMOSPHERE

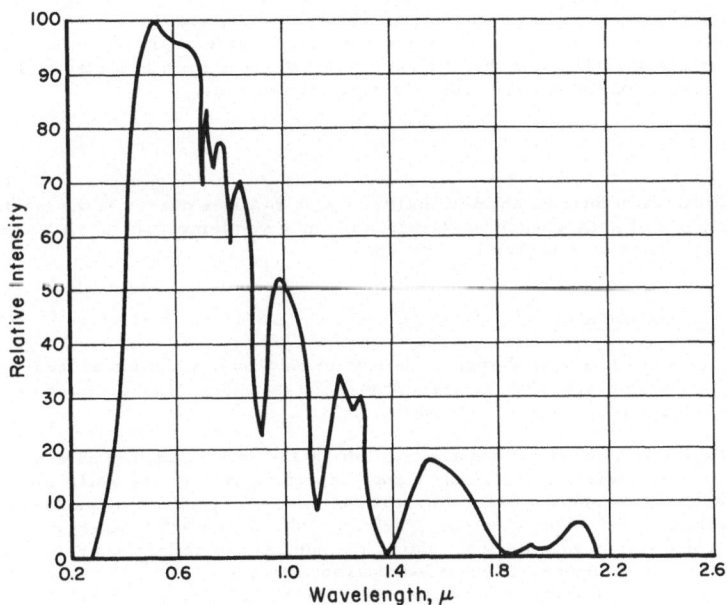

FIGURE 3. SOLAR–SPECTRAL–ENERGY DISTRIBUTION
AT SEA LEVEL

The terms normal and directional are also used to define the direction of propagation, relative to a surface, of radiation incident upon the surface. The term diffuse is used to describe radiation incident upon a surface from all possible directions.

Blackbody Radiation

The emissive power of a blackbody is proportional to the fourth power of its absolute temperature. For other bodies, the rate of radiation varies, depending on the material, surface conditions, and temperature. The rate of emission of energy per unit area for nonblackbody materials is never greater than the rate of energy emission per unit area from a blackbody. For this reason the blackbody is used as a standard or reference and emission from other bodies is compared with it.

Definition

A blackbody is defined as an ideal emitter which radiates energy at the maximum possible rate per unit area at each wavelength for any given temperature. A blackbody also absorbs all the radiant energy incident upon it.

Blackbody Approximations

There are some fairly good approximations of blackbody radiation sources but there are no perfect blackbody materials. Among the materials which come closest are carbon black, zinc black, and platinum black.

Good blackbody approximations are possible by the construction of an enclosure or cavity of uniform temperature containing a small aperture through the wall. Multiple reflections inside the enclosure assure a source of energy very close to blackbody conditions regardless of the material used. Worthing and Halliday[1]* have shown that, to obtain an aperture that is 99.9 per cent black, the numbers of reflections needed for cavities constructed of copper, tungsten, and carbon are 45, 12, and 3, respectively, while if a 99.5 per cent blackbody is acceptable, only 33, 10, and 2 reflections are needed for the same respective materials.

Definition of Terms

The Need for Precise Definitions

The word emit means to send out; therefore, as we might expect, an emissivity or emittance for a given material is a measure of the ability of the material to send out radiant energy. Some authors use emissivity in the sense of the rate of emission per unit area of surface. General usage now terms this ability as radiance and defines emissivity or emittance as the ratio of emission from a nonblackbody to the rate of emission

---

*References for this section are listed on page 17.

from a blackbody at the same temperature. Obviously, since the rate of emission from a nonblackbody can never exceed the rate of emission from a blackbody, this ratio can never exceed 1.

For normal materials (nonblackbodies) the rate of absorption will be reduced by reflection at the surface for opaque bodies and by transmission through the material for transparent or translucent bodies. Surface conditions also alter the rate of emission or absorption. Roughness and oxidation of a surface generally increase the rate of emission. For example, the emittance of a copper specimen can be increased from about 0.02 for a polished surface to a value of approximately 0.8 for copper with a thick oxide layer.[2]

Since surface conditions and degree of opaqueness affect the rate of emission, specified conditions must be considered in any definitions. Our definitions will conform in terminology and symbolic representation to those established by Worthing[3] and most widely used in modern literature.[4-7]

Definitions and Symbols

Radiance (R). The rate of radiant-energy emission from a unit area of a source in all the radial directions of the overspreading hemisphere.

Steradiance (B). The rate of radiant emission per unit solid angle and per unit of projected area of a source, in a stated angular direction from the surface (usually normal).

Emittance (ϵ). The ratio of the rate of radiant emission from a body, as a consequence of its temperature only, to the corresponding rate of emission from a blackbody at the same temperature.

Emissivity (ϵ'). The term emissivity is reserved for the case of an opaque material having an optically smooth surface composed of the material. It follows that, for a body composed of an opaque material, the emissivity of the material is the lower limiting value approached by the emittance of the body as its surface is made more and more optically smooth.

Hemispherical Emittance (ϵ_h). The ratio of the radiance from a body as a consequence of its temperature to that of a blackbody at the same temperature.

Directional Emittance (ϵ_θ). The ratio of the steradiance from a body as a consequence of its temperature to that from a blackbody at the same temperature. Ordinarily directional emittance is in the direction normal to the surface and is called normal emittance and designated by (ϵ_n).

Total Emittance (ϵ_t). The ratio of the total radiance from a body as a consequence of its temperature to that of a blackbody at the same temperature.

Figure 4 shows the spectral-radiance-distribution curves for tungsten and for a blackbody, both at a temperature of 2450 K. The tungsten curve is seen to lie below the blackbody curve at all wavelengths. If the range of wavelengths were extended to infinity, the ratio of the area beneath the tungsten curve to the area beneath the blackbody curve would be the total emittance of the tungsten at 2450 K.

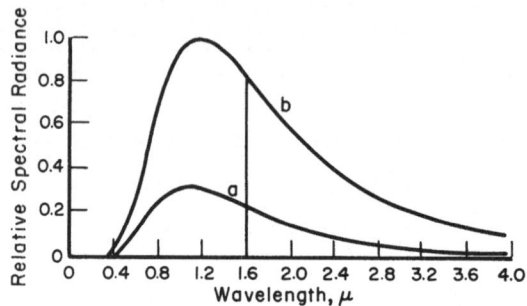

FIGURE 4. SPECTRAL-RADIANCE-DISTRIBUTION CURVES FOR
(a) TUNGSTEN AT 2450 K AND (b) A BLACKBODY AT
2450 K

Spectral Emittance (ϵ_λ). The ratio of the spectral radiance (or monochromatic radiance at a given wavelength) from a body as a consequence of its temperature to that of a blackbody at the same temperature and the same wavelength.

In Figure 4, an ordinate has been drawn vertically through the two curves at a wavelength of 1.6 microns. The ratio of the ordinate (or height) of the tungsten curve (a) to that of the blackbody curve (b) at 1.6 microns is the spectral emittance of the particular sample of tungsten at 2450 K and 1.6 microns.

All of the above emittances have corresponding emissivities for the optically smooth, opaque materials.

Absorptance (α). The ratio of the radiation absorbed by a body to that incident upon it.

For a body at a given temperature, the spectral absorptance is equal to the corresponding spectral emittance ($\alpha_\lambda = \epsilon_\lambda$). Since spectral absorptances of actual bodies usually vary with wavelength, total absorptances usually depend on the temperature of the source.

If the incident radiation has the spectral distribution of that from a blackbody source at the temperature of the specimen, the total absorptance equals the total emittance ($\alpha_t = \epsilon_t$). This does not hold true, however, if the source is not a blackbody (except for graybodies, whose spectral emittance is the same at all wavelengths) and if the source is not at the temperature of the specimen.

Reflectance (ρ). The ratio of the radiant flux reflected by a body to that incident upon it. For an opaque body, the sum of the reflectance and the absorptance for the incident radiation is unity ($\rho + \alpha = 1$).

Transmittance (τ). The ratio of the radiant flux transmitted through a body to that incident upon it. For a transparent or translucent material, the sum of the transmittance, reflectance, and absorptance for the incident radiation is unity ($\tau + \rho + \alpha = 1$).

As with emittances, absorptances, reflectances, and transmittances are of various types such as spectral, normal, and hemispherical. The term "diffuse" is ordinarily used instead of hemispherical, since it refers to energy incident upon a surface from all possible directions. The term "total", when applied to absorptance, reflectance, or transmittance, is not meaningful unless the spectral distribution of the source is specified, and usually the term "total" is not used.

The reflectance and transmittance that are complements to the normal emittance or absorptance are for conditions of either normal illumination and hemispherical viewing, or, vice versa, diffuse illumination and normal viewing. Those that are complements to hemispherical emittance or diffuse absorptance are for conditions of diffuse illumination and hemispherical viewing.

When an emittance or one of the other properties described above is measured, the measurement must be concerned with the total radiation or with radiation at a specified wavelength, and it must also differentiate between radiation in all directions and radiation measured in a certain direction relative to the surface. As an example, let us suppose that an emittance has been obtained by a method which takes into consideration radiation in all directions from a flat surface of a body, but that the energy receiver used in the measurement was receptive to only one wavelength. We would then combine the appropriate emittance terms given above to give us the hemispherical spectral emittance ($\epsilon_{h\lambda}$) of the body.

The types of emittances ordinarily found in the literature today include the hemispherical total emittance (ϵ_{ht}), normal total emittance (ϵ_{nt}), hemispherical spectral emittance ($\epsilon_{h\lambda}$), and the normal spectral emittance ($\epsilon_{n\lambda}$). Absorptance, reflectance, and transmittance all have total, spectral, hemispherical, and normal relationships similar to the above.

Radiation Laws

Total-Radiation Laws

Lambert's Cosine Law. This law states that the directional radiance of a plane source of radiation varies as the cosine of the angle of emission.

The apparent visible brightness, or steradiance, of a radiating surface depends on the energy emitted by a unit area of the radiating body. Essentially, this law states that a radiating blackbody has the same apparent brightness regardless of the angle from which it is viewed.

True surfaces vary from this law depending upon the material. When the directional radiance of a surface follows the cosine law, the directional emittance is independent of the angle of emission and is identical with the hemispherical emittance. Actually, all true surfaces exhibit a certain degree of dependence of the emittance on the angle of emission.

For most metals, as the angle between the direction of emission from an element of surface and the normal to the surface is increased, the emittance first increases and then decreases rapidly to zero at angles near 90 degrees with the normal. Although many substances vary with angle in a similar manner, others, such as carbon and the metal oxides, vary in just the reverse manner. For example, a sphere whose surface radiates as a blackbody, appears as a disk of uniform brightness; a sphere of tungsten appears with a brighter rim since its emittance increases as the angle of emission increases from the normal, and a carbon sphere will appear to have a brighter center due to its higher emittance at angles near the normal.

Most dielectric materials conform closely to the cosine law out to angles of about 60 to 65 degrees from the normal. Their directional emittances decrease rapidly with the angle at angles greater than 75 degrees from the normal.[9]

Prevost's Theory of Exchange. In the late 18th century, Pierre Prévost stated that all bodies at all times are emitting radiation into the space about them and are absorbing, partially at least, those radiations originating elsewhere and incident upon them.

Experience tells us that when a body at one temperature is suspended by a fine thread in an evacuated chamber whose walls are maintained at another temperature, the body will eventually attain the same temperature as the walls. Prévost's theory accounts for this reaching of equilibrium, assuming no conduction along the thread.

Boltzmann's Fourth-Power Law. This law, established by Ludwig Boltzmann in 1884, shows that the heat radiated by a blackbody is proportional to the fourth power of its absolute temperature. In equation form the law becomes

$$R = \sigma T^4 ,$$

where

R is blackbody radiance, watts/cm^2

σ is Boltzmann's constant, 5.673×10^{-12} watts cm^{-2} K^{-4}

T is the absolute temperature, K.

For nonblackbodies, since the emittance is the ratio of the nonblackbody to blackbody radiances, the nonblackbody radiance equals the blackbody radiance multiplied by the emittance or

$$R_n = \epsilon \sigma T^4 ,$$

where

R_n is nonblackbody radiance, watts/cm^2

σ is the Boltzmann's constant

T is the absolute temperature

ϵ is the hemispherical total emittance of the nonblackbody.

Kirchoff's Law. In 1858, the German physicist, G. R. Kirchoff, stated this law quantitatively, relating the radiating and absorbing properties of bodies. Essentially, the law states that, for a blackbody placed in a completely enclosed cavity whose opaque walls are maintained at some constant temperature, and when the body is in thermal equilibrium with the cavity walls, the rate of incidence of radiant energy per unit area of its surface from the surrounding walls is equal to its emission of radiant energy per unit area due to its temperature. In equation form

$$E_b = R_b \ ,$$

where E_b = the irradiance of the surface from its surroundings and R_b = the radiance of the body. The subscript "b" indicates blackbody conditions. From this relationship was derived the fact that the total emittance of a nonblackbody at any given temperature is equal to its total absorptance of radiation from a blackbody at the same temperature, or

$$\epsilon = \alpha .$$

Kirchoff's law applied equally well to spectral radiation and becomes $E_\lambda = R_\lambda$, where the irradiance E_λ and the radiance R_λ are taken at the same wavelength and also the spectral emittance equals the spectral absorptance, or

$$\epsilon_\lambda = \alpha_\lambda .$$

Spectral-Radiation Laws

Wien's Displacement Law. In 1896, Wilhelm Wien found that, when the temperature of a radiating blackbody increases, the wavelength corresponding to the maximum energy decreases (see Figure 1) in such a way that the product of the absolute temperature and wavelength is a constant.

$$\lambda \max T = \text{constant} \ (= 2897.9 \ \mu K) .$$

In obtaining the above, Wien also determined that if the wave of length λ_2 at T_2 is displaced from that of length λ_1 at T_1, such that $\lambda_2 T_2 = \lambda_1 T_1$, the monochromatic radiances at these two wavelengths are directly proportional to the fifth powers of the absolute temperature, or

$$\frac{R_{\lambda_1}}{R_{\lambda_2}} = \frac{T_1^{\ 5}}{T_2^{\ 5}}$$

Wien's Distribution Law. Wien also obtained an expression for the spectral distribution of radiant energy at a given temperature.

$$R_\lambda = \frac{C_1 \lambda^{-5}}{e^{C_2/\lambda T}} \quad .$$

This relation, however, holds true only for the short-wavelength region of the spectrum.

For nonblackbody relationship, the spectral radiance R_λ becomes

$$R_\lambda = \frac{\epsilon_\lambda C_1 \lambda^{-5}}{e^{C_2/\lambda T}} \quad ,$$

where ϵ_λ = spectral emittance and e = base of natural logarithms. C_1 and C_2 are radiation constants.

Rayleigh – Jean's Distribution Law. This law gives the spectral-radiation distribution applicable to the longer wavelengths, but not to shorter ones.

$$R_\lambda = \frac{C_1 \lambda^{-4} T}{C_2} \quad ,$$

where the units are the same as before and whose values are given for Planck's law below.

For nonblackbodies, the right side of the equation should again be multiplied by the spectral emittance at the given wavelength.

Planck's Distribution Law. Max Planck, in 1900, published the following expression for the radiance distribution at a given temperature.

$$R_\lambda = \frac{C_1 \lambda^{-5}}{e^{C_2 \lambda_T^{-1}}} \quad ,$$

where C_1 and C_2 are often referred to as the first and second radiation constants.

C_1 = 3.7403 x 10^{-12} watts cm^{-2} for 0.01μ zone of spectrum

C_2 = 1.4388 cm K (= 14,388 μK.

This law holds for all wavelengths.

It is interesting to note that, in seeking an explanation for an apparently correct expression for a law of spectral distribution, Planck arrived at his Nobel Prize-winning quantum theory, which revolutionized modern physics.

λ is the wavelength in microns

e is the base of natural logarithms.

As in the foregoing, for the nonblackbody case, the relationship becomes

$$R_\lambda = \frac{\epsilon_\lambda C_1 \, \lambda^{-5}}{e^{\,C_2/\lambda_T} - 1} \quad ,$$

where ϵ_λ is the spectral emittance.

General Relationships

All real materials are nonblack, and all real bodies are nonblackbodies, excluding, of course, so-called blackbody cavities or enclosures. As a general case, any radiation incident upon a body must be reflected, absorbed, or transmitted by the body and, therefore, the reflectance plus the absorptance plus the transmittance equals unity, or

$$\alpha + \rho + \tau = 1.$$

For a body in equilibrium with its surroundings, in an opaque walled cavity, Kirchoff's law shows that the absorptance is equal to the emittance or $\alpha = \epsilon$ and we define $\alpha = \epsilon = R/R_b$. However, if the temperature of the body and its surroundings are different, the absorptance varies with the spectral distribution of the incident radiation. In general, the total emittance varies with temperature.

If the nonblackbody is optically smooth and opaque, we have the reflectivity, absorptivity, and emissivity relationships similar to the above.

For some materials, the emittance is nearly constant for all wavelengths and temperatures and we can say that

$$\epsilon_\lambda = \alpha_\lambda = \epsilon = \alpha \ .$$

These materials are called graybodies. Although no materials are strictly gray, for some materials it is possible to use the above relationships as a close approximation in the interest of ease of calculation.

It should be mentioned that some materials are opaque to certain wavelengths of radiation and transparent to others, and that some materials become more or less transparent to a varying range of wavelengths as the temperature is changed. Theoretically, no specimen of a material is completely opaque except at "infinite thickness". A specimen may be considered to be opaque, however, when sufficiently thick that additional thickness produces no observable difference in opacity between that thickness and an infinite thickness.

For a body of infinite thickness all energy will be either reflected or absorbed. The thickness which can be taken as infinite for practical purposes depends on the absorption coefficient and is given by

$$X \cong \frac{3.5}{a} ,$$

where X is the infinite thickness and a is the absorption coefficient.

The absorption coefficient depends strongly on wavelength. For most glasses it varies from the order of 0.1 per cm for wavelengths below 2.5 μ to about 10 per cm for wavelengths above 2.5 μ. This means that a 0.3-cm-thick sheet is effectively infinite at room temperature where most of the radiation is at long wavelengths, but that a 35-cm thickness is necessary at elevated temperatures.[10] Infinite thickness for most metals is in the order of a few hundred to a few thousand atomic diameters.

In many cases it is necessary to obtain the emission properties of surfaces covered by special films or coatings of a different material. Many coatings are applied in a thickness at which they transmit significant amounts of radiation. In such a case, the emittance of a specimen will be influenced by the substrate material as well as the coating material. Usually the emittance of a metal specimen coated with a nonmetallic coating is much greater than that of the base metal, but often it is less than the emissivity of the coating material.[5]

For most of the above we have assumed solid bodies and materials. All of the above relationships hold equally well for liquids. Gases, however, are different in their behavior from solid and liquid bodies in that they radiate and absorb radiation only within certain limited wavelength ranges. Gases also, in contrast to solids and liquids, need great thicknesses to absorb the major part of incident radiation.

Figure 5 shows the absorption distribution for water vapor. As with many gases, the radiation and absorption take place mainly at wavelengths longer than 1 μ and are, therefore, invisible.

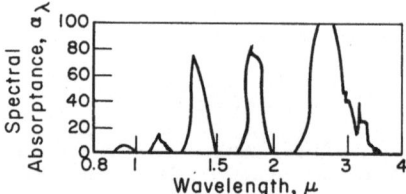

FIGURE 5. ABSORPTION COEFFICIENT VERSUS WAVELENGTH
FOR WATER VAPOR 109 CM THICK AT 127 C

From Eckert and Drake, Heat and Mass Transfer,
McGraw-Hill Book Company, New York (1959), p 383.

Figure 4 shows the change in the spectral-distribution curve for solar radiation due to absorption in the earth's atmosphere.

The elementary gases such as hydrogen, oxygen, and nitrogen radiate no measurable heat and are perfectly transparent to incident radiation for all practical purposes.

REFERENCES

(1)   Worthing and Halliday, Heat, John Wiley and Sons, Inc., New York (1948), p 435.

(2)   Brown and Marco, Introduction to Heat Transfer, McGraw-Hill Book Company, New York (1958), p 54.

(3)   Temperature – Its Measurement and Control in Science and Industry, American Institute of Physics, Reinhold Publishing Corporation, "Temperature Radiation Emissivities and Emittances" (A. G. Worthing) (1941), pp 1164-1187.

(4)   Kingery, W. D., Property Measurements at High Temperatures, John Wiley and Sons, Inc., New York (1959), p 92.

(5)   Richmond, J. C., "Coatings for Space Vehicles", First Symposium – Surface Effects on Spacecraft Materials, John Wiley and Sons, Inc., New York (1960, p 94.

(6)   A.S.T.M. Standards, Supplement for 1959, Revision to C-168, American Society for Testing Materials.

(7)   Harrison, Richmond, Plyler, Stair, and Skramstad, "Standardization of Thermal Emittance Measurements", WADC Technical Report 59-510, March 1960.

(8)   Eckert and Drake, Heat and Mass Transfer, 2nd Edition, McGraw-Hill Book Company, Inc., New York (1959), p 37.

(9)   Jakob, Max, Heat Transfer, Volume 1, John Wiley and Sons, Inc., New York (1959), p 42.

(10)   Kingery, W. D., Property Measurements at High Temperatures, John Wiley and Sons, Inc., New York (1959), p 93.

METHODS OF MEASUREMENT

Measurements Needed

Emittance was previously defined as the ratio of the rate of radiant emission from a body to that from a blackbody at the same temperature. From this definition we can see that, in order to measure the emittance from a body, we must measure three things – the rate of radiation from the body, its true (or blackbody) temperature, and the rate of radiation from a blackbody at the same temperature. The rate of radiation from a blackbody can be calculated by using Boltzmann's fourth-power law, $R = \sigma T^4$, where R is the radiance (watts/cm^2), σ is Boltzmann's constant (5.673×10^{-12} watts cm^{-2} K^{-4}), and T is the absolute temperature (K). The rate of radiation and the temperature of the body must be determined experimentally. For nonblackbodies, Boltzmann's fourth-power law shows the rate of radiation from the body to be $R = \epsilon \sigma T^4$, where ϵ is the emittance. Radiation detectors and temperature-measuring devices are described at the end of this section.

Methods for Making Emittance Measurements

Let us consider a few of the methods used to measure emittance. Detailed descriptions of these methods and their related apparatus and equipment can be found in the references given for each method.

Total-Emittance Measurements

Hemispherical Total Emittance. One of the most common methods of measuring the hemispherical total emittance of electrically conducting materials is the so-called hole-in-tube method described by Worthing and Halliday.[1]* Measurements can be made at temperatures almost as high as the melting points of the materials. The specimen material is formed into a long, thin-walled tube, and a small hole, is drilled through the wall of the tube near its center. Current electrodes are clamped at each end and it is suspended in a chamber whose walls are maintained at a known temperature that is below the temperature at which the measurements are to be made. Leads of fine wire are fastened to the tube equidistant from the hole so that the voltage drop in the central (evenly heated) portion of the tube can be measured. The tube is heated, by resistance heating, to the desired temperature. An optical pyrometer is sighted through a suitably placed window and through the small hole in the tube wall to measure the temperature of the inside of the tube. Since the inside of the tube is essentially in blackbody conditions, the temperature thus measured is the blackbody or true temperature. If the wall is thin enough, the inside and outside temperatures can be assumed to be the same. Sighting the optical pyrometer on the outside wall of the tube will give not true temperature but an apparent or brightness temperature, which is a function of the emittance of the outside wall. Emittance and the apparent or brightness temperature of the outside wall are discussed below in the section "Normal Spectral Emittance". For thick-walled tubes, the outside temperature will be lower than the inside because of radiation to the chamber wall. An extrapolation from the inside temperature to the outside surface temperature of the tube can be made if the thermal conductivity of the material is known. If a window is used, the pyrometer must be calibrated through the window material to correct for any absorption by the window. If the chamber wall is maintained at a temperature, T_0, such that the temperature, T, of the tube is much greater than T_0, the hemispherical total emittance can be calculated from the relationship:

$$\frac{IV}{2\pi r l} = \epsilon_{ht}\, \sigma T^4 \quad,$$

where

    I = current through the tube, amperes

    V = voltage drop between potential leads, volts

    r = the radius of the tube, cm

    l = the distance between potential leads, cm

*References for this section are listed on pages 29, 30, and 31.

ϵ_{ht} = the hemispherical total emittance

σ = Boltzmann's radiation constant

T = the blackbody temperature of the tube, K

T_o = the temperature of the chamber wall, K.

Another method of determining the hemispherical total emittance of materials having a high thermal conductivity is described by Drummeter and Goldstein[2] and by Shaw[3]. This method uses a small spherical specimen containing an internal heater, suspended in the center of a large, water-cooled, spherical chamber. The temperatures in this case are measured by suitably placed thermocouples. The same general principles hold for this method as for the hole-in-tube method.

J. H. Cairns[4] describes an apparatus for measuring the hemispherical total emittance of polished metals and alloys at temperatures from 100 to 900 C. The assembly consists, essentially, of a highly polished, hollow, cylindrical sample supported within an evacuated enclosure of only slightly larger dimensions. The temperature of the sample is raised above that of the enclosure by an internal heater within the sample. Observations are made of the rate of heat loss from the sample after the heater has been switched off. Heat losses other than by radiation are determined experimentally. Using the mass and specific heat of the sample and the rate of heat loss, Cairns has derived a special equation, based upon Boltzmann's fourth power law, from which the emittance can be obtained. Temperatures are measured with thermocouples. A method is also given for determining the specific heat, using the same initial conditions.

With an apparatus similar in design to that described above for the hole-in-tube method, Allen, Glasier, and Jordan[5] have measured the hemispherical total and normal spectral emittances and the thermal conductivity of molybdenum, tantalum, and tungsten above 2300 K.

The specimens, in this case, are polished 1/8-inch-diameter rods. The electrical power input to the central section of the rod is determined, and the surface temperature is measured with a disappearing-filament optical pyrometer. An equation has been derived whereby the heat-flow rate and brightness temperature can be used to obtain the spectral emittance (ϵ_λ). From this, the true temperature of the surface can be calculated from Wien's equation,

$$\ln \epsilon_\lambda = C_2/\lambda \left(\frac{1}{T} - \frac{1}{T'} \right) ,$$

where

C_2 = Planck's second radiation constant

λ = wavelength of transmission of the pyrometer

T = true temperature

T' = apparent (brightness) temperature.

With the true temperature known, the total emittance can be calculated from Boltzmann's equation and the fact that the power input must equal the power radiated from the surface at equilibrium conditions, or

$$EI = \epsilon_{ht} \, A \, \sigma T^4 \, ,$$

where

      E   = voltage drop in the measured section

      I   = current through the rod

      ϵ_{ht} = total hemispherical emittance

      σ = Boltzmann's constant

      T   = true temperature.

Richmond[6] used essentially the same method, but at lower temperatures. A rod specimen is placed in a water-cooled vacuum envelope. The specimen is heated by passing a current through it, and the power input required to maintain a uniform temperature in a known length of the specimen near its midlength is measured. Temperatures of the specimen and the water-cooled shell are measured with thermocouples. The familiar relationship given above,

$$EI = \epsilon_{ht} \, A \, \sigma \, T^4 \, ,$$

is used to calculate the hemispherical total emittance.

A method for measuring the hemispherical total emittance of small, thin-disk specimens of metals and metal alloys was developed by Butler and Inn[7]. The specimen is suspended by thermocouple wires inside a blackened, water-cooled vacuum chamber. A carbon arc is arranged outside a window in the vacuum chamber so that its radiation can be focused upon the specimen. When the specimen has been heated to the desired temperature by the converging rays from the arc, a shutter is placed between the arc and the specimen, and the rate of cooling of the specimen is determined by thermocouple readings. Knowing the mass and specific heat of the specimen, the emittance can be calculated from the relationship,

$$\epsilon_{ht} = \frac{mc \, dT/dt}{A \, \sigma \, (T^4 - T_w^4)} \, ,$$

where

      m   = mass of the specimen

      c   = specific heat of the specimen material

      dT/dt = rate of cooling of the specimen

      A   = surface area of the specimen

σ   = Boltzmann's constant

T   = specimen temperature, K

T_w = wall temperature, K.

This method assumes no heat loss from the specimen by conduction through the thermocouple leads and no molecular convection losses within the vacuum chamber. It also assumes negligible absorption or reflection by the specimen of radiation from the chamber walls. Suitable design and techniques must be used to minimize these possible errors.

Normal Total Emittance. Wade[8] has developed a method for measuring the normal total emittance of electrically heated strip specimens of metals and metal alloys. A total-radiation detector of the thermopile type is used to obtain the rate of radiation from a strip specimen and from a blackbody furnace at the same temperature. Suitably placed thermocouples are used to measure temperatures. The ratio of these two measurements is assumed to be the normal total emittance of the specimen. The rate of radiation is also measured at angles of from 0 to 60 degrees from normal to the specimen surface. When no change in the emittance can be found as the angle varies, the specimen surface is considered to be diffuse and to obey Lambert's cosine law. The hemispherical total emittance is, therefore, equal to the normal total emittance. For a specimen whose surface is not diffuse, that is, in which the emittance varies with the angle of incidence, a double integration method developed by O'Sullivan and Wade[9] is used to calculate the hemispherical total emittance from the measured data.

A method for measuring the normal total emittance of opaque, nonconducting materials has been developed by Olson and Katz.[10] Basically, this method consists of a tubular furnace in which the upright disk specimen moves back and forth past an opening in the wall through which the specimen surface can be measured. A total-radiation detector and its associated amplifying and indicating system measures the radiation alternately from the specimen surface and from a specially designed blackbody cavity which also moves past the opening. The ratio of the two readings is the normal total emittance of the specimen surface.

A method devised by McMahon[11] has been used to measure, simultaneously, emittance, reflectance, and transmittance of transparent or semitransparent materials and emittance and reflectance of opaque materials. It consists of a small furnace enclosure having a small water-cooled aperture through one wall. A semicircular disk of the material to be measured is mounted on a shaft so that it can be rotated past, and close to, the aperture. A blackened, water-cooled shutter is positioned so that it can be lowered into the furnace, covering the inside of the aperture but with room for the specimen disk to rotate between the aperture and the shutter. In operation, the radiation detector is adjusted to view the interior of the furnace through the water-cooled aperture. Assuming blackbody conditions in the furnace cavity, this first reading then is a measure of the blackbody radiation at that temperature. The specimen disk is then rotated past the aperture. A reading of blackbody transmission through the material plus that radiation emitted by the material itself is obtained. With the disk covering the aperture and the shutter inserted behind it, the radiation emitted by the material itself is measured.

The emittance, reflectance, and transmittance of the disk can be determined, provided blackbody conditions exist within the furnace cavity, and a small enough area of the disk is exposed to the cooled aperture to insure negligible cooling of the disk surface as it passes across the opening.

The measured properties can be either normal total or normal spectral, depending upon the type of radiation detector used.

Spectral-Emittance Measurements

Normal Spectral Emittance. The hole-in-tube method described above for hemispherical-total-emittance measurements can also be used to obtain the normal spectral emittance.[12] The optical pyrometer (or other spectral detector) is sighted upon the inner tube surface through the small hole to obtain the blackbody temperature, T, and on the external tube wall beside the hole to obtain the spectral or apparent temperature, S_λ. The following relationship is then used to obtain the normal spectral emittance:

$$\ln \epsilon_{n\lambda} = \frac{C_2}{\lambda} \left(\frac{1}{T} - \frac{1}{S_\lambda}\right) ,$$

where

$\epsilon_{n\lambda}$ = the normal spectral emittance at the measuring wavelength

C_2 = the second radiation constant

λ = the wavelength at which the detector measures, microns

T = the blackbody temperature, K

S_λ = the spectral or apparent temperature, K.

The methods of Olson and Katz[10] and McMahon[11], as mentioned above in connection with normal-total-emittance measurements, can also be used to measure normal spectral emittance merely by using a spectral detector instead of the total detector described.

A modification of Worthing's[12] hole-in-tube method was used by Krishman and Jain[13] in conjunction with high-temperature thermal-conductivity measurements. A graphite tube with a small hole through its wall is used in a manner similar to that described by Worthing[1,12]. A single turn of wire of the material whose emittance is to be measured is placed in a groove in the surface of the graphite tube near the sight hole in the constant-temperature section of the tube. Temperatures of the inside of the tube (blackbody temperature), the outside of the tube, and the surface of the wire are measured with an optical pyrometer. The thermal conductivity of the graphite is used with the true temperature inside the tube to obtain the true temperature of the outside of the tube. This temperature is assumed to be the true temperature of the wire. Using this true temperature and the brightness temperature of the wire obtained with the pyrometer, the normal spectral emittance of the wire can be calculated from Wien's Law, as shown above for the hole-in-tube method.

McDonough[14] has used a spectrophotometer and a single-beam optical system to measure the normal spectral emittance of materials, near room temperature, in the range of 4 to 13.5 μ. The infrared spectrophotometer is used as a monochromator. The radiation from a blackbody at 360 K and that from another blackbody at room temperature are alternated on the entrance slit of the monochromator. A bolometer at the exit of the monochromator gives a signal proportional to the energy difference of the two blackbodies, which is amplified and measured. For a sample run, the 360 K blackbody is replaced by a sample heated to 360 K. The radiation from the room-temperature blackbody is used to correct for energy reflected by the sample from the room itself, and is subtracted from the 360 K blackbody and the 360 K sample readings. The ratio of these two net readings then becomes the emittance of the sample at 360 K.

Two methods for measuring the normal spectral emittance have been described by Richmond.[15,16] The first method measures the emittance of specimens, heated by passing a current through them, in the wavelength range 2 to 15 μ.[15] The specimen and a blackbody furnace are mounted behind a blackened shield so that they can be moved to radiate alternately through an aperture to the measuring system. The measuring system consists of focusing mirrors and chopper and an infrared spectrometer. The output of the spectrometer is focused on a thermocouple detector, and suitably amplified and recorded. The ratio of the reading from this sample to that from the blackbody at the same temperature is the normal spectral emittance of the sample.

The second method[16] uses a double-beam, ratio-recording, infrared spectrometer which has been modified to record normal spectral emittance directly. Two sources, the blackbody furnace and a sample at the same temperature, are each situated so that their radiation is fed through a chopper which alternately passes one beam and then the other. This effectively produces an alternating split beam, with the two sources 180 degrees out of phase. This split beam then passes through a system of combining mirror optics, where the beams follow the same path but are separated in time. The combined beam passes through the monochromator, the spectrometer, and to the thermocouple detector. The a-c signal from the detector is preamplified and fed to commutators on the chopper shaft where it is separated into two a-c signals proportional to the intensities of the respective beams. These two a-c signals are then amplified and rectified to produce two d-c potentials whose ratio, recorded on a potentiometer, is equal to the normal spectral emittance of the specimen.

Reflectance Measurements

Emittance is sometimes determined from reflectance methods. If the material is opaque, Kirchoff's law shows that the emittance equals unity minus the reflectance, or

$$\epsilon = 1 - r.$$

Ried and McAlister[17] used a reflectance method to measure the normal spectral emittance in the range from 2 to 15 μ. A Gier-Dunkle reflectometer[18] was constructed so that a double-beam spectrometer could "look" through an entrance port normally at two different sections of the inner wall. The specimen, maintained at a constant temperature, is mounted flush with the inner wall of the blackbody cavity, at one of the viewing areas. The ratio of the readings from the sample and from the wall of the cavity yields the reflectance of the sample at a given wavelength.

A discussion of various methods for determining spectral reflectance is given by R. F. Dunkle[18]. He individually evaluates the most used types of reflectometers, including the Coblentz hemispherical reflectometer[19-21], the paraboloidal reflectometer, and the integrating-sphere reflectometer. The Gier-Dunkle reflectometer is described in considerable detail. The only difference between its use by Dunkle and by Ried and McAllister[17] is that Dunkle uses a single-beam spectrometer instead of the double beam used by Ried and McAllister. Dunkle rotates the reflectometer to obtain alternate readings of the wall and the sample.

General Considerations

In the foregoing discussion it has been tacitly assumed that equilibrium conditions have prevailed. That is, whenever temperatures have been changed, measurements were not made until all bodies were again in equilibrium with their surroundings. The study of radiant emission under transient conditions is beyond the scope of this review.

No consideration has been given to absorption or emission from gases. It is assumed that corrections will be made if absorbing and radiating gases are present or that the measurements will be made in a vacuum.

Optical systems used with the various detectors or pyrometers have been largely ignored, with the assumption that all possible care will be taken to insure against any error in measurement resulting from their use.

Only a few of the methods and corresponding apparatus are included in this review. A more detailed discussion of these can be found in the cited references. Other methods can be found in the publications cited in References 1, 2, 12, 22, and 23.

Radiation Detectors

Depending on circumstances, we are sometimes interested in the radiating effects of all wavelengths taken collectively; at other times only the effects of spectral or monochromatic radiation are of interest. There are, therefore, two general groups of radiation detectors used in conjunction with radiation-measuring devices: the nonselective or total detectors and the selective or spectral detectors. A brief discussion of a few of each type will be given. A more detailed description of their construction and operation is given in References 22-25.

Total-Radiation Detectors

Total, or nonselective, radiation detectors are those whose efficiencies depend but little on the spectral characteristics of the radiation being measured. Included in these are the thermocouple, the thermopile, the bolometer, and the pyroheliometer.

The Thermocouple. A thermocouple consists of two wires of different metals joined to each other at their ends. When a temperature difference between these two junctions is produced, current flows in the circuit and a voltage known as a thermoelectric emf is produced. This voltage may be measured directly with a potentiometer or a millivoltmeter. Radiation from a source of radiant energy focused upon one junction of such a thermocouple would heat this junction in a manner proportional to the amount of radiation being received. The emf output of a single thermocouple is sometimes quite small, depending upon the energy of radiation being received, and must be suitably amplified for use. A beam chopper and a-c amplifier are ordinarily used for this purpose. The thermocouple detector, although one of the oldest, is still one of the best in use today.

Many of the total-radiation detectors commercially available use thermocouple construction. Ordinarily, the thermocouple is so arranged that a collecting lens, or focusing front-surface mirror system, focuses the incoming radiation onto the hot junction. It should be noted that any lens or system of optics will absorb or scatter some of the incident radiation and will be opaque or only semitransparent to some wavelengths. When lenses are used with a total-radiation detector, therefore, the radiation received by the detector is no longer total radiation. The optical system in most commercial total-radiation detectors, however, transmits radiation over a wide range of wavelengths. Front-surface mirrors, in general, are much less selective in the wavelengths reflected than is a lens through which the radiation must pass.

The Thermopile. A number of thermocouples are sometimes connected so that the emf's produced are additive. This is known as a "thermopile". All of the "hot" junctions are placed so as to receive the radiation from the source, while the "cold" junctions are shielded from the incoming radiation. If the temperature and emittance are high enough, it is possible to measure the output directly with a meter or potentiometer. Although sensitivity and response may be less than that of a single thermocouple with an amplifier, the added convenience of operation and the simplification of circuitry are sometimes desirable.

The Bolometer. The bolometer consists, essentially, of two thin metal strips blackened on one side. These strips form two arms of a Wheatstone bridge. Radiation is allowed to fall on one strip only, thereby raising its temperature and increasing its electrical resistance. This increase in resistance unbalances the Wheatstone bridge, which is indicated by a galvanometer deflection.

The bolometer can be calibrated in terms of a known source of radiant energy. This instrument frequently is used in connection with a spectrometer, since the metal strips can be designed to correspond to a desired width of slit image. In recent years a number of bolometers have been designed, using thermistors in place of the metal strips.

The Pyroheliometer. The pyroheliometer is used almost exclusively to measure solar radiation. There are two major types, absolute instruments and instruments which must be calibrated by comparing their indications with those of absolute instruments. Only the water-flow, absolute type will be discussed here. The water-flow type consists, essentially, of two identical cylindrical chambers with blackened interior walls closed completely except for an opening in one end of each chamber through which radiation may enter. The exterior walls of each chamber are formed about a water-flow channel which absorbs heat from the incident radiation. A thermopile with its hot junctions at the exit of the water channel of one chamber and the cold junctions at the exit of the other detects differences in temperature between the two heat-absorbing streams. A heater is wound about each chamber. In operation, both chambers are immersed in a Dewar flask to maintain a constant external temperature. Radiation is allowed to enter one chamber, and the other is heated electrically until the output of the thermopile registers zero on a galvanometer. It is then assumed that the rate of heat produced by the entering radiation in one chamber is equal to the rate of heat produced electrically in the other. In the case of solar measurements, the irradiation of the orifice of the one chamber by the sun (ϵ_s) multiplied by the area of the orifice (A) is then equated to the electrical power input to the other chamber (P), or:

$$\epsilon_s = P/A \; ,$$

from which the irradiance per unit area of the earth's surface (watts/cm^2) can be determined,

and where

ϵ_s is the irradiance, watts/cm^2

A is the orifice area, cm^2

P is the electrical power, watts.

Spectral or Selective Radiation Detectors

Spectral or selective detectors are those whose efficiencies depend upon the spectral character of the incident radiation. These include the human eye, the photographic plate, the photovoltaic cell, and the photoelectric cell. The sensitivity of spectral detectors varies greatly with wavelength. They must, therefore, be calibrated in terms of a known radiation intensity at the wavelength desired, depending on the characteristics of the source and the instrumentation used, or calibrated by comparison with a total-radiation detector.

The Human Eye. The eye is a radiation detector of unparalled sensitivity and, although limited to a small range of wavelengths and obviously incapable of accurate calibration, it can readily pick up small differences in color and intensity. It is extensively used in conjunction with color- and intensity-matching devices such as the optical pyrometer, which will be discussed under Temperature-Measuring Devices.

Photographic Materials. Photographic plates or films are sensitive radiation detectors of the selective or spectral type. Therefore, they must be calibrated against some type of total-radiation detector. Exposure of the plate or film to incident radiation gives a response (the amount of developed silver, usually expressed as density) proportional in a complex manner to the intensity of the radiation and to the exposure time. In conjunction with the photographic material, some means of measuring the density of the developed film must be available. Ordinarily an instrument called a densitometer is used. This instrument is designed to measure the amount of light transmitted by the developed film or plate.

This method of measuring radiant intensities can be used with precision only if a complete knowledge of the characteristics of the photographic material used is available and the proper handling, developing, and measuring techniques are known. It is common practice to expose the same photographic emulsion to known amounts of radiation to furnish a calibrated density-exposure relationship. Usually, photographic materials are used for approximate determinations. Photographic emulsions are now available for sensitivities over a wide range of wavelengths.

The Photoelectric Tube. The photoelectric tube consists of two metal electrodes in a vacuum or in a low-pressure gas. The positive electrode (anode) is maintained at a positive potential with respect to the negative electrode (cathode). Radiation, allowed to fall on the cathode, causes electrons to be emitted which are drawn across the space to the anode, producing a flow of current in an external measuring circuit. This current is proportional complexly to the amount of incident radiation.

Most phototubes are sensitive only to a limited spectral range. They reach a high peak or maximum photocurrent at or near a certain wavelength and fall off sharply for longer or shorter wavelengths. Special tubes are commercially available from which one can choose maximum sensitivities over a wide range of wavelengths.

Phototubes must be calibrated against a known source.

The Photovoltaic Cell. One type of photovoltaic, or photo emf, cell (now generally considered obsolete) consists of a metallic surface covered with a thin film of a semiconductor which is, in turn, covered by a translucent film of another metal. Exposure of the cell to radiant energy generates an emf of a few millivolts, which increases with increasing radiation. The internal resistance decreases with increasing radiation. When used with the proper value of external resistance, the response is nearly proportional to the incident energy. At present, photovoltaic cells based on silicon of controlled purity and with a diffused impurity to produce a p-n junction are largely supplanting the barrier-type photocells formerly based on selenium.

The main advantage of these cells is that no external source of voltage is required, although ordinarily the low voltage obtained is suitably amplified for use. Modern silicon cells have a much higher voltage output than older types of photovoltaic cells.

The spectral characteristics of the photovoltaic cell are quite similar to those of photoelectric cells.

The Photoresistive Cell. Recent developments in solid-state physics have led to the construction of many radiation-sensing devices based on changes of electrical resistance of cadmium sulfide crystals and films. Capable of extremely wide changes in electrical resistance between the dark state and when illuminated, such photoresistive cells offer a new tool for measuring radiant energy.

Temperature-Measuring Devices

The measurement of low temperatures is usually called thermometry, whereas the measurement of high temperatures is called pyrometry. The normally accepted dividing line between thermometry and pyrometry is considered to be "dull red" heat, or about

1000 F. A number of temperature-measuring devices will be considered here. For further information the reader is referred to References 26-30.

Low-Temperature Devices

Several devices exist for the measurement of low temperatures, including liquid-in-glass thermometers; bimetal expansion thermometers; Bourdon-tube-type, filled-system thermometers; thermocouples; and resistance thermometers. However, in the discussion of the measurement of the radiation properties of materials, only the last two, thermocouples and resistance thermometers, will be discussed since they are the most widely used for radiant-energy measurements.

The Thermocouple. Thermocouples were previously discussed in reference to total-radiation detectors. A number of standard thermocouple materials are manufactured, and for each type, the relation of temperature difference to thermal emf is well known. With proper care, thermocouples can be used to measure temperatures ranging from near the temperature of liquid nitrogen (-320 F) to above 3000 F, depending upon the type of thermocouple used and the conditions under which the measurements are made.

The Resistance Thermometer. Basically, a resistance thermometer consists of a resistance coil or spiral, suitably shielded from contamination, mechanical damage, and strain, and with suitable lead wires to the measuring system. Temperature measurement consists of measuring the resistance of the thermometer coil, which changes with a change in temperature. Temperatures measurable with this type of instrument range from near absolute zero to 1300 F or higher, depending on the resistance-coil material.

The platinum resistance thermometer is used as the International Standard between -190 C (-310 F) and 660 C (1220 F). Its resistance is usually measured with a Wheatstone, Mueller, or other sensitive bridge. The operation of a resistance thermometer and that of a bolometer are similar.

High-Temperature Devices

The Human Eye. The eye is surprisingly sensitive to temperature changes between 1000 F and 2750 F. As a body is heated (in a darkened room), the first visible radiation occurs for most individuals near 950 F. A dark-adapted eye, in a darkened room, may see visible radiation as a colorless glow at a temperature as low as 750 F. As the temperature is raised, more of the radiation is in the visible region, and the color changes from dull red to deep red at about 1200 F, and to bright red at about 1400 F. A trained eye can readily detect temperature changes of the order of 50 F, but actual temperature measurement by the unaided eye can be in error by 200 to 250 F.

The Optical Pyrometer. The instrument most used in industry for the measurement of high temperatures is the optical pyrometer.

This instrument uses the ability of the eye to distinguish brightness differences equivalent to a change of a few degrees of temperature when the two sources are close together in the field of view.

Although many types of optical pyrometers are in use today, probably most used is the disappearing-filament type. The operator looks through the eyepiece and sees the filament of the pyrometer lamp super-imposed on the body whose temperature is being measured. He then varies the current through the filament, which changes its brightness until it matches that of the background and the filament seems to disappear. To minimize errors in color matching, a nearly monochromatic red filter is in the light path. Ordinarily the filament-control rheostat is graduated directly in degrees. To extend the temperature range of the instrument, a gray filter (one whose absorptance does not change with wavelength) is rotated into the optical path in front of the filament.

Optical pyrometers are usually calibrated against a standard tungsten lamp.

It must be remembered that, when using an optical pyrometer, blackbody conditions must be present for the temperature reading to indicate true temperature. When measuring nonblackbodies in a cooler environment, an apparent temperature lower than the true temperature will be indicated, because the emittance is less than one. Conversely, in a hotter environment, the observed temperature will be higher than the true temperature of the target.

The Total-Radiation Pyrometer. Total-radiation pyrometers are many and varied. Those most frequently seen are of the thermopile or the bolometer type, similar to the total-radiation detectors but calibrated in temperature units. They can be of the "open" type or they can be sealed and use windows, lenses, and mirrors for focusing the incoming radiation and thereby keep out undesirable contaminants such as dust and dirt. To determine which instrument to use, a survey of the manufacturer's literature is suggested. A selection should be made for the specified conditions under which it will be used.

REFERENCES

(1)   Worthing and Halliday, Heat, John Wiley and Sons, Inc., New York (1948), p 436.

(2)   Drummeter and Goldstein, "Vanguard Emittance Studies at NRL", First Symposium – Surface Effects on Spacecraft Materials, John Wiley and Sons, Inc., New York (1960), p 153.

(3)   Shaw, C. C., "Apparatus for the Measurement of Spectral and Total Emittance of Opaque Solids", First Symposium – Surface Effects on Spacecraft Materials, John Wiley and Sons, Inc., New York (1960), p 228.

(4)   Cairns, J. H., "Apparatus for Investigating Total Hemispherical Emissivity", Journal of Scientific Instruments, 37 (3), 84-87 (March, 1960).

(5)   Allen, Glasier, and Jordan, "Spectral Emissitivity, Total Emissivity, and Thermal Conductivity of Molybdenum, Tantalum, and Tungsten Above 2300°K", Journal of Applied Physics, 31 (8), 1382-87 (August, 1960).

(6)   Richmond, J. C., "Some Methods Used at NBS for Measuring Thermal Emittance at High Temperatures", First Symposium – Surface Effects on Spacecraft Materials, John Wiley and Sons, Inc., New York (1960), 182-192.

(7)   Butler and Inn, "A Method for Measuring Total Hemispherical Emissivity of Metals", First Symposium – Surface Effects on Spacecraft Materials, John Wiley and Sons, Inc., New York (1960), pp 195-200.

(8)   Wade, W. R., "Measurements of Total Hemispherical Emissivity of Several Stably Oxidized Metals and Some Refractory Oxide Coatings", National Aeronautics and Space Administration Memorandum 1-20-59L (January, 1959).

(9)   O'Sullivan and Wade, "Theory and Apparatus for Measurement of Emissivity for Radiative Cooling of Hypersonic Aircraft With Data for Inconel and Inconel-X", National Aeronautics and Space Administration TN-4121, 1957.

(10) Olson and Katz, "Emissivity, Absorptivity, and High-Temperature Measurements at Armour Research Foundation", First Symposium – Surface Effects on Spacecraft Materials, John Wiley and Sons, Inc., New York (1960), p 169.

(11) McMahon, H. O., "Thermal Radiation Characteristics of Some Glasses", Journal American Ceramic Society, 34 (3), 91 (1951).

(12) Worthing, A. G., "Temperature Radiation Emissivities and Emittances", Temperature – Its Measurement and Control in Science and Industry, Reinhold Publishing Company (1941), p 1171.

(13) Krishman and Jain, "Determination of Thermal Conductivities at High Temperatures", Brit. J. Appl. Phys., 5, 426-430 (December, 1954).

(14) McDonough, R., "Emissivity of Materials Near Room Temperature", First Symposium – Surface Effects on Spacecraft Materials, John Wiley and Sons, Inc., New York (1960), pp 142-145.

(15) Richmond, J. C., "Some Methods Used at NBS for Measuring Thermal Emittance at High Temperatures", First Symposium – Surface Effects on Spacecraft Materials, John Wiley and Sons, Inc., New York (1960), pp 182-185.

(16) Richmond, J. C., "Some Methods Used at NBS for Measuring Thermal Emittance at High Temperatures", First Symposium – Surface Effects on Spacecraft Materials, John Wiley and Sons, Inc., New York (1960), pp 186-191.

(17) Ried and McAllister, "Measurement of Spectral Emissivity From 2μ to 15μ", J. Opt. Soc. Am., 49 (1), 78-82 (January, 1959).

(18) Dunkle, R. V., "Spectral Reflectance Measurements", First Symposium – Surface Effects on Spacecraft Materials, John Wiley and Sons, Inc., New York (1960), pp 117-137.

(19) Coblentz, W. W., "Radiometric Investigations of Infrared Absorption and Reflection Spectra", Bull. Bur. Stds., 2, 457 (1906).

(20) Coblentz, W. W., "Selective Radiation From Various Substances", Bull. Bur. Stds., 7, 243 (1911).

(21) Coblentz, W. W., "The Reflecting Power of Monel Metal, Stellite, and Zinc", Bull. Bur. Stds., 16, 249 (1920).

(22) Worthing and Halliday, Heat, John Wiley and Sons, New York (1948), pp 421-429.

(23) Forsythe, W. E., Measurement of Radiant Energy, McGraw-Hill Book Company (1937), pp 189-354.

(24) Behar, M. F., "Thermometry", and "Pyrometry", Handbook of Measurement and Control, Instruments Publishing Company (1951), pp 107-113.

(25) Giedt, W. H., Principles of Engineering Heat Transfer, D. Van Nostrand Company, Inc. (1957), pp 267-272.

(26) Forsythe, W. E., Measurement of Radiant Energy, McGraw-Hill Book Company (1937), pp 355-387.

(27) Behar, M. F., "Thermometry" and "Pyrometry", Handbook of Measurement and Control, Instruments Publishing Company (1951), p 82.

(28) Forsythe, W. E., "Optical Pyrometry", Temperature – Its Measurement and Control in Science and Industry, Reinhold Publishing Company (1941), pp 1115-1131.

(29) Stauffer and Hunter, "Alloys of Iron and Nickel in Resistance Thermometry", Temperature – Its Measurement and Control in Science and Industry, Reinhold Publishing Company (1941), pp 1236-1237.

(30) Gier and Boelter, "The Silver-Constantan Plated Thermopile", Temperature – Its Measurement and Control in Science and Industry, Reinhold Publishing Company (1941), pp 1284-1292.

RADIATIVE PROPERTY DATA

Titanium and Titanium Alloys

TABLE OF CONTENTS

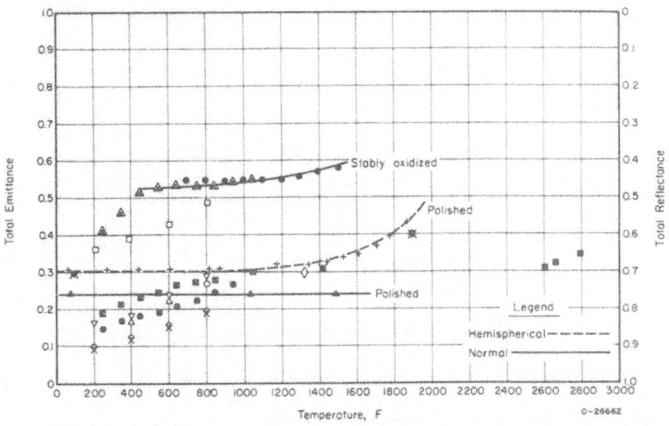

NORMAL AND HEMISPHERICAL TOTAL EMITTANCE OF TITANIUM

Normal and Hemispherical Total Emittance of Titanium as a Function of Temperature-Reference Information

Reference	Investigator	Symbol	Composition and Surface Condition	Test Method	Remarks
1	Bevans, Gier, and Dunkle	⌀ △ ▽ ○ □ ✕	Ti-75A titanium 300 hours at 600 F 100 hours at 810 F 306 hours at 820 F 303 hours at 871 F 303 hours at 1003 F No thermal treatment	Normal total emittance. Calibrated thermopile detector. Temperatures measured with thermocouples.	Various oxidation treatments. Data taken from curves.
2	Carpenter and Mair	■ ▾	Magnesium reduced Iodine reduced	Hemispherical total emittance. Power dissipation of wire to concentric cylinder.	Measured in vacuum.
3	Carpenter and Reavel	◇	Alpha phase, iodine reduced	Hemispherical total emittance. Power dissipation of wire to concentric cylinder.	Measured in vacuum.
4	Dobbins	● ■ ▲	Purity - 99+ per cent Polished Oxidized blue Oxidized gray (100 hours at 1200 F)	Normal total emittance. Calibrated thermopile detector. Temperatures measured with thermocouples. Blackbody radiation calculated from known emittance of "secondary standard".	Measured in air. Data taken from curves.
5	Michaels and Wilford	+	Purity - 95.5 per cent (commercial purity)	Hemispherical total emittance. Power dissipation of wire to concentric cylinder. Temperatures measured with calibrated optical pyrometer.	Measured in vacuum. Data taken from curves.
6	Skinner, Johnson, and Beckett	✖ ✳	Alpha phase Beta phase	Hemispherical total emittance. Hole-in-tube method.	Measured in vacuum.
7	Wade	▲ ●	Ti-75A titanium Polished Oxidized, stable	Normal total emittance. Calibrated thermopile detector. Comparison blackbody. Temperatures measured with thermocouples.	Measured in air. Data taken from curves.

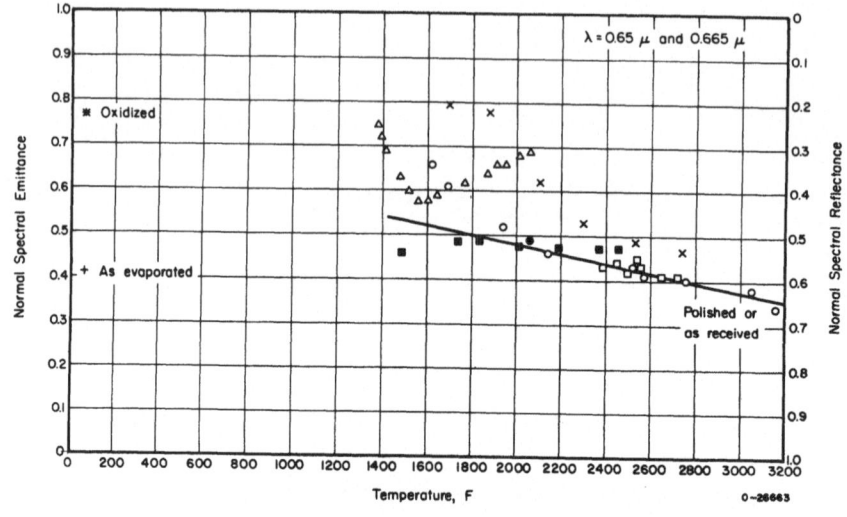

NORMAL SPECTRAL EMITTANCE OF TITANIUM

Normal Spectral Emittance of Titanium--Reference Information

Reference	Investigator	Symbol	Composition and Surface Condition	Test Method	Remarks
8	Blocher and Campbell	●		Normal spectral emittance. Optical pyrometer. Comparison blackbody.	Measured in air (λ = 0.65μ).
9	Bradshaw	■	Purity - 99.5 per cent, magnesium reduced Polished	Normal spectral emittance. Induction-heated cylinder. Drilled blackbody hole.	Measured in vacuum (λ = 0.65μ).
10	Edwards, Johnston, and Ditmars	□	Purity - 99.88 per cent	Normal spectral emittance. Hole-in-tube method.	Measured in vacuum (λ = 0.65μ).
11	Hass and Bradford	 + ✻	Purity - 99.9 per cent iodine reduced As evaporated - 0.15 thick Oxidized	Normal spectral reflectance. Double beam, recording spectrophotometer.	Titanium film evaporated on quartz in vacuum of 5 x 10^{-6} mm of Hg (λ = 0.665μ).
5	Michaels and Wilford	△	Purity - 99.5 per cent (commercial purity)	Normal spectral emittance. Surface brightness compared with blackbody hole. Calibrated optical pyrometer.	Data taken from curves (λ = 0.665μ).
12	Powers and Wilhelm	 ○ ✕	Polished Crystal-bar titanium Du Pont titanium	Normal spectral emittance. Surface brightness compared with blackbody hole. Optical pyrometer. Resistance-heated strip specimens.	Measured in vacuum (λ = 0.665μ). Data taken from curves.

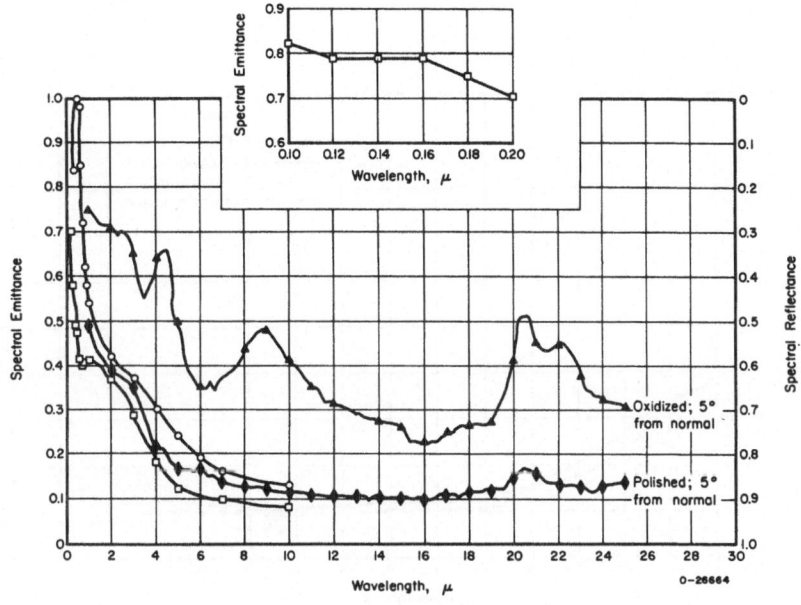

SPECTRAL EMITTANCE OF TITANIUM AS A FUNCTION OF WAVELENGTH

Spectral Emittance of Titanium as a Function of Wavelength--Reference Information

Reference	Investigator	Symbol	Composition and Surface Condition	Test Method	Remarks
11	Hass and Bradford	□ ○	Purity 99.9 per cent, iodine reduced Freshly deposited opaque film Heated in air at 400 C for 3 hours (oxide film of about 0.03 μ)	Spectral reflectance. Double-beam recording spectrophotometer. Measurements made at "close to normal incidence".	Film evaporated on quartz in vacuum of 5 x 10⁻⁶ mm of Hg. Measured in air at room temperature. Data taken from reflectance curves.
1	Bevans, Gier, and Dunkle	◆ ▲	No thermal treatment 303 hours at 1003 F.	Spectral reflectance at 5° from normal. Gier-Dunkle reflectometer. Temperatures measured with thermocouples. Diffuse illumination - normal viewing.	Measured in air at room temperature. Data taken from reflectance curves.

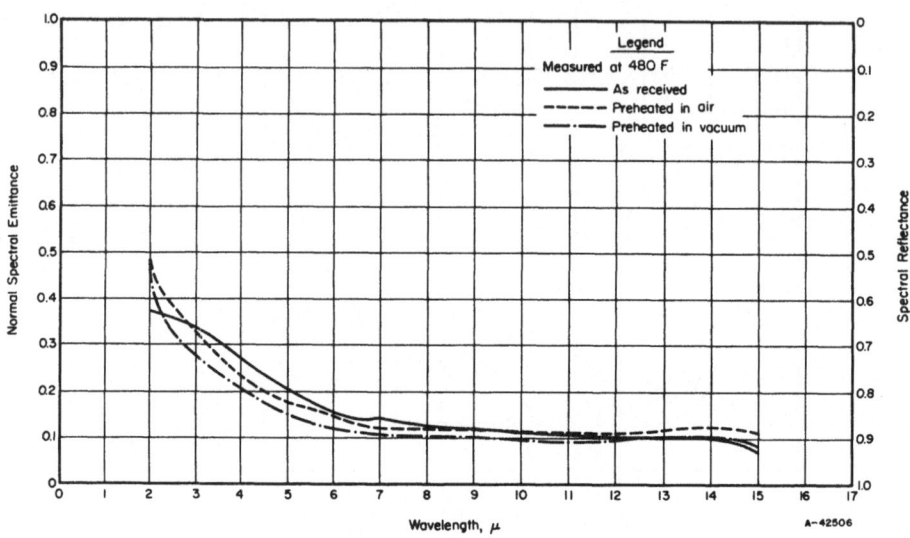

NORMAL SPECTRAL EMITTANCE OF TITANIUM AT 480 F

NORMAL SPECTRAL EMITTANCE OF TITANIUM AT 480 F--REFERENCE INFORMATION

Reference	Investigator	Symbol	Composition and Surface Condition	Test Method	Remarks
15	Adams, J. G.		As received Heated 30 minutes in air at 800 F Heated 30 minutes in 2.8 x 10^{-5} mm Hg pressure at 800 F	Normal spectral emittance. Furnace-heated disk specimen. Comparison blackbody (Hohlraun). Spectrometer-monochromator with photomultiplier, lead sulphide, and thermocouple detectors. Temperatures measured with thermocouples.	Measured in air.

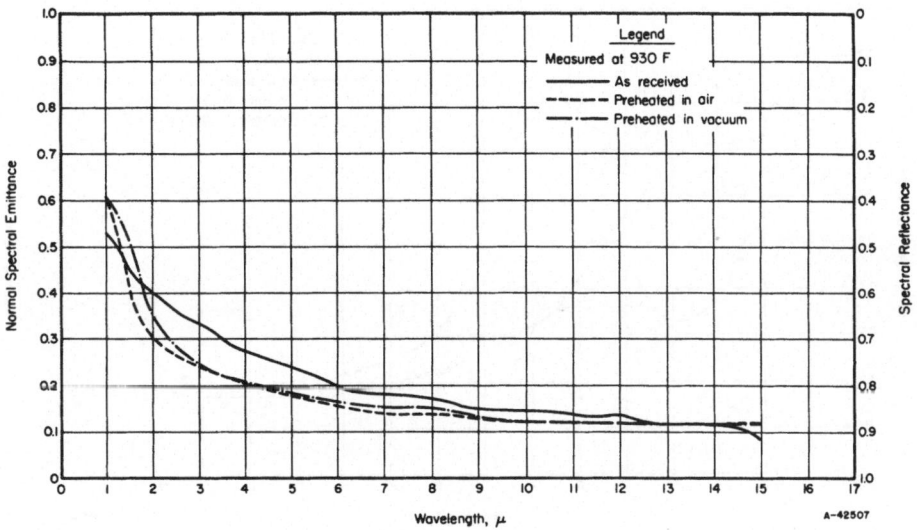

NORMAL SPECTRAL EMITTANCE OF TITANIUM AT 930 F

NORMAL SPECTRAL EMITTANCE OF TITANIUM AT 930 F--REFERENCE INFORMATION

Reference	Investigator	Symbol	Composition and Surface Condition	Test Method	Remarks
15	Adams, J. G.		As received Heated 30 minutes in air at 800 F Heated 30 minutes in 2.8 x 10^{-5} mm Hg pressure at 800 F	Normal spectral emittance. Furnace-heated disk speci-men. Comparison blackbody (Hohlraun). Spectrometer-mono-chromator with photo-multiplier, lead sulphide, and thermocouple de-tectors. Temperatures measured with thermocouples.	Measured in air.

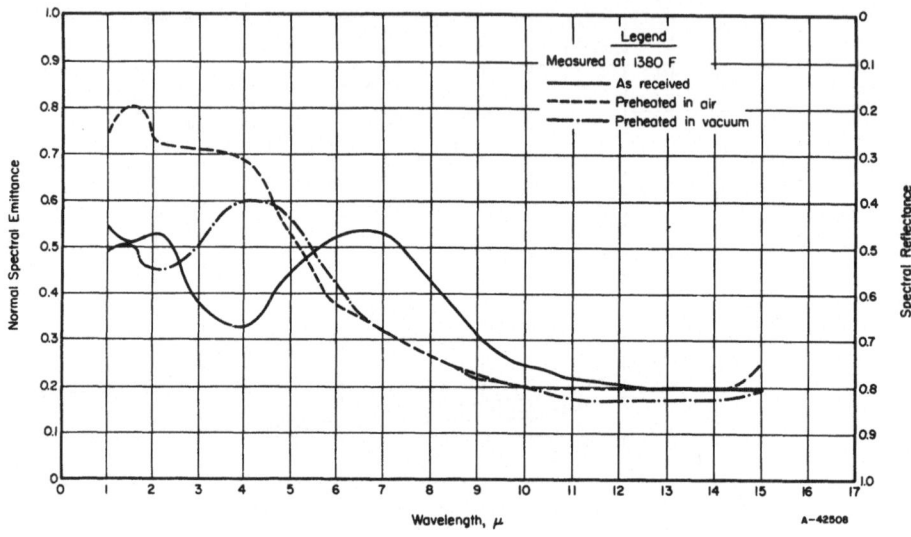

NORMAL SPECTRAL EMITTANCE OF TITANIUM AT 1380 F

NORMAL SPECTRAL EMITTANCE OF TITANIUM AT 1380 F--REFERENCE INFORMATION

Reference	Investigator	Symbol	Composition and Surface Condition	Test Method	Remarks
15	Adams, J. G.		As received Heated 30 minutes in air at 800 F Heated 30 minutes at 2.8 x 10⁻⁵ mm Hg at 800 F	Normal spectral emittance. Furnace-heated disk specimen. Comparison blackbody (Hohlraun). Spectrometer-monochromator with photomultiplier, lead sulphide, and thermocouple detectors. Temperatures measured with thermocouples.	Measured in air.

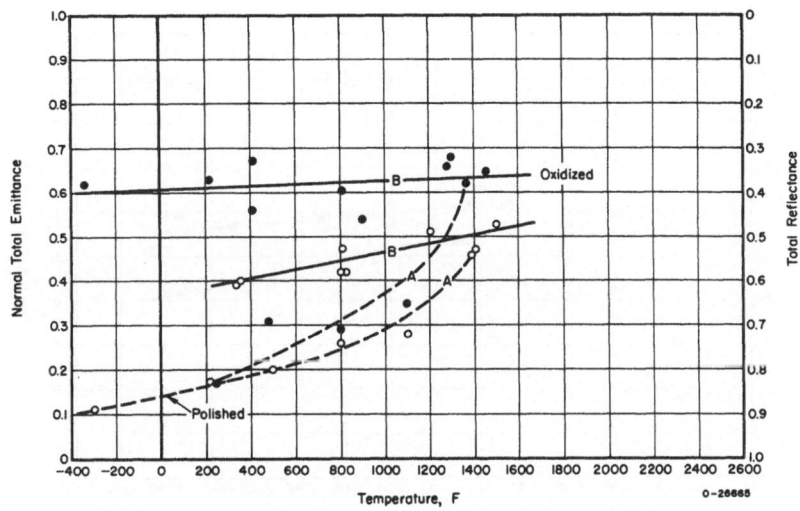

NORMAL TOTAL EMITTANCE OF ALLOY Ti-5Al-2.5Sn

Normal Total Emittance of Alloy Ti-5Al-2.5Sn--Reference Information

Reference	Investigator	Symbol	Composition and Surface Condition	Test Method	Remarks
13	Olson and Morris	● ○	Polished Oxidized 30 minutes at red heat	Normal total emittance. Resistance-heated specimen. Thermistor-bolometer detector. Temperatures measured with thermocouples.	Measured in air; three cycles. Curve A shows first cycle; Curve B shows combined second and third cycles.

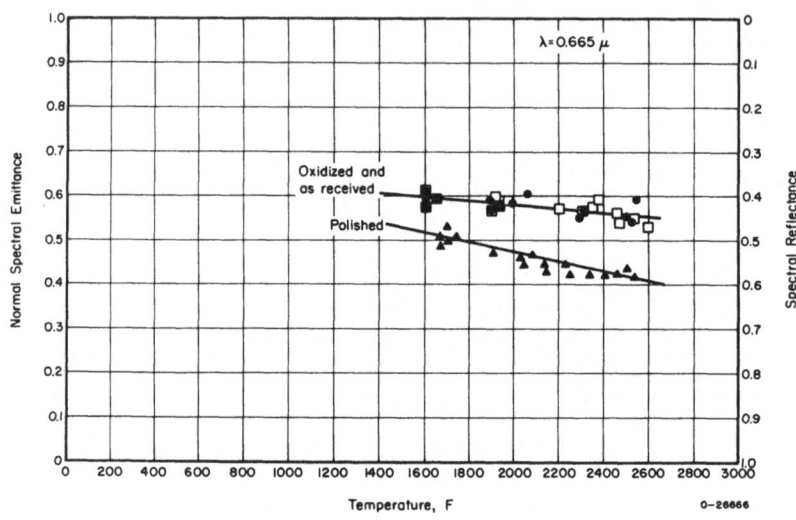

NORMAL SPECTRAL EMITTANCE OF ALLOY Ti-5Al-2.5Sn

Normal Spectral Emittance of Alloy Ti-5Al-2.5Sn--Reference Information

Reference	Investigator	Symbol	Composition and Surface Condition	Test Method	Remarks
14	Betz, Olson, Schurin, and Morris	▲ ● ▫	Polished As received or wiped clean Oxidized	Normal spectral emittance. Modified hole-in-tube method. Drilled black-body hole. Temperatures measured with thermo-couples.	Measured in vacuum (λ = 0.665μ). Data taken from curves.

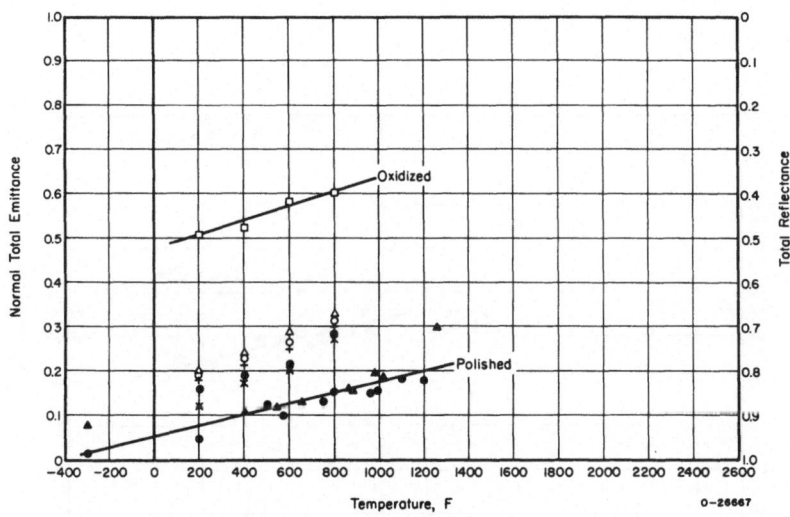

NORMAL TOTAL EMITTANCE OF ALLOY Ti-8Mn

Normal Total Emittance of Alloy Ti-8Mn--Reference Information

Reference	Investigator	Symbol	Composition and Surface Condition	Test Method	Remarks
14	Betz, Olson, Schurin, and Morris	●	As received, cleaned, and polished	Normal total emittance. Resistance-heated specimen. Thermistor-bolometer detector. Comparison blackbody Temperatures measured with thermocouples.	Measured in vacuum. Data taken from curves.
		▲	Oxidized		
1	Bevans, Gier, and Dunkle	●	300 hours at 600 F	Normal total emittance. Calibrated thermopile detector. Temperatures measured with thermo-couples.	Various oxidation treatments. Measured in air. Data taken from curves.
		+	100 hours at 810 F		
		△	306 hours at 820 F		
		○	303 hours at 871 F		
		□	303 hours at 1003 F		
		×	No thermal treatment		

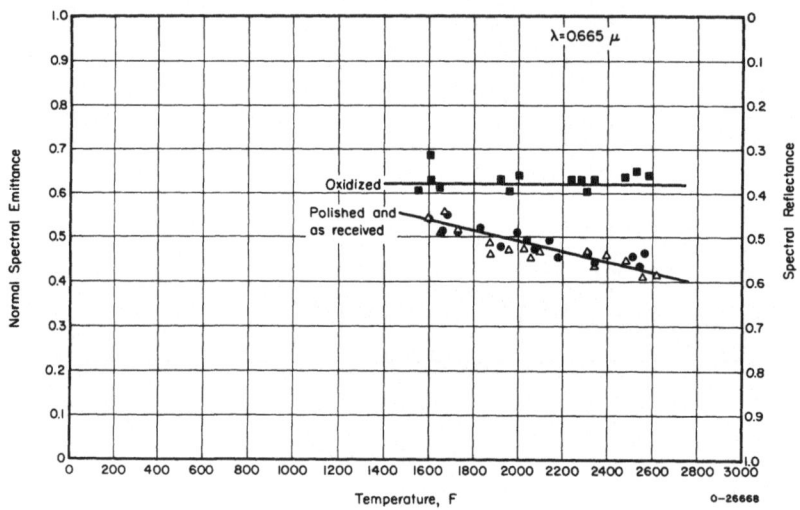

NORMAL SPECTRAL EMITTANCE OF ALLOY Ti-8Mn

Normal Spectral Emittance of Alloy Ti-8Mn--Reference Information

Reference	Investigator	Symbol	Composition and Surface Condition	Test Method	Remarks
14	Betz, Olson, Schurin, and Morris	● △ ■	As received, or wiped clean Polished Oxidized	Normal spectral emittance. Self-resistance heating. Modified hole-in-tube method.	Measured in vacuum. (λ = 0.665μ). Data taken from curves.

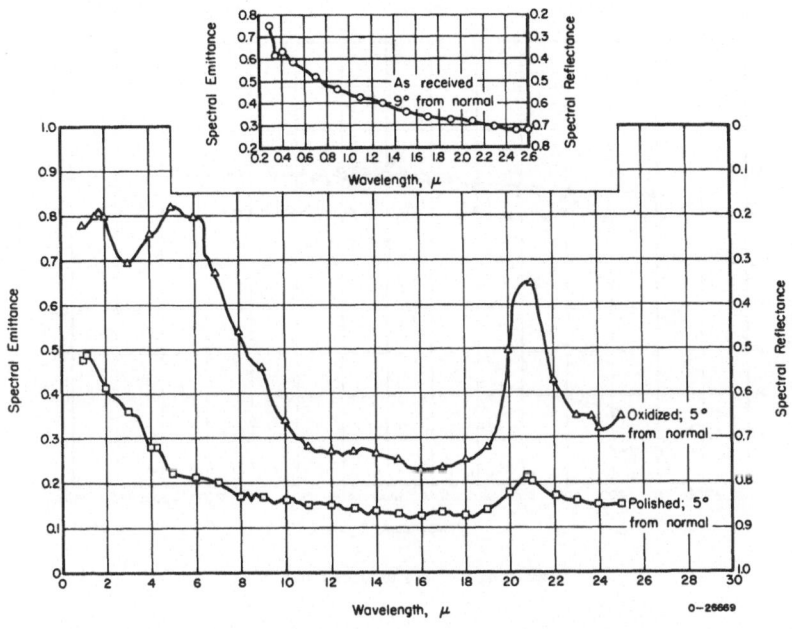

SPECTRAL EMITTANCE OF ALLOY Ti–8Mn AS A FUNCTION OF WAVELENGTH

Spectral Emittance of Alloy Ti–8Mn as a Function of Wavelength--Reference Information

Reference	Investigator	Symbol	Composition and Surface Condition	Test Method	Remarks
14	Betz, Olson, Schurin, and Morris	O	As received	Spectral reflectance. Monochromator, integrating sphere, lead sulfide detector, and MgO standard. Near normal (9°) irradiation – diffuse viewing.	Measured in air at room temperature. Data taken from reflectance curves.
1	Bevans, Gier, and Dunkle	△ □	303 hours at 1003 F No thermal treatment	Spectral reflectance. Gier, Dunkle reflecto-meter. Diffuse illumination – near normal (5°) viewing.	Measured in air at room temperature. Data taken from reflectance curves.

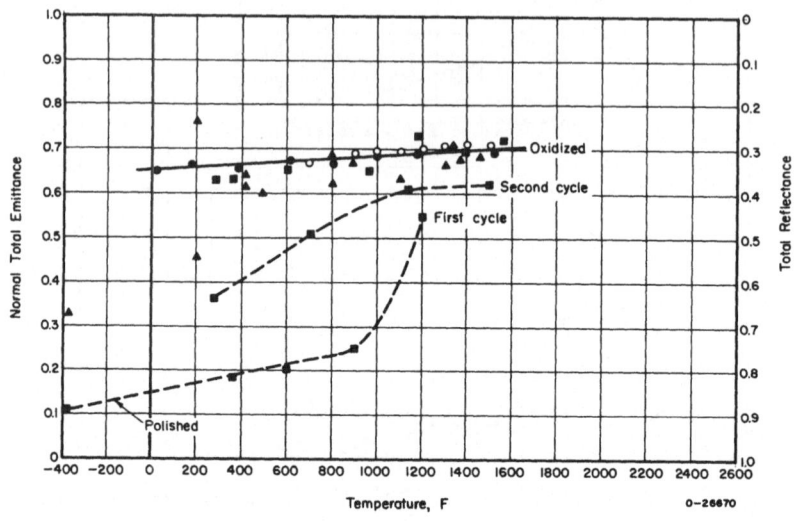

NORMAL TOTAL EMITTANCE OF ALLOY Ti-6Al-4V

Normal Total Emittance of Alloy Ti-6Al-4V—Reference Information

Reference	Investigator	Symbol	Composition and Surface Condition	Test Method	Remarks
13	Olson and Morris	▲ ■ ●	Oxidized Polished Smoothed data after three cycles in air	Normal total emittance. Resistance-heated specimen. Thermistor-bolometer detector. Comparison blackbody. Temperatures measured with thermocouples.	Measured in air. First and second cycles for the polished specimen are indicated. Data taken from curves.
7	Wade	○	Stably oxidized at 1500 F	Normal total emittance. Resistance-heated ribbon specimen. Total-radiation pyrometer. Comparison blackbody. Temperatures measured with thermocouples.	Measured in air. Data taken from curves.

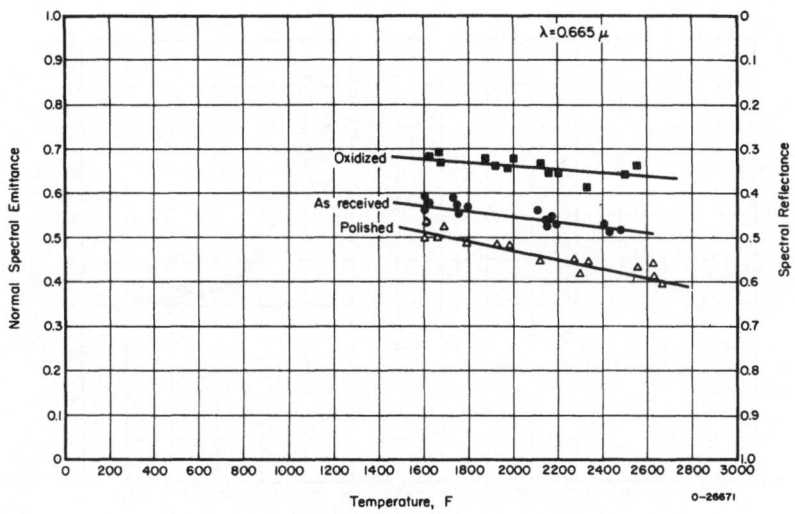

NORMAL SPECTRAL EMITTANCE OF ALLOY Ti-6Al-4V

Normal Spectral Emittance of Alloy Ti-6Al-4V--Reference Information

Reference	Investigator	Symbol	Composition and Surface Condition	Test Method	Remarks
14	Betz, Olson, Schurin, and Morris	△ ● ■	Polished As received or wiped clean Oxidized	Normal spectral emittance. Modified hole-in-tube method. Drilled black-body hole. Temperatures measured with thermo-couples.	Measured in vacuum (λ = 0.665 μ). Data taken from curves.

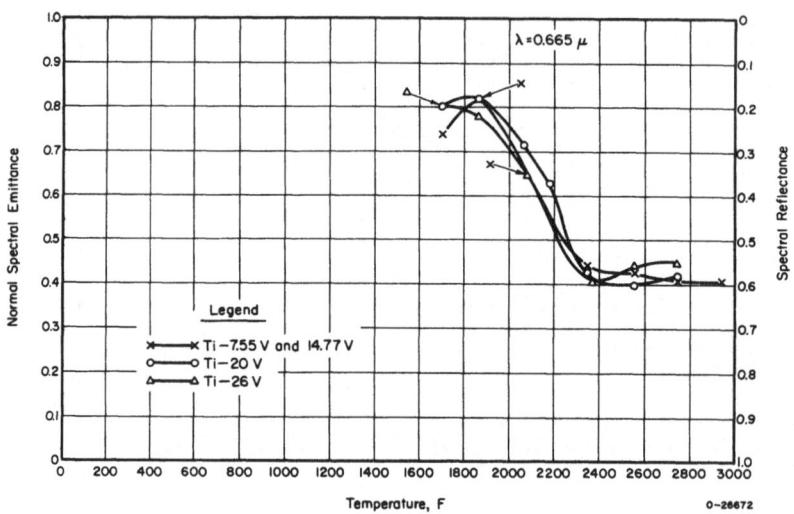

NORMAL SPECTRAL EMITTANCE OF TITANIUM-VANADIUM ALLOYS

Normal Spectral Emittance of Titanium-Vanadium Alloys--Reference Information

Reference	Investigator	Symbol	Composition and Surface Condition	Test Method	Remarks
12	Powers and Wilhelm	× ○ △	Polished specimens Ti-7.55V and Ti-14.77V Ti-20V Ti-26V	Normal spectral emittance. Resistance-heated strip specimens. Drilled blackbody hole. Optical pyrometer.	Measured in vacuum of 0.1 micron (λ = 0.665μ). Data taken from curves.

REFERENCES

(1) Bevans, J. T. Gier, J. T. , and Dunkle, R. V. , "Comparison of Total Emittances with Values Computed from Spectral Measurements", Trans. ASME, 80 (2), 1405-1414 (1958).

(2) Carpenter, L. G. , and Mair, W. N. , "The Evaporation of Titanium", Proc. Phys. Soc. (London), B64, 57 (1951).

(3) Carpenter, L. G. , and Reavel, F. W. , "Vaporization of Titanium", Nature, 163, 527 (1949).

(4) Dobbins, J. P. , "Emittances of Titanium, Aluminum, and Stainless Steel", North American Aviation Report No. NA-49-238 (March, 1949).

(5) Michaels, W. C. , and Wilford, S. E. (Maur College), "The Physical Properties of Titanium, Emissivity, and Resistivity of Commercial Metal", J. Appl. Phys. , 20, 1223 (1949).

(6) Skinner, G. , Johnston, H. L. , and Beckett, C. , "Titanium and Its Compounds", Herrick L. Johnston Enterprises, Columbus, Ohio (1954).

(7) Wade, W. R. , "Measurement of Total Hemispherical Emissivity of Various Oxidized Metals at High Temperatures", Langley Research Center, NACA TN 4206 (1958).

(8) Blocher, J. M. , Jr. , and Campbell, I. E. , "Vapor Pressure of Titanium", JACS, 71, 4040 (1949).

(9) Bradshaw, F. I. , "The Optical Emissivity of Titanium and Zirconium", Proc. Phys. Soc. (London), p 573 (1950).

(10) Edwards, J. W. , Johnson, H. L. , and Ditmars, W. , "Vapor Pressures of Inorganic Substances", JACS, 75, 2467 (1953).

(11) Hass, G. , and Bradford, A. P. , "Optical Properties and Oxidation of Evaporated Titanium Films", J. Opt. Soc. of America, 47, 125-129 (1957).

(12) Powers, R. M. , and Wilhelm, H. A. , "The Titanium Vanadium System", Iowa State College, ISC-228, pp 91-103 (September, 1952).

(13) Olson, H. O. , and Morris, J. C. , "Determination of Emissivity and Reflectivity Data on Aircraft Structural Materials", WADC TR 56-222, Part II, Supp. I. (October, 1958).

(14) Betz, H. T. , Olson, H. Q. , Schurin, B. D. , and Morris, J. C. , "Determination of Emissivity and Reflectivity Data on Aircraft Structural Materials", WADC TR 56-222, Part II (October, 1958).

(15) Adams, J. G. , "The Determination of Spectral Emissivities, Reflectivities, and Absorptivities of Materials and Coatings", Northrup Corporation Report No. NOR-61-189 (August 3, 1961).

RADIATIVE PROPERTY DATA

Stainless Steels

TABLE OF CONTENTS

TABLE OF CONTENTS
(Continued)

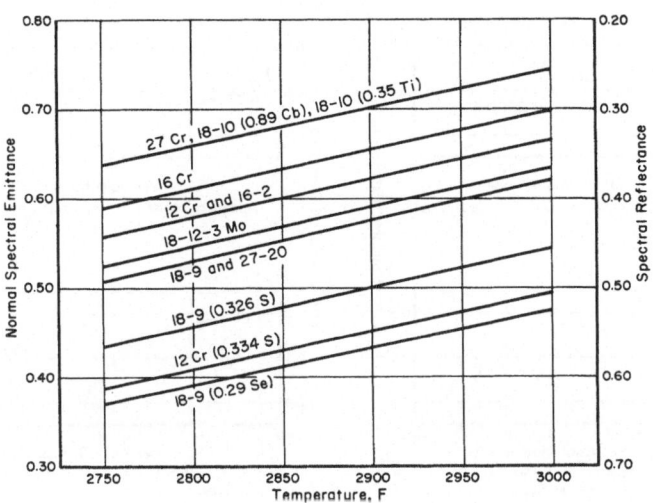

NORMAL SPECTRAL EMITTANCE OF MOLTEN STAINLESS STEELS

NORMAL SPECTRAL EMITTANCE OF MOLTEN STAINLESS STEELS--REFERENCE INFORMATION

Reference	Investigator	Symbol	Composition and Surface Condition	Test Method	Remarks
1	Goller, G. N.	None	Molten. Measured at lip while pouring. (Analysis of each type given below.)	Normal spectral emittance. Calibrated optical pyrometer. Immersion thermocouple.	Measured in air. Data actually fell in flattened s-shaped curves, but were reported as straight lines for illustration purposes. (No wavelength given for the Pyro optical pyrometer used - probably near 0.65μ.)

Analysis of steels studied:

Steel	C	Mn	P	S	Si	Cr	Ni	Mo	Other
27 Cr	0.126	0.43	0.026	0.021	0.67	26.80	0.33	--	----
18-10 (0.89 Cb)	0.070	1.27	0.020	0.020	0.55	17.96	11.01	--	0.89 Cb
18-10 (0.35 Ti)	0.054	1.32	0.025	0.018	0.79	18.34	10.87	--	0.35 Ti
16 Cr	0.067	0.47	0.012	0.018	0.34	16.27	0.40	--	----
12 Cr	0.093	0.39	0.012	0.021	0.25	12.00	0.21	0.44	----
16-2	0.144	0.41	0.018	0.022	0.45	16.44	1.82	--	----
18-12-3 Mo	0.049	1.58	0.019	0.010	0.52	17.05	12.42	2.43	----
18-9	0.056	0.48	0.016	0.019	0.38	17.98	9.01	--	----
27-20	0.106	1.60	0.023	0.016	0.48	26.47	21.67	--	----
18-9 (0.326 S)	0.071	0.65	0.016	0.326	0.38	18.54	9.59	0.38	----
12 Cr (0.334 S)	0.089	0.39	0.015	0.334	0.27	12.21	0.19	0.40	----
18-9 (0.29 Se)	0.071	0.87	0.156	0.021	0.60	18.50	8.74	--	0.29 Se

54

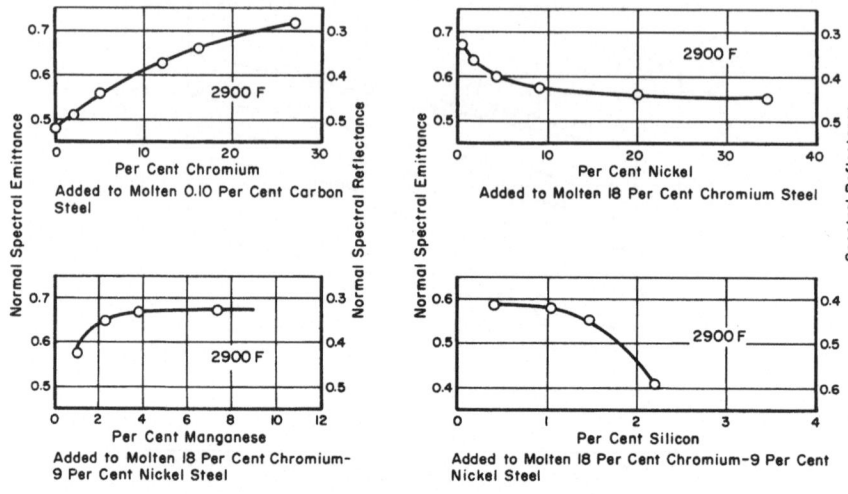

VARIATION IN THE NORMAL SPECTRAL EMITTANCE OF MOLTEN STEELS WITH ADDITIONS OF SELECTED ELEMENTS

VARIATION IN THE NORMAL SPECTRAL EMITTANCE OF MOLTEN STEELS WITH ADDITIONS
OF SELECTED ELEMENTS--REFERENCE INFORMATION

Reference	Investigator	Symbol	Composition and Surface Condition	Test Method	Remarks
1	Goller, G. N.	O	Molten. Measured at lip while pouring. (Analysis given on previous information sheet.)	Normal spectral emittance. Calibrated optical pyrometer. Immersion thermo- couple.	Measured in air. (No wavelength given for the Pyro optical pyrometer used.)

Note: The author shows the following effects
of increasing concentrations of various
alloying elements upon the emittance.

Element	Emittance
Cr	Increases
Mn	Increases
Ti	Increases
Cb	Increases
Ni	Decreases
Si	Decreases
S	Decreases
Se + P	Decreases
C	No appreciable effect
Mo	No appreciable effect

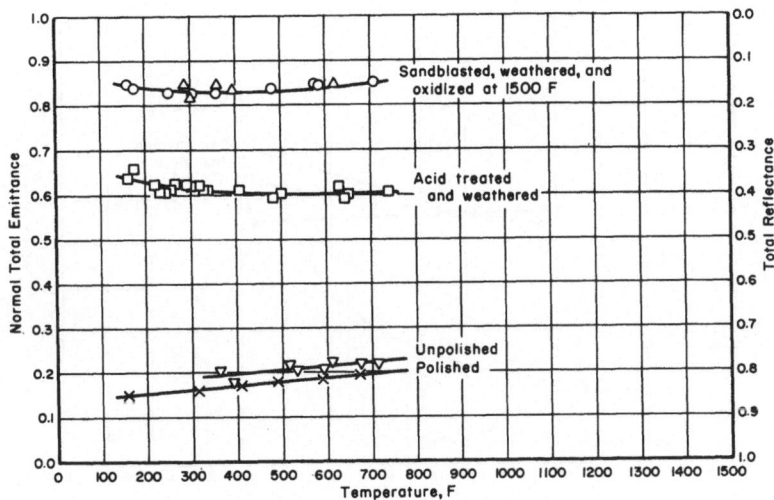

NORMAL TOTAL EMITTANCE OF STAINLESS STEEL TYPE 18-8

NORMAL TOTAL EMITTANCE OF STAINLESS STEEL TYPE 18-8--REFERENCE INFORMATION

Reference	Investigator	Symbol	Composition and Surface Condition	Test Method	Remarks
2	Snyder, Gier, and Dunkle	O	Nominal composition. Oxidized at 1500 F and weathered	Normal total emittance. Thermopile detector. Temperatures measured with thermocouples.	Measured in air. Data taken from curves.
		△	Sandblasted and weathered		
		□	Chromic and sulfuric acid treated		
		▽	Unpolished		
		×	Polished		

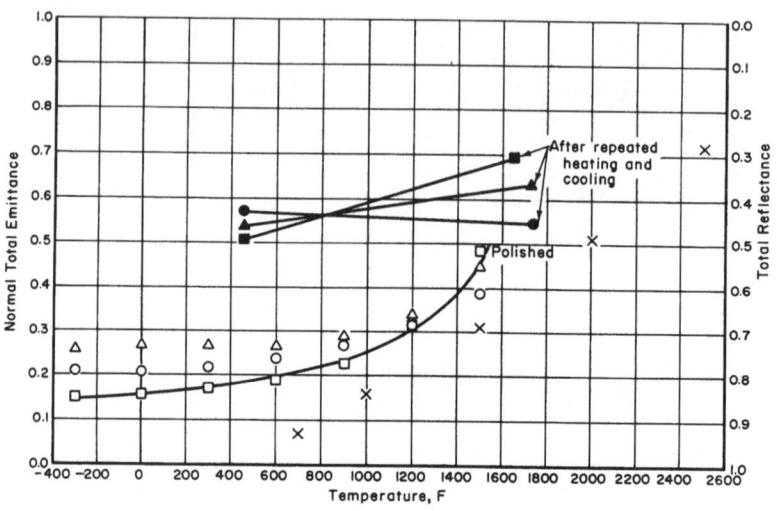

NORMAL TOTAL EMITTANCE OF STAINLESS STEEL TYPE 301

NORMAL TOTAL EMITTANCE OF STAINLESS STEEL TYPE 301--REFERENCE INFORMATION

Reference	Investigator	Symbol	Composition and Surface Condition	Test Method	Remarks
3	Wilkes, G. B.	O △ □	First Heating: As received Clean and smooth Polished	Normal total emittance. Total radiation detector. Comparison blackbody. Temperature measured with thermocouples.	Measured in 10 micron pressure of helium. Data taken from table.
		● ▲ ■	After Repeated Heating and Cooling: As received Clean and smooth Polished		
4	Anthony and Pearl	×	As received (Surface oxidation indicated after test)	Normal total emittance. Calibrated thermopile detector. Comparison blackbody. Temperatures measured with thermocouples.	Measured in flow of helium gas. Data taken from table.

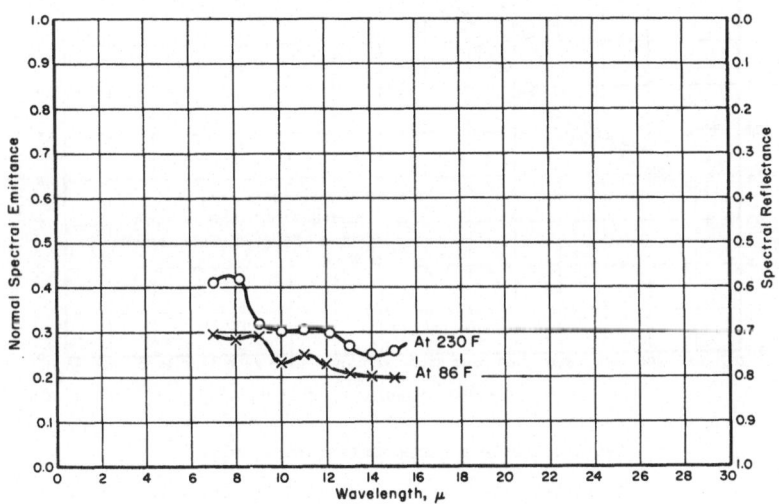

NORMAL SPECTRAL EMITTANCE OF STAINLESS STEEL TYPE 301

NORMAL SPECTRAL EMITTANCE OF STAINLESS STEEL TYPE 301--REFERENCE INFORMATION

Reference	Investigator	Symbol	Composition and Surface Condition	Test Method	Remarks
5	Weber, D.	O X	Measured at 230 F Measured at 86 F (Samples in as-received condition)	Normal spectral emittance. Infrared spectrometer. Comparison blackbody. Temperatures measured with thermocouples.	Measured in air at 230 and 86 F. Data taken from curves.

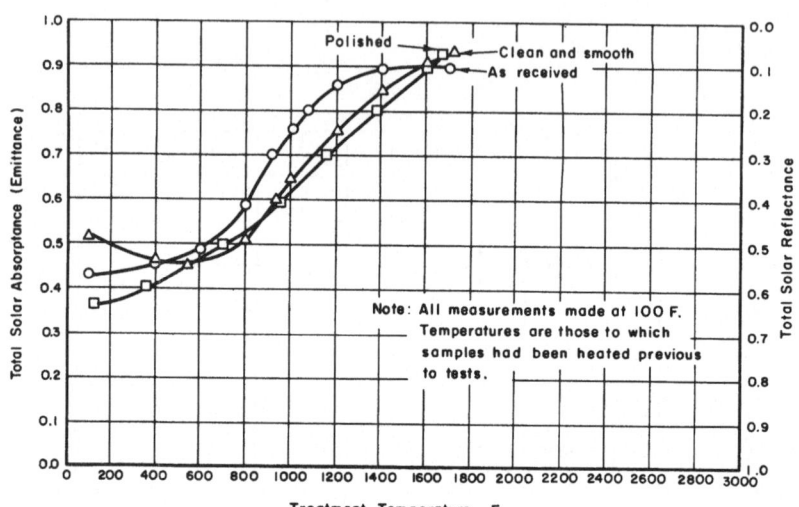

TOTAL SOLAR ABSORPTANCE OF STAINLESS STEEL TYPE 301 AT 100 F

TOTAL SOLAR ABSORPTANCE OF STAINLESS STEEL TYPE 301 AT 100 F--REFERENCE INFORMATION

Reference	Investigator	Symbol	Composition and Surface Condition	Test Method	Remarks
3	Wilkes, G. B.	□ △ ○	Polished Clean and smooth As received	Total solar absorptance. Comparison standards. Comparison pyroheliometer. Output measured with thermocouples.	Measured in air at 100 F. Temperatures shown are those to which samples had been heated previous to tests. Data taken from curves.

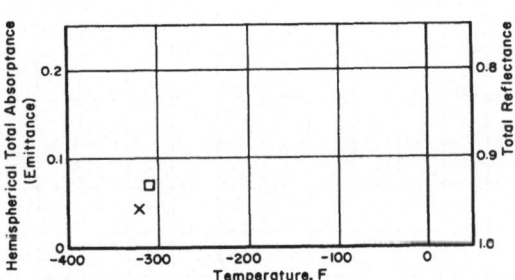

HEMISPHERICAL TOTAL ABSORPTANCE OF STAINLESS STEEL TYPE 302

HEMISPHERICAL TOTAL ABSORPTANCE OF STAINLESS STEEL TYPE 302--REFERENCE INFORMATION

Reference	Investigator	Symbol	Composition and Surface Condition	Test Method	Remarks
7	Fulk and Reynolds	□ X	Commercial ball type 0.005-inch-thick sheet	Hemispherical total absorptance. Calibrated calorimeter. Heat transfer measured by liquid nitrogen boil-off.	Measured in vacuum.

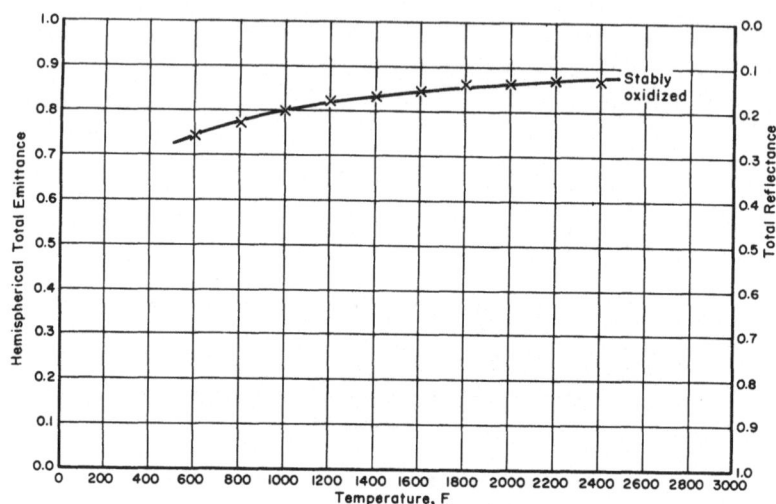

HEMISPHERICAL TOTAL EMITTANCE OF STAINLESS STEEL TYPE 303

HEMISPHERICAL TOTAL EMITTANCE OF STAINLESS STEEL TYPE 303--REFERENCE INFORMATION

Reference	Investigator	Symbol	Composition and Surface Condition	Test Method	Remarks
7	Wade, W. R.	✗	Stably oxidized 60 minutes at 2000 F	Normal total emittance. Total radiation pyrometer. Comparison blackbody. Temperatures measured with thermocouples.	Measured in air at various angles to the normal. Normal emittance equals hemispherical emittance for this specimen.

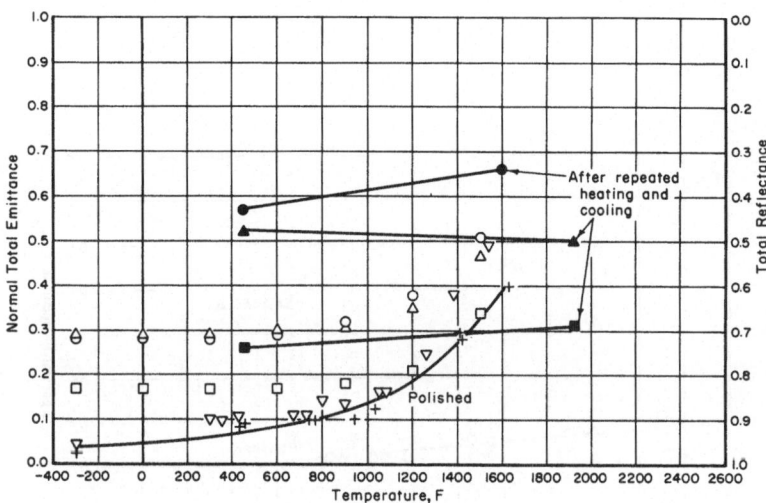

NORMAL TOTAL EMITTANCE OF STAINLESS STEEL TYPE 316

NORMAL TOTAL EMITTANCE OF STAINLESS STEEL TYPE 316--REFERENCE INFORMATION

Reference	Investigator	Symbol	Composition and Surface Condition	Test Method	Remarks
3	Wilkes, G. B.	O △ □	First Heating: As received Clean and smooth Polished	Normal total emittance. Total radiation de- tector. Comparison blackbody. Temperatures measured with thermocouples.	Measured in 10 micron pressure of helium. Data taken from table.
		● ▲ ■	After Repeated Heating and Cooling: As received Clean and smooth Polished		
8	Betz, Olson, Schurin, and Morris	▽ +	Rms finish of approxi- mately 15 microinches Rms finish of approxi- mately 2 microinches.	Normal total emittance. Thermistor-bolometer detector. Resistance heated strip specimens. Comparison blackbody. Temperatures measured with thermocouples.	Measured in vacuum. Data taken from curves.

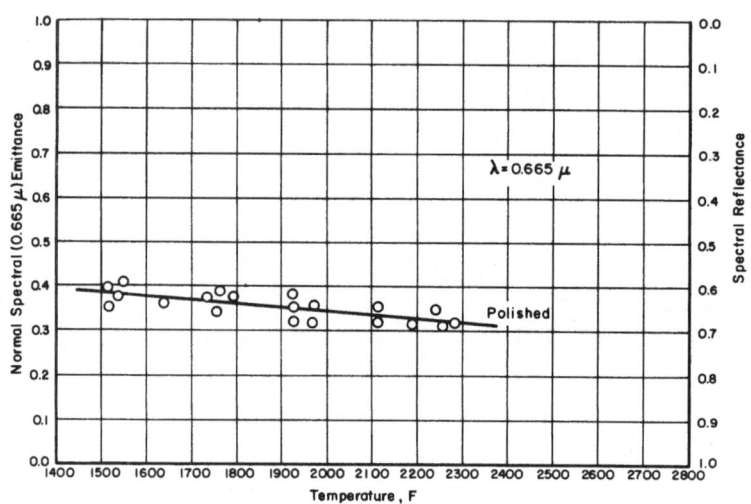

NORMAL SPECTRAL EMITTANCE OF STAINLESS STEEL TYPE 316

NORMAL SPECTRAL EMITTANCE OF STAINLESS STEEL TYPE 316--REFERENCE INFORMATION

Reference	Investigator	Symbol	Composition and Surface Condition	Test Method	Remarks
8	Betz, Olson, Schurin, and Morris	O	Polished Combined 15 and 2 rms microinch finishes	Normal spectral emittance. Hole-in-tube method. Optical pyrometer.	Measured in vacuum. Data taken from curves. (λ = 0.665 μ)

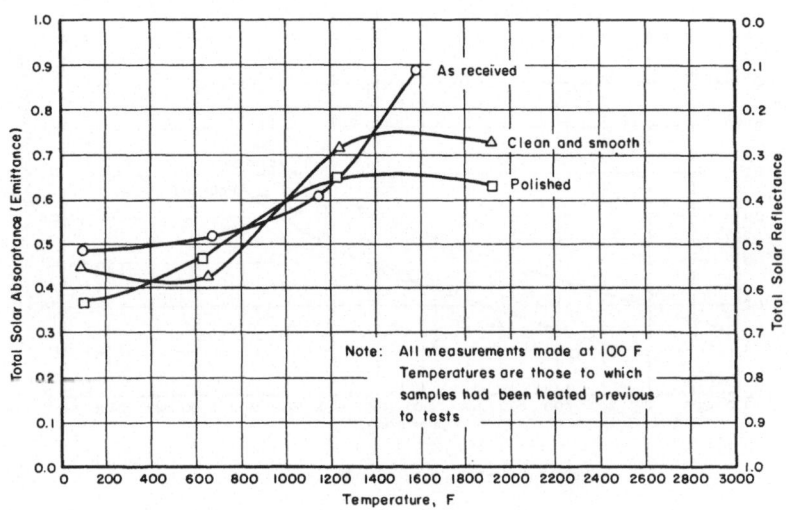

TOTAL SOLAR ABSORPTANCE OF STAINLESS STEEL TYPE 316 AT 100 F

TOTAL SOLAR ABSORPTANCE OF STAINLESS STEEL TYPE 316 AT 100 F--REFERENCE INFORMATION

Reference	Investigator	Symbol	Composition and Surface Condition	Test Method	Remarks
3	Wilkes, G. B.	□ △ ○	Polished Clean and smooth As received	Total solar absorptance. Comparison standards. Comparison pyro- heliometer. Output measured with thermocouples.	Measured in air at 100 F. Temperatures shown are those to which samples had been heated previous to tests. Data taken from table.

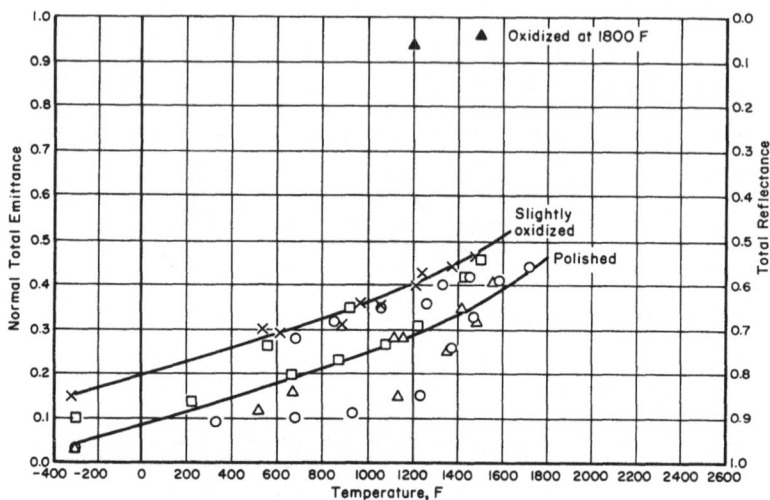

NORMAL TOTAL EMITTANCE OF STAINLESS STEEL TYPE 321

NORMAL TOTAL EMITTANCE OF STAINLESS STEEL TYPE 321--REFERENCE INFORMATION

Reference	Investigator	Symbol	Composition and Surface Condition	Test Method	Remarks
8	Betz, Olson, Schurin, and Morris	O △ ☐ X	Finish – 2 microinches rms Finish – not given, except (No. 2 bright) Finish – 6 microinches rms Finish – 6 microinches rms Oxidized 30 min at red heat	Normal total emittance. Thermistor-bolometer detector. Resistance heated strip specimens. Comparison blackbody. Temperatures measured with thermocouples.	Measured in vacuum. Data taken from curves.
9	Douglas	▲	Oxidized 15 minutes at 1800 F	Normal total emittance. Rotating sample in blackbody furnace. Total radiation pyro- meter. Temperatures measured with thermocouples.	Measured in air. Data taken from table.

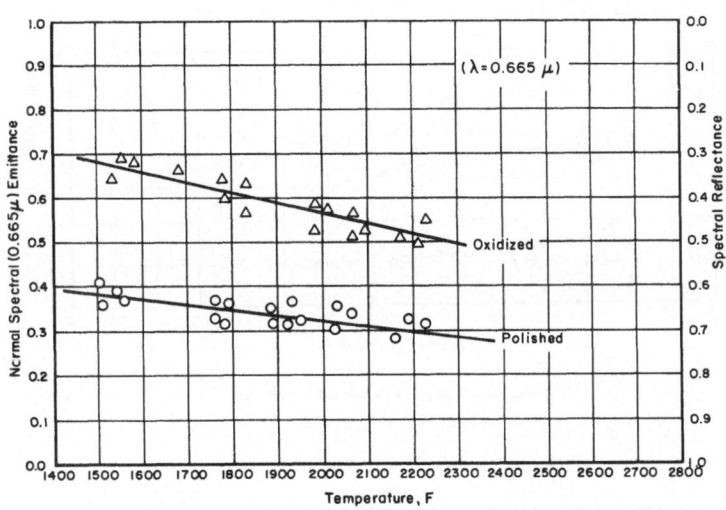

NORMAL SPECTRAL EMITTANCE OF STAINLESS STEEL TYPE 321

NORMAL SPECTRAL EMITTANCE OF STAINLESS STEEL TYPE 321--REFERENCE INFORMATION

Reference	Investigator	Symbol	Composition and Surface Condition	Test Method	Remarks
8	Betz, Olson, Schurin, and Morris	O	2 to 6 microinches rms finish	Normal spectral emittance.	Measured in vacuum.
		△	6 microinch rms finish oxidized 30 min at red heat	Hole-in-tube method. Optical pyrometer.	Data taken from curves. ($\lambda = 0.665\mu$)

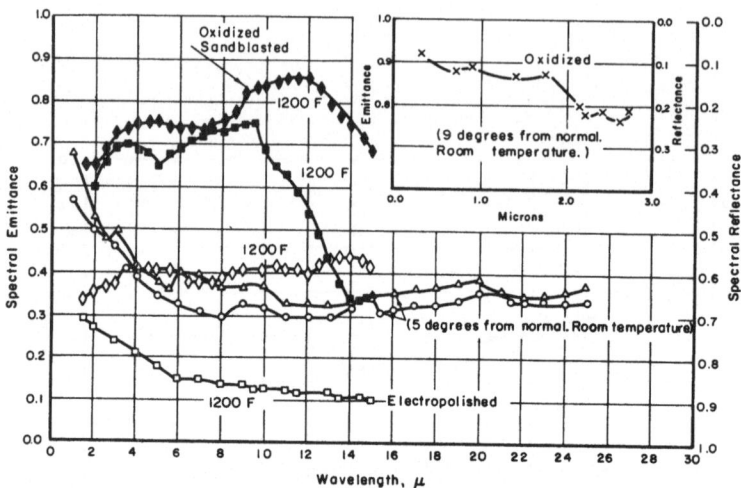

SPECTRAL EMITTANCE OF STAINLESS STEEL TYPE 321

SPECTRAL EMITTANCE OF STAINLESS STEEL TYPE 321--REFERENCE INFORMATION

Reference	Investigator	Symbol	Composition and Surface Condition	Test Method	Remarks
10	Olson and Morris	X	Oxidized	Spectral reflectance at 9 degrees from the normal. Monochromator, integrating sphere reflectometer, and lead sulphide detector. "Normal" illumination, hemispherical viewing.	Measured in air at room temperature. Data taken from reflectance curve.
11	Bevans, Gier, and Dunkle	△ ○	No thermal treatment 1000 hours at 705 F	Spectral reflectance at 5 degrees from the normal. Gier-Dunkle reflectometer. Monochromator. Temperatures measured with thermocouples. Diffuse illumination, "normal" viewing.	Measured in air. Data taken from reflectance curves.
12	Richmond and Stewart	□ ◇ ■ ◆	Electropolished Sandblasted Electropolished, oxidized 1/2 hour at 1800 F Sandblasted, oxidized 1/2 hour at 1800 F (All measurements at 1200 F)	Normal spectral emittance. Recording, double-beam spectrophotometer. Comparison blackbody. Temperatures measured with thermocouples.	Measured in air. Data taken from table. Measured at 1200 F.

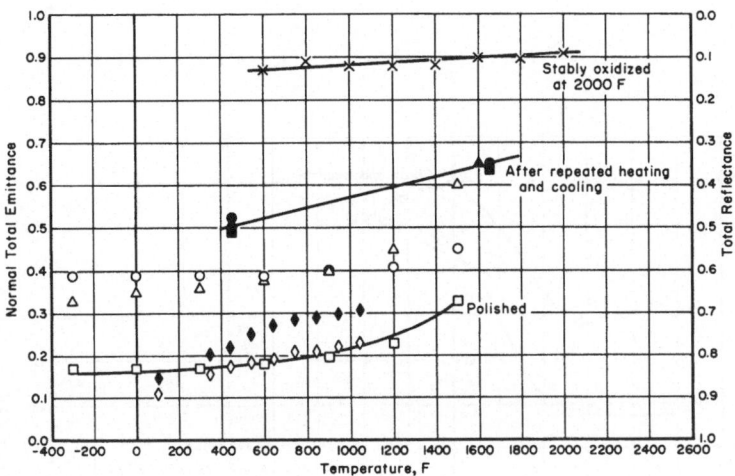

NORMAL TOTAL EMITTANCE OF STAINLESS STEEL TYPE 347

NORMAL TOTAL EMITTANCE OF STAINLESS STEEL TYPE 347—REFERENCE INFORMATION

Reference	Investigator	Symbol	Composition and Surface Condition	Test Method	Remarks
3	Wilkes, G. B.	O △ □ ● ▲ ■	First Heating: As received Clean and smooth Polished After Repeated Heating and Cooling: As received Clean and smooth Polished	Normal total emittance. Total radiation detector. Comparison blackbody. Temperatures measured with thermocouples.	Measured in 10 micron pressure of helium. Data taken from table.
7	Wade, W. R.	X	Stably oxidized at 2000 F	Normal total emittance. Thermopile detector. Comparison blackbody. Temperatures measured with thermocouples.	Measured in air. Measurements were taken normally and at various angles with the normal. Normal total equals hemispherical total emittance for this specimen. Data taken from curves.
13	Dobbins, J. P.	◊ ♦	Bare (polished) Oxidized blue (100 hours at 1200 F)	Normal total emittance. Calibrated thermopile detector. Temperatures measured with thermocouples. Blackbody radiation calculated from known emittance of "secondary standard".	Measured in air. Data taken from curves.

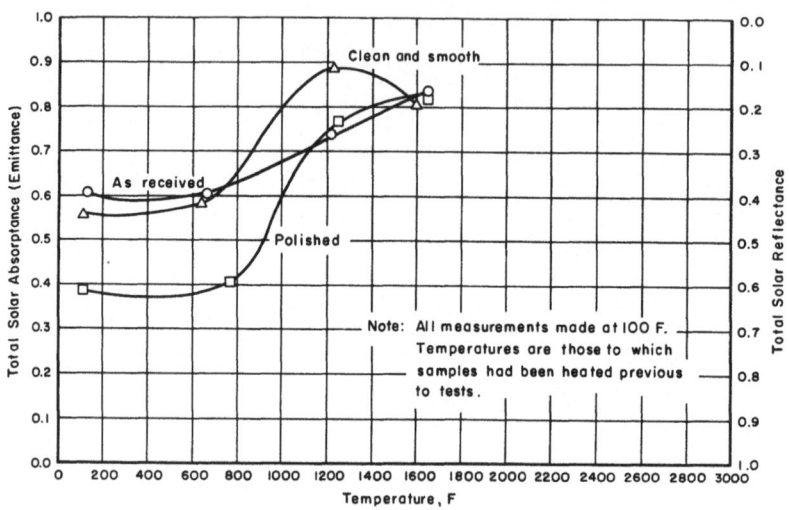

TOTAL SOLAR ABSORPTANCE OF STAINLESS STEEL TYPE 347 AT 100 F

TOTAL SOLAR ABSORPTANCE OF STAINLESS STEEL TYPE 347 AT 100 F--REFERENCE INFORMATION

Reference	Investigators	Symbol	Composition and Surface Condition	Test Method	Remarks
3	Wilkes, G. B.	○ ● △ □	As received Clean and smooth Polished	Total solar absorptance. Comparison standards. Comparison pyro- heliometer. Output measured with thermocouples.	Measured in air at 100 F. Temperatures shown are those to which samples had been heated previous to tests. Data taken from table.

Surface Treatment	Absorptance , per cent	
	Solar Radiation	Terrestrial Radiation
Untreated	74.2	40 I
Fine sandblasted	72.4	49.8
Fine sandblasted , heated in air to 600 F	80.7	43.2

TOTAL SOLAR AND TERRESTRIAL ABSORPTANCE OF STAINLESS STEEL TYPE 410

TOTAL SOLAR AND TERRESTRIAL ABSORPTANCE OF STAINLESS STEEL TYPE 410—REFERENCE INFORMATION

Reference	Investigator	Symbol	Composition and Surface Condition	Test Method	Remarks
14	Shipley and Thostesen		As shown in table	Calculated from reflectance data and the known energy distribution curve for solar radiation (or for a 250 K blackbody for terrestrial radiation) outside the earth's atmosphere. Graphical integration. Reflectance data were extrapolated with relatively small error.	Reflectance data obtained from the University of California (Berkeley).

Surface Treatment	Absorptance, per cent	
	Solar Radiation	Terrestial Radiation
Etched	62.3	17.0
Fine sandblasted, heated in air to 600 F	73.2	39.3
Etched and heated in air to 600 F	63.9	15.9
Lightly abraded with 600-mesh aluminum, by hand	59.9	14.3

TOTAL SOLAR AND TERRESTRIAL ABSORPTANCE OF STAINLESS STEEL TYPE 430

TOTAL SOLAR AND TERRESTRIAL ABSORPTANCE OF STAINLESS STEEL TYPE 430--REFERENCE INFORMATION

Reference	Investigator	Symbol	Composition and Surface Condition	Test Method	Remarks
14	Shipley and Thostesen		As shown in table	Calculated from reflectance data and the known energy distribution curve for solar radiation (or terrestrial radiation) outside the earth's atmosphere. Graphical integration. Reflectance data were extrapolated with relatively small error.	Reflectance data obtained from the University of California (Berkeley).

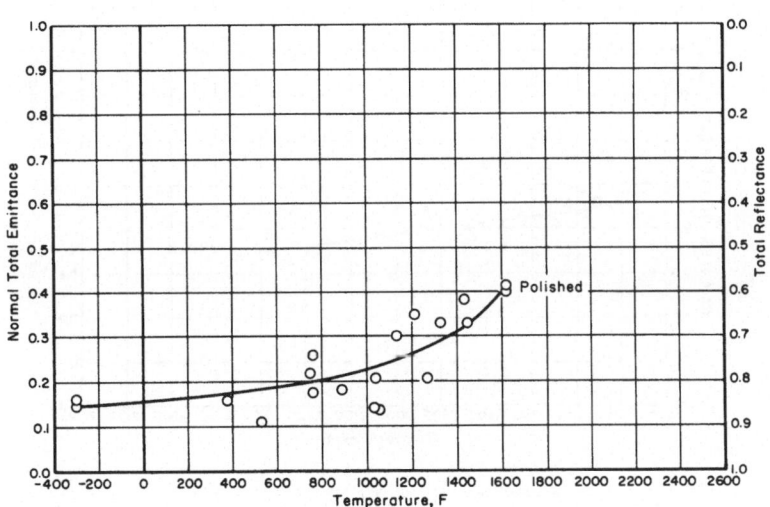

NORMAL TOTAL EMITTANCE OF STAINLESS STEEL TYPE 446

NORMAL TOTAL EMITTANCE OF STAINLESS STEEL TYPE 446—REFERENCE INFORMATION

Reference	Investigator	Symbol	Composition and Surface Condition	Test Method	Remarks
8	Betz, Olson, Schurin, and Morris	O	Polished: surface finished to 15 and 2 microinches rms	Normal total emittance. Thermistor-bolometer detector. Resistance-heated strip specimens. Comparison black-body. Temperatures measured with thermocouples.	Measured in vacuum. Data taken from curves.

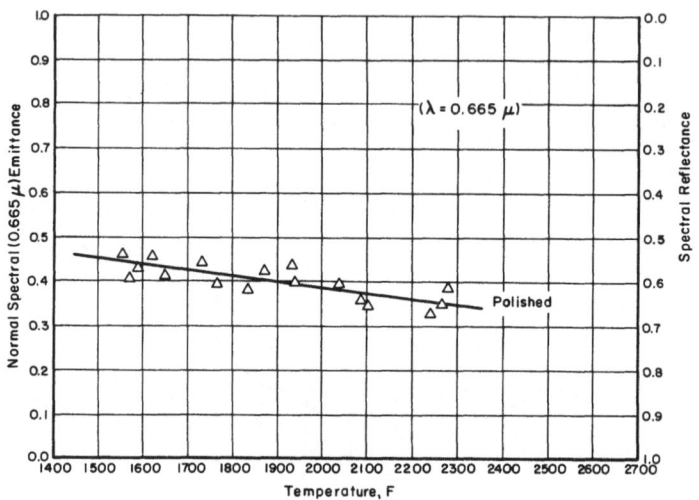

NORMAL SPECTRAL EMITTANCE OF STAINLESS STEEL TYPE 446

NORMAL SPECTRAL EMITTANCE OF STAINLESS STEEL TYPE 446--REFERENCE INFORMATION

Reference	Investigator	Symbol	Composition and Surface Condition	Test Method	Remarks
8	Betz, Olson, Schurin, and Morris	△	Polished: surface finished to 2 and 15 microinches rms	Normal spectral emittance. Hole-in-tube method. Optical pyrometer.	Measured in vacuum. ($\lambda = 0.665 \mu$) Data taken from curves.

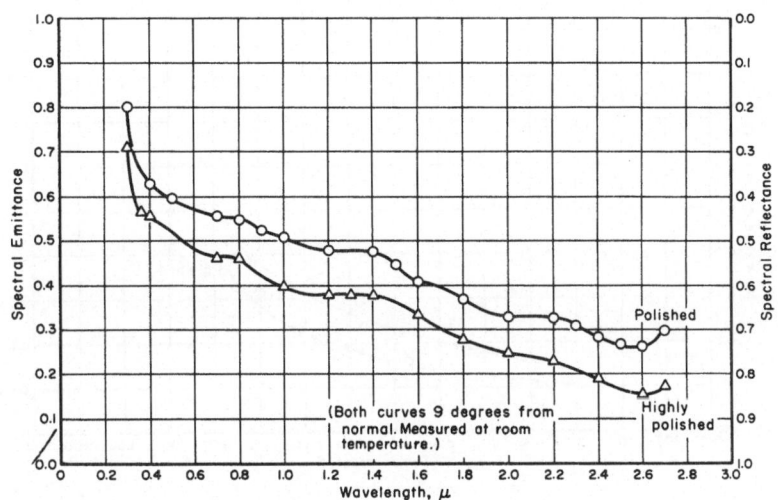

SPECTRAL EMITTANCE OF STAINLESS STEEL TYPE 446

SPECTRAL EMITTANCE OF STAINLESS STEEL TYPE 446—REFERENCE INFORMATION

Reference	Investigator	Symbol	Composition and Surface Condition	Test Method	Remarks
8	Betz, Olson, Schurin, and Morris	O △	Polished: 15 microinches rms Polished: 2 microinches rms	Spectral reflectance at 9 degrees to the normal. Monochromator, integrating sphere reflectometer, and lead sulfide detector. "Normal" illumination, hemispherical viewing.	Measured in air at room temperature. Data taken from reflectance curves.

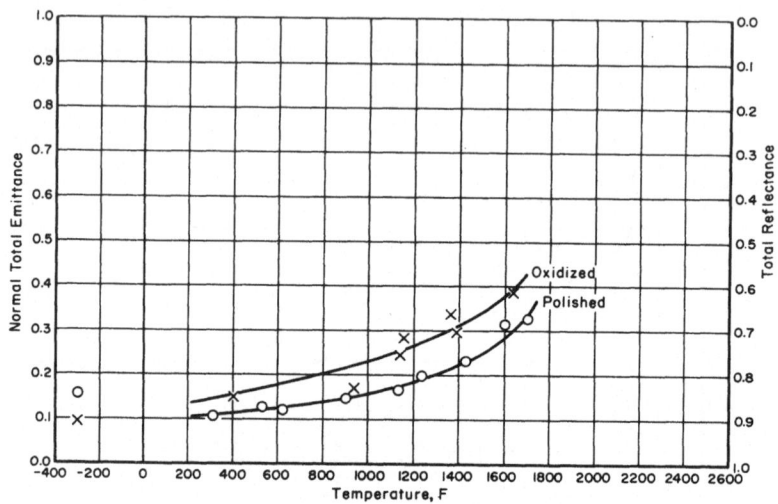

NORMAL TOTAL EMITTANCE OF STAINLESS STEEL TYPE AM-350

NORMAL TOTAL EMITTANCE OF STAINLESS STEEL TYPE AM-350--REFERENCE INFORMATION

Reference	Investigator	Symbol	Composition and Surface Condition	Test Method	Remarks
8	Betz, Olson, Schurin, and Morris	O	Polished: 2 microinches rms	Normal total emittance.	Measured in vacuum.
		X	2 microinches rms - oxidized	Thermistor-bolometer detector. Resistance-heated strip specimens. Comparison blackbody. Temperatures measured with thermocouples.	Data taken from curves.

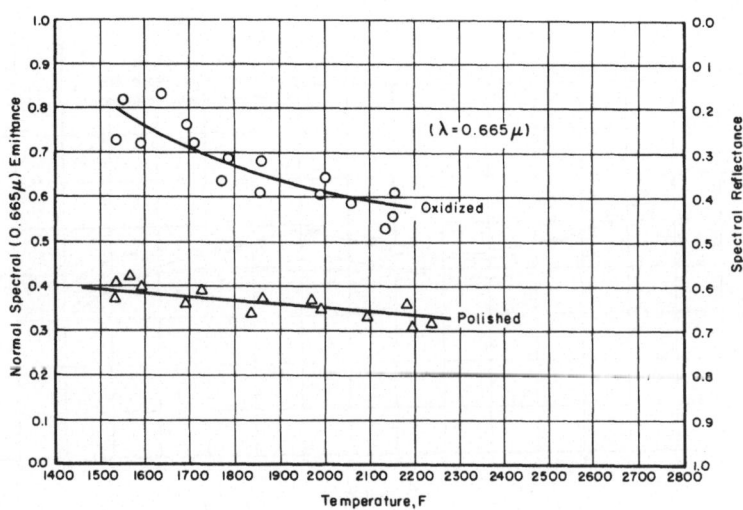

NORMAL SPECTRAL EMITTANCE OF STAINLESS STEEL TYPE AM-350

NORMAL SPECTRAL EMITTANCE OF STAINLESS STEEL TYPE AM-350—REFERENCE INFORMATION

Reference	Investigator	Symbol	Composition and Surface Condition	Test Method	Remarks
8	Betz, Olson, Schurin, and Morris	△	Polished: surface finished to 15 and 2 microinches rms	Normal spectral emittance.	Measured in vacuum.
		○	2 microinch surfaces oxidized	Hole-in-tube method. Optical pyrometer.	(λ = 0.665 μ) Data taken from curves.

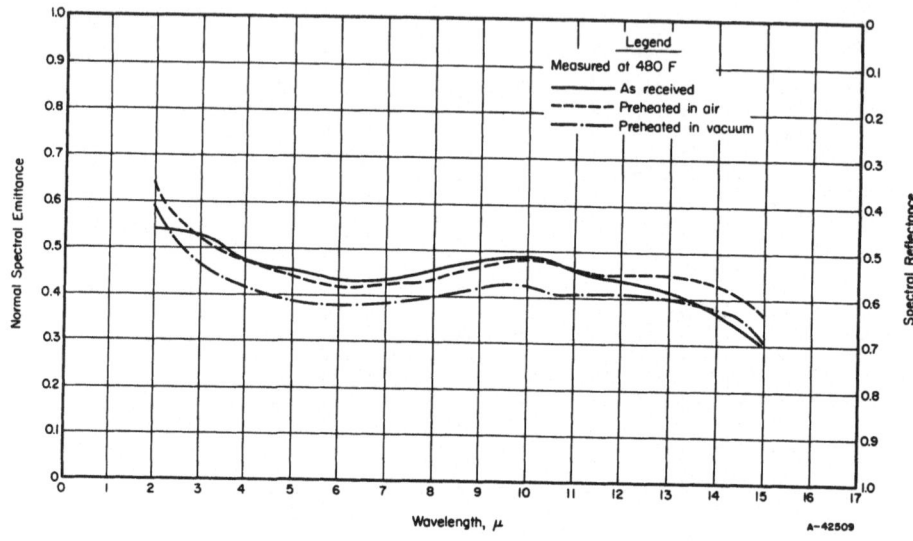

NORMAL SPECTRAL EMITTANCE OF STAINLESS STEEL TYPE AM-350 AT 480 F

NORMAL SPECTRAL EMITTANCE OF STAINLESS STEEL TYPE AM-350 AT 480 F--REFERENCE INFORMATION

Reference	Investigator	Symbol	Composition and Surface Condition	Test Method	Remarks
16	Adams, J. G.		As received Heated 30 minutes in air at 900 F Heated 30 minutes in 4.4 x 10^{-5} mm Hg pressure at 900 F	Normal spectral emittance. Furnace-heated disk specimen. Comparison blackbody (Hohlraun). Spectrometer-monochromator with photomultiplier, lead sulphide, and thermocouple detectors. Temperatures measured with thermocouples.	Measured in air.

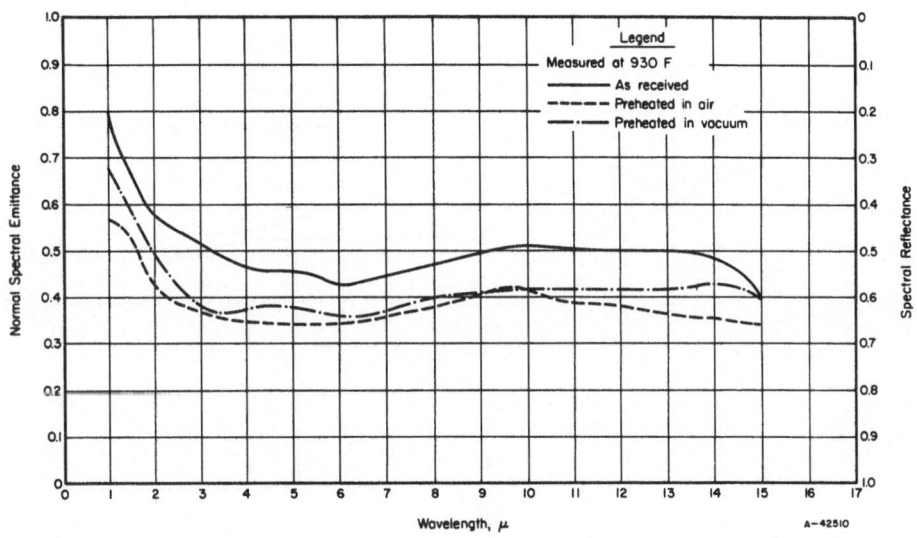

NORMAL SPECTRAL EMITTANCE OF STAINLESS STEEL TYPE AM-350 AT 930 F

NORMAL SPECTRAL EMITTANCE OF STAINLESS STEEL TYPE AM-350 AT 930 F--REFERENCE INFORMATION

Reference	Investigator	Symbol	Composition and Surface Condition	Test Method	Remarks
16	Adams, J. G.		As received Heated 30 minutes in air at 900 F Heated 30 minutes in 4.4 x 10^{-5} mm Hg pressure at 900 F	Normal spectral emittance. Furnace-heated disk specimen. Comparison blackbody (Hohlraun). Spectrometer-monochromator with photomultiplier, lead sulphide, and thermocouple detectors. Temperatures measured with thermocouples.	Measured in air.

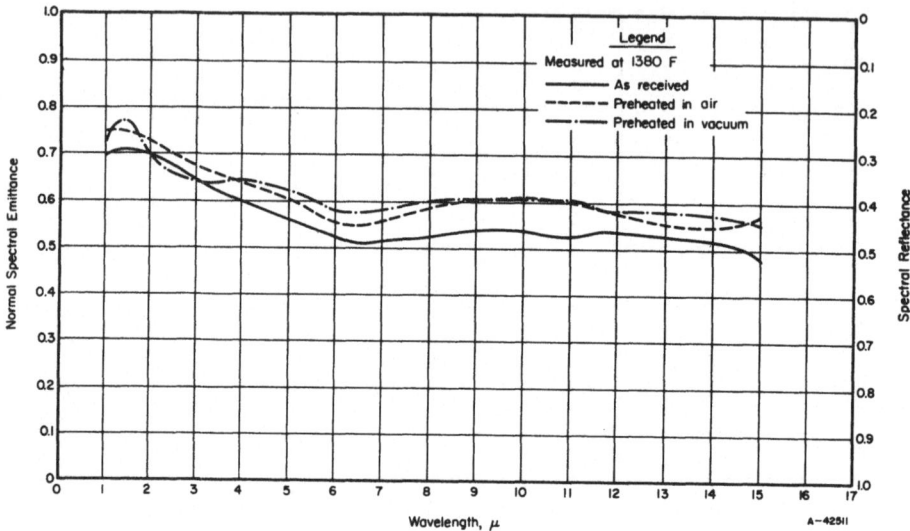

NORMAL SPECTRAL EMITTANCE OF STAINLESS STEEL TYPE AM-350 AT 1380 F

NORMAL SPECTRAL EMITTANCE OF STAINLESS STEEL TYPE AM-350 AT 1380 F--REFERENCE INFORMATION

Reference	Investigator	Symbol	Composition and Surface Condition	Test Method	Remarks
16	Adams, J. G.		As received Heated 30 minutes in air at 900 F Heated 30 minutes in 4.4 x 10^{-5} mm Hg pressure at 900 F	Normal spectral emittance. Furnace-heated disk specimen. Comparison blackbody (Hohlraun). Spectrometer-monochromator with photomultiplier, lead sulphide, and thermocouple detectors. Temperatures measured with thermocouples.	Measured in air.

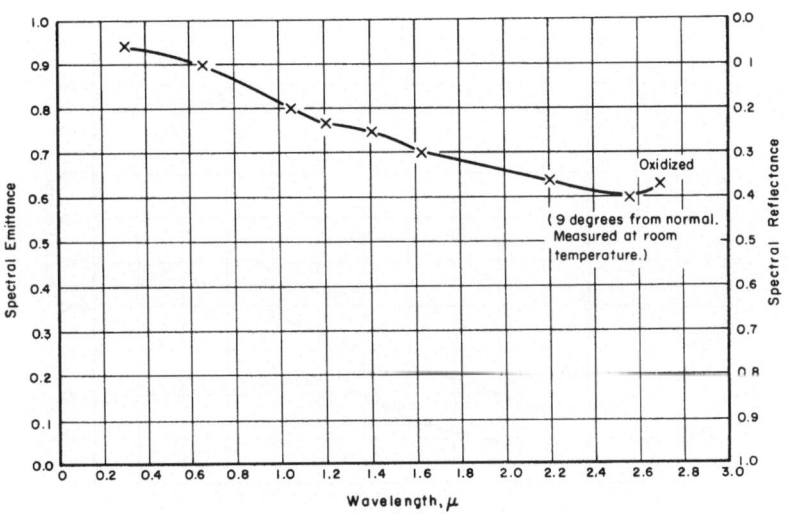

SPECTRAL EMITTANCE OF STAINLESS STEEL TYPE AM-350

SPECTRAL EMITTANCE OF STAINLESS STEEL TYPE AM-350--REFERENCE INFORMATION

Reference	Investigator	Symbol	Composition and Surface Condition	Test Method	Remarks
10	Olson and Morris	X	Oxidized	Spectral reflectance at 9 degrees from the normal. Monochromator, integrating sphere reflectometer, and lead sulfide detector. "Normal" illumination, hemispherical viewing.	Measured in air at room temperature. Data taken from reflectance curve.

80

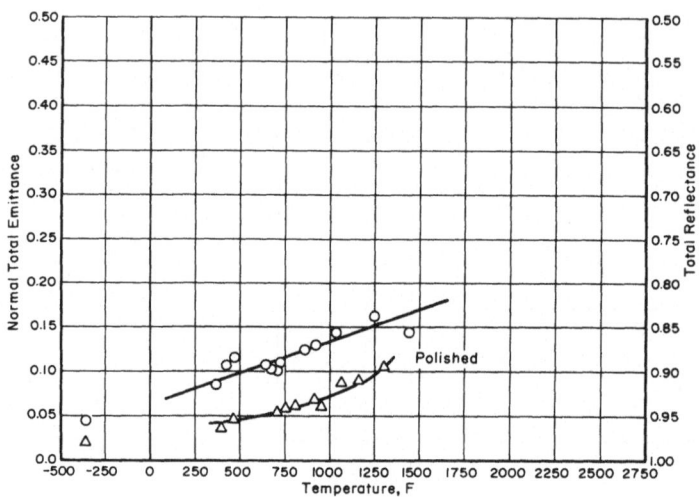

NORMAL TOTAL EMITTANCE OF STAINLESS STEEL TYPE 17-7 PH

NORMAL TOTAL EMITTANCE OF STAINLESS STEEL TYPE 17-7 PH--REFERENCE INFORMATION

Reference	Investigator	Symbol	Composition and Surface Condition	Test Method	Remarks
8	Betz, Olson, Schurin, and Morris	△	Highly polished: 2 micro-inches rms	Normal total emittance.	Measured in vacuum.
		○	Polished: 15 microinches rms	Thermistor-bolometer detector. Resistance-heated strip specimens. Comparison black-body. Temperatures measured with thermocouples.	Data taken from curves.

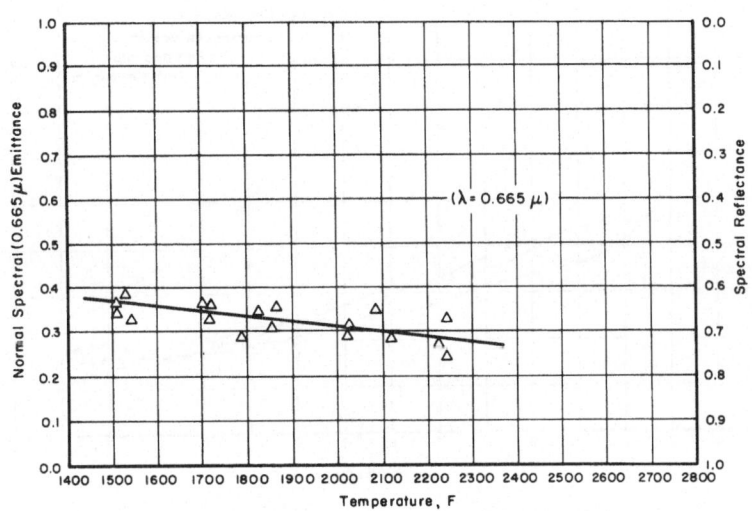

NORMAL SPECTRAL EMITTANCE OF STAINLESS STEEL TYPE 17-7 PH

NORMAL SPECTRAL EMITTANCE OF STAINLESS STEEL TYPE 17-7 PH--REFERENCE INFORMATION

Reference	Investigator	Symbol	Composition Surface Condition	Test Method	Remarks
8	Betz, Olson, Schurin, and Morris	△	Polished: either 2 or 15 microinches rms	Normal spectral emittance. Hole-in-tube method. Optical pyrometer.	Measured in vacuum. (λ = 0.665μ) Data taken from curves.

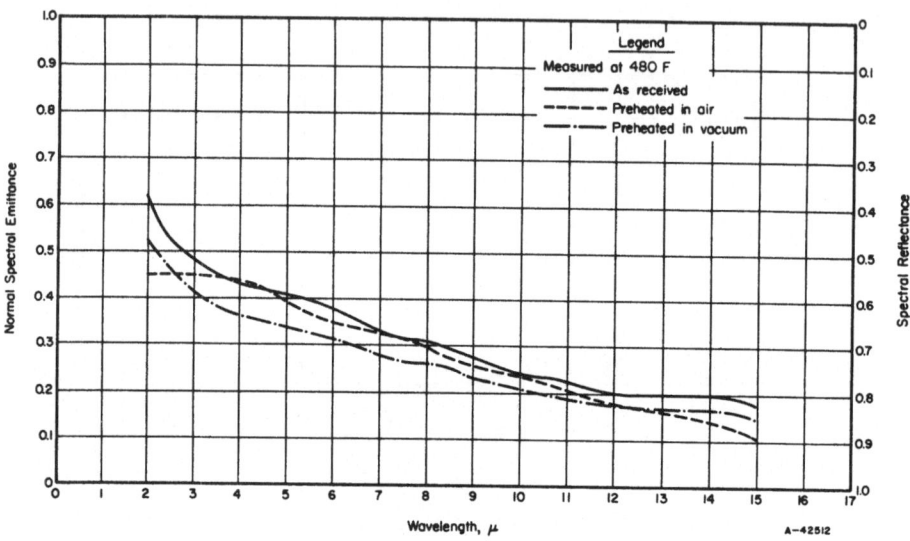

NORMAL SPECTRAL EMITTANCE OF STAINLESS STEEL TYPE 17-7PH AT 480 F

NORMAL SPECTRAL EMITTANCE OF STAINLESS STEEL TYPE 17-7PH AT 480 F--REFERENCE INFORMATION

Reference	Investigator	Symbol	Composition and Surface Condition	Test Method	Remarks
16	Adams, J. G.		As received Heated 30 minutes in air at 900 F Heated 30 minutes in 4.4 x 10^{-5} mm Hg pressure at 900 F	Normal spectral emittance. Furnace-heated disk specimen. Comparison blackbody (Hohlraun). Spectrometer-monochromator with photomultiplier, lead sulphide, and thermocouple detectors. Temperatures measured with thermocouples.	Measured in air.

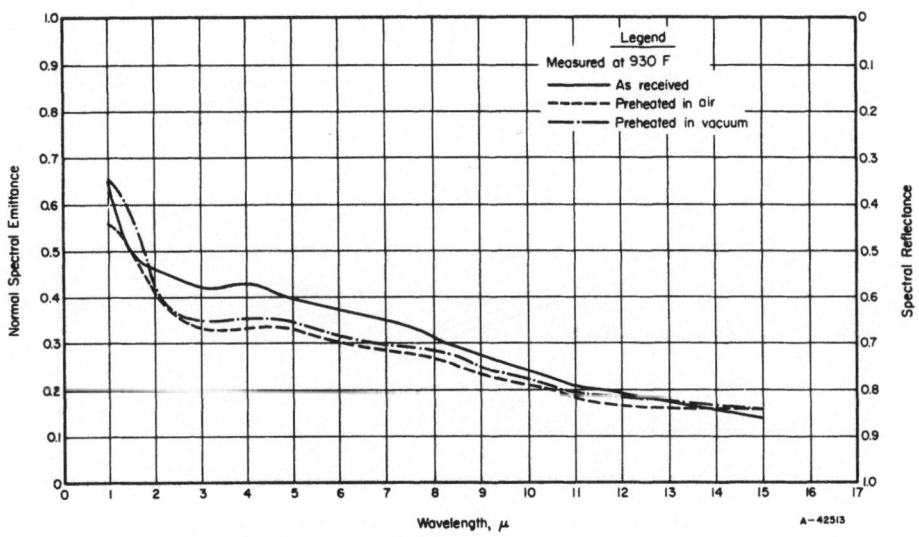

NORMAL SPECTRAL EMITTANCE OF STAINLESS STEEL TYPE 17-7PH AT 930 F

NORMAL SPECTRAL EMITTANCE OF STAINLESS STEEL TYPE 17-7PH AT 930 F--REFERENCE INFORMATION

Reference	Investigator	Symbol	Composition and Surface Condition	Test Method	Remarks
16	Adams, J. G.		As received Heated 30 minutes in air at 900 F Heated 30 minutes in 4.4×10^{-5} mm Hg pressure at 900 F	Normal spectral emittance. Furnace-heated disk specimen. Comparison blackbody (Hohlraun). Spectrometer-monochromator with photomultiplier, lead sulphide, and thermocouple detectors. Temperatures measured with thermocouples.	Measured in air.

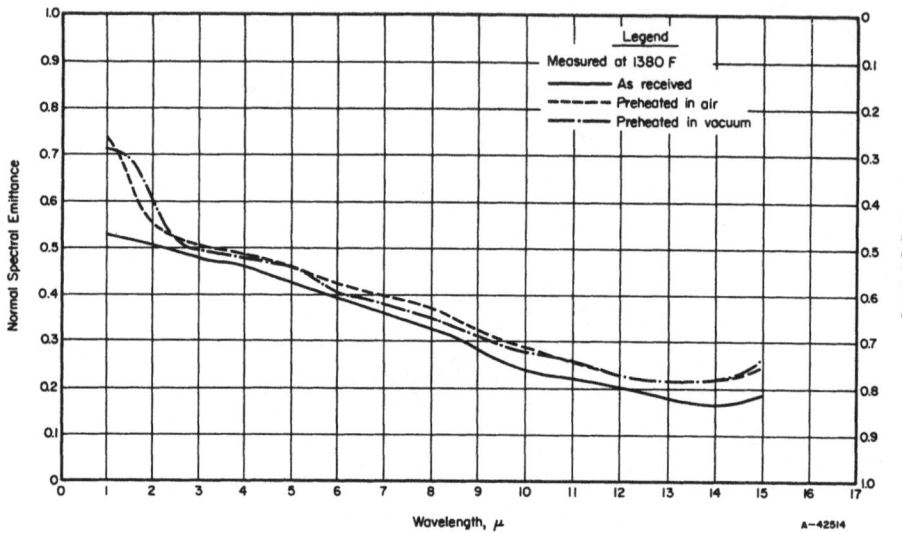

NORMAL SPECTRAL EMITTANCE OF STAINLESS STEEL TYPE 17-7PH AT 1380 F

NORMAL SPECTRAL EMITTANCE OF STAINLESS STEEL TYPE 17-7PH AT 1380 F—REFERENCE INFORMATION

Reference	Investigator	Symbol	Composition and Surface Condition	Test Method	Remarks
16	Adams, J. G.		As received Heated 30 minutes in air at 900 F Heated 30 minutes in 4.4 x 10^{-5} mm Hg pressure at 900 F	Normal spectral emittance. Furnace-heated disk specimen. Comparison blackbody (Hohlraun). Spectrometer-monochromator with photomultiplier, lead sulphide, and thermocouple detectors. Temperatures measured with thermocouples.	Measured in air.

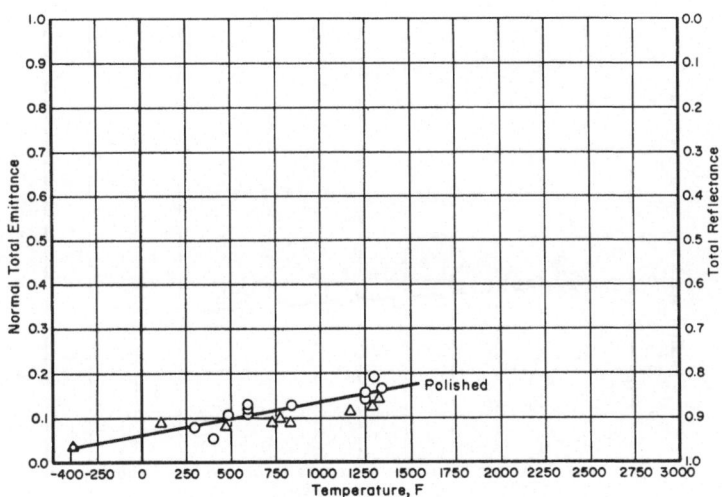

NORMAL TOTAL EMITTANCE OF STAINLESS STEEL TYPE PH 15-7 Mo

NORMAL TOTAL EMITTANCE OF STAINLESS STEEL TYPE PH 15-7 Mo—REFERENCE INFORMATION

Reference	Investigator	Symbol	Composition and Surface Condition	Test Method	Remarks
8	Betz, Olson, Schurin, and Morris	△ ○	Highly polished: 2 microinches rms Polished: 15 microinches rms	Normal total emittance. Thermistor-bolometer detector. Resistance-heated strip specimens. Comparison blackbody. Temperatures measured with thermocouples.	Measured in vacuum. Data taken from curves.

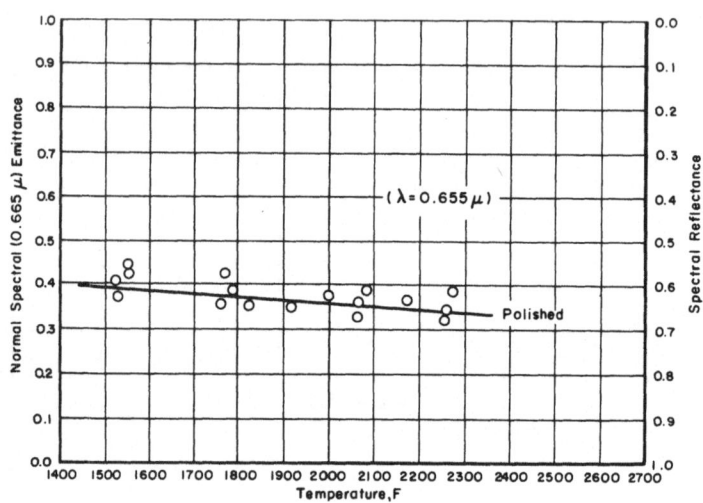

NORMAL SPECTRAL EMITTANCE OF STAINLESS STEEL TYPE PH 15-7 Mo

NORMAL SPECTRAL EMITTANCE OF STAINLESS STEEL TYPE PH 15-7 Mo--REFERENCE INFORMATION

Reference	Investigator	Symbol	Composition and Surface Condition	Test Method	Remarks
8	Betz, Olson, Schurin, and Morris	O	Polished: either 2 or 15 microinches rms	Normal spectral emittance. Hole-in-tube method. Optical pyrometer.	Measured in vacuum. ($\lambda = 0.665\mu$) Data taken from curves.

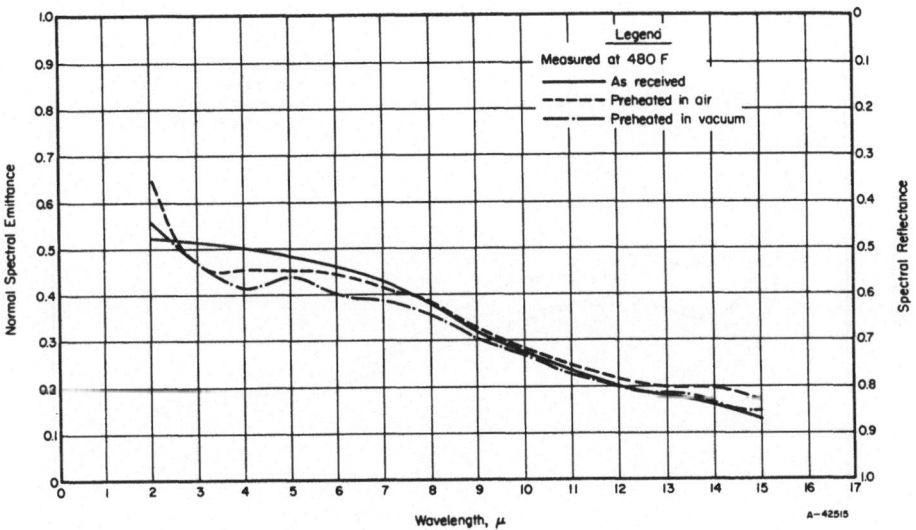

NORMAL SPECTRAL EMITTANCE OF PH 15-7 Mo STEEL AT 480 F

NORMAL SPECTRAL EMITTANCE OF STAINLESS STEEL TYPE PH 15-7 Mo AT 480 F--REFERENCE INFORMATION

Reference	Investigator	Symbol	Composition and Surface Condition	Test Method	Remarks
16	Adams, J. G.		As received Heated 30 minutes in air at 900 F Heated 30 minutes in 4.4 x 10^{-5} mm Hg pressure at 900 F	Normal spectral emittance. Furnace-heated disk specimen. Comparison blackbody (Hohlraun). Spectrometer-monochromator with photomultiplier, lead sulphide, and thermocouple detectors. Temperatures measured with thermocouples.	Measured in air.

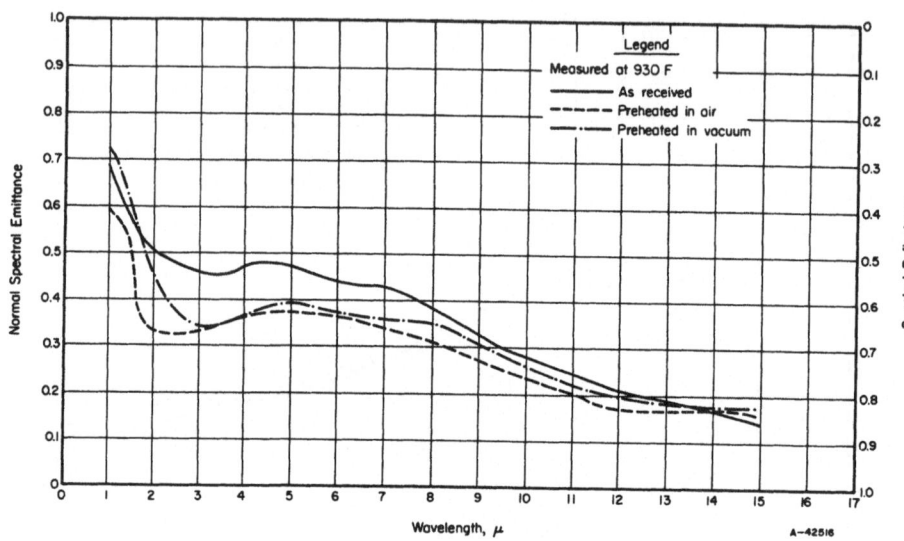

NORMAL SPECTRAL EMITTANCE OF STAINLESS STEEL TYPE PH 15-7 Mo AT 930 F

NORMAL SPECTRAL EMITTANCE OF STAINLESS STEEL TYPE PH 15-7 Mo AT 930 F—REFERENCE INFORMATION

Reference	Investigator	Symbol	Composition and Surface Condition	Test Method	Remarks
16	Adams, J. G.		As received Heated 30 minutes in air at 900 F Heated 30 minutes in 4.4 x 10⁻⁵ mm Hg pressure at 900 F	Normal spectral emittance. Furnace-heated disk specimen. Comparison blackbody (Hohlraun). Spectrometer-monochromator with photomultiplier, lead sulphide, and thermocouple detectors. Temperatures measured with thermocouples.	Measured in air.

89

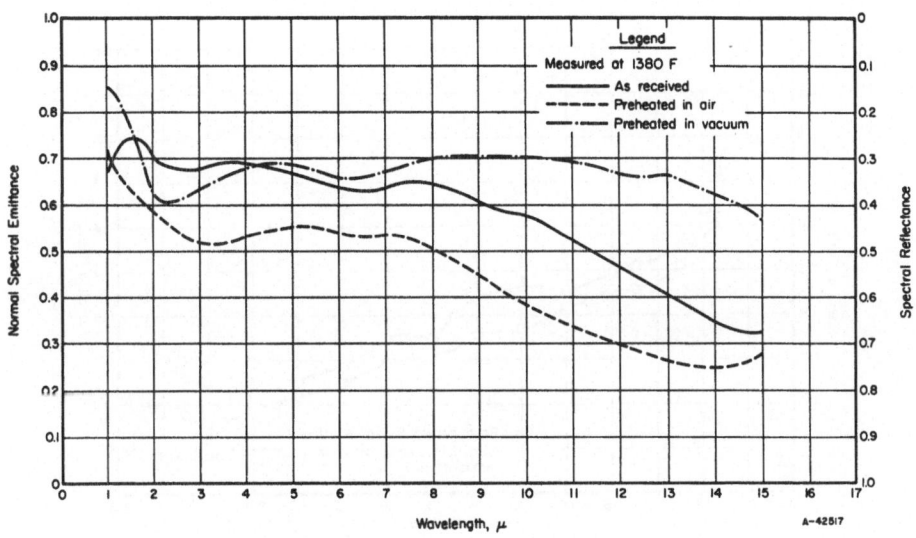

NORMAL SPECTRAL EMITTANCE OF STAINLESS STEEL TYPE PH 15-7 Mo AT 1380 F

NORMAL SPECTRAL EMITTANCE OF STAINLESS STEEL TYPE PH 15-7 Mo AT 1380 F--REFERENCE INFORMATION

Reference	Investigator	Symbol	Composition Surface Condition	Test Method	Remarks
16	Adams, J. G.		As received Heated 30 minutes in air at 900 F Heated 30 minutes in 4.4 x 10⁻⁵ mm Hg pressure at 900 F	Normal spectral emittance. Furnace-heated disk specimen. Comparison blackbody (Hohlraun). Spectrometer-mono-chromator with photo-multiplier, lead sulphide, and thermo-couple detectors. Temperatures measured with thermocouples.	Measured in air.

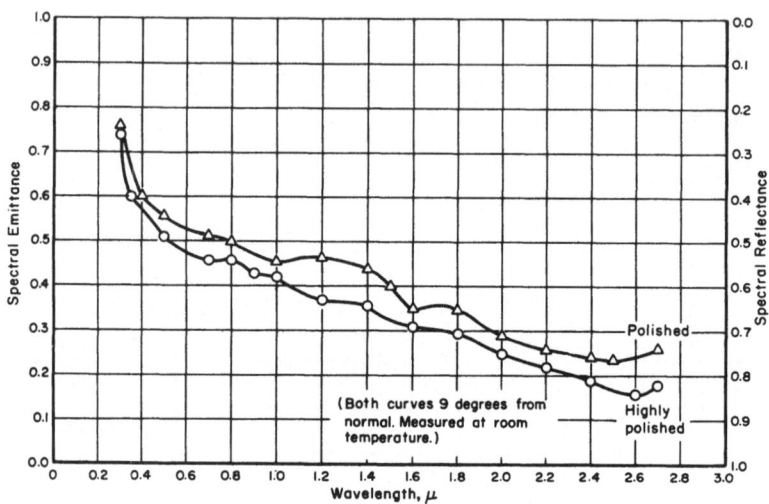

SPECTRAL EMITTANCE OF STAINLESS STEEL PH 15-7 Mo

SPECTRAL EMITTANCE OF STAINLESS STEEL PH 15-7 Mo--REFERENCE INFORMATION

Reference	Investigator	Symbol	Composition and Surface Condition	Test Method	Remarks
8	Betz, Olson, Schurin, and Morris	O △	Highly polished: 2 micro-inches rms Polished: 15 microinches rms	Spectral reflectance at 9 degrees to the normal. Monochromator, inte-grating sphere re-flectometer, and lead sulfide de-tector. "Normal" illumi-nation, hemispherical viewing.	Measured in air at room temper-ature. Data taken from reflectance curves.

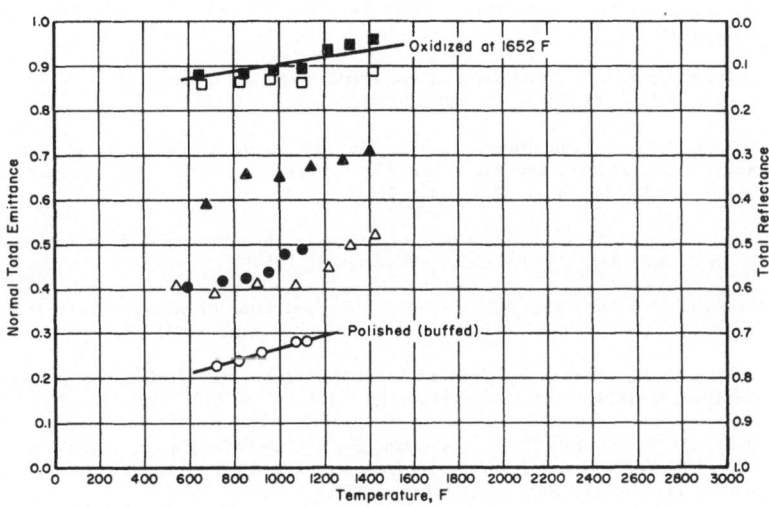

NORMAL TOTAL EMITTANCE OF STAINLESS STEEL TYPE VICKERS F.D.P.

NORMAL TOTAL EMITTANCE OF STAINLESS STEEL TYPE VICKERS F.D.P.--REFERENCE INFORMATION

Reference	Investigator	Symbol	Composition and Surface Condition	Test Method	Remarks
15	Sully, Brandes, and Waterhouse	O ● △ ▲ □ ■	Unoxidized Oxidized at 1112 F Oxidized at 1652 F Open symbols – buffed surface Solid symbols – shotblasted surface	Normal total emittance. Thermopile detector. Comparison blackbody. Temperatures measured with thermocouples. Resistance-heated strip specimens.	Measured in air.

REFERENCES

(1)   Goller, G. N. , "The Emissivity of Molten Stainless Steels", Trans. ASM, 32-33, 239-254 (1944).

(2)   Snyder, N. W. , Gier, J. T. , and Dunkle, R. V. , "Total Normal Emissivity Measurements on Aircraft Materials Between 100 and 800 F", J. ASME, 77, 1011-1019 (1955).

(3)   Wilkes, G. B. , "Total Normal Emissivities and Solar Absorptivities of Materials", WADC TR 54-42 (March, 1954).

(4)   Anthony, F. M. , and Pearl, H. A. , "Investigation of Feasibility of Utilizing Available Heat Resistant Materials for Hypersonic Leading Edge Applications", WADC TR 59-744, Part III (July, 1960).

(5)   Weber, D. , "Spectral Emissivity of Solids in the Infrared at Low Temperatures", J. Opt. Soc. Am. , 49 (8), 815-820 (August, 1959).

(6)   Fulk, M. M. , and Reynolds, M. M. , "Emissivities of Metallic Surfaces at 76°K", J. Appl. Phys. , 28 (12), 1464-1467 (December, 1957).

(7)   Wade, W. R. , "Measurements of Total Hemispherical Emissivity of Various Oxidized Metals at High Temperatures", NACA TN 4206 (October, 1957).

(8)   Betz, H. T. , Olson, H. O. , Schurin, B. D. , and Morris, J. C. , "Determination of Emissivity and Reflectivity Data on Aircraft Structural Materials", WADC TR 56-222, Part II (October, 1958).

(9)   Douglas, E. A. , "Investigation Directed Toward the Development of Ceramic Coatings with High Reflectivities and Emissivities for Use in Aircraft Power Plants", WADC TR 56-110 (February, 1956).

(10)   Olson, O. H. , and Morris, J. C. , "Determination of Emissivity and Reflectivity Data on Aircraft Structural Materials", WADC TR 56-222, Part II, Suppl. I (October, 1958).

(11)   Bevans, J. T. , Gier, J. T. , and Dunkle, R. V. , "Comparison of Total Emittances with Values Computed from Spectral Measurements", Trans. ASME, 80, Part II, 1405-1412 (October, 1958).

(12)   Richmond, J. C. , and Stewart, J. E. , "Spectral Emittance of Uncoated and Ceramic-Coated Inconel and Type 321 Stainless Steel", NASA Memo 4-9-59 W (April, 1959).

(13)   Dobbins, J. P. , "Emittances of Titanium, Aluminum, and Stainless Steel", North American Aviation, Inc. , Report No. NA-49-238 (March 25, 1949).

(14)   Shipley, W. S. , and Thostesen, T. O. , "Radiative Properties of Surfaces Considered for Use on the Explorer Satellites and Pioneer Space Probes", JPL Memo No. 20-194 (February, 1960).

(15) Sully, A. H. , Brandes, E. A. , and Waterhouse, R. B. , "Some Measurements of the Total Emissivity of Metals and Pure Refractory Oxides and the Variation of Emissivity with Temperature", Brit. J. Appl. Phys. , $\underline{3}$, 97- 101 (March, 1952).

(16) Adams, J. G. , "The Determination of Spectral Emissivities, Reflectivities, and Absorptivities of Materials and Coatings, Northrup Corporation Report No. NOR-61-189 (August 3, 1961).

RADIATIVE PROPERTY DATA

Iron, Nickel, and Cobalt and Their Alloys

TABLE OF CONTENTS

TABLE OF CONTENTS
(Continued)

TABLE OF CONTENTS
(Continued)

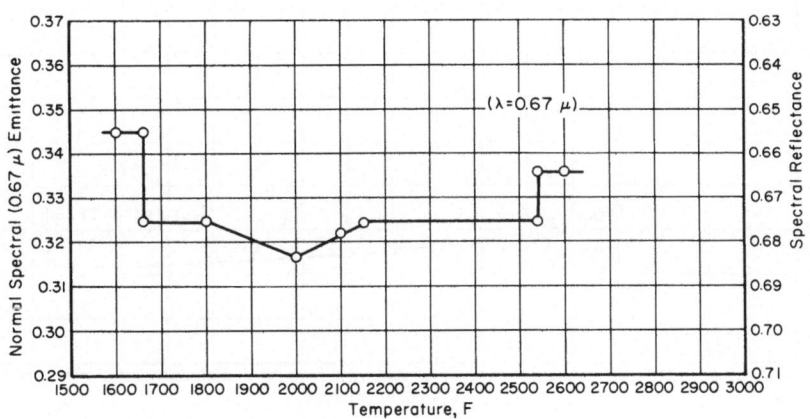

NORMAL SPECTRAL EMITTANCE OF IRON

NORMAL SPECTRAL EMITTANCE OF IRON--REFERENCE INFORMATION

Reference	Investigator	Symbol	Composition and Surface Condition	Test Method	Remarks
1	Wahlin, Zentner, and Martin	O	As received	Normal spectral emittance. Hole-in-tube method.	Measured in reducing atmosphere ($\lambda = 0.67\mu$). Data taken from curve.

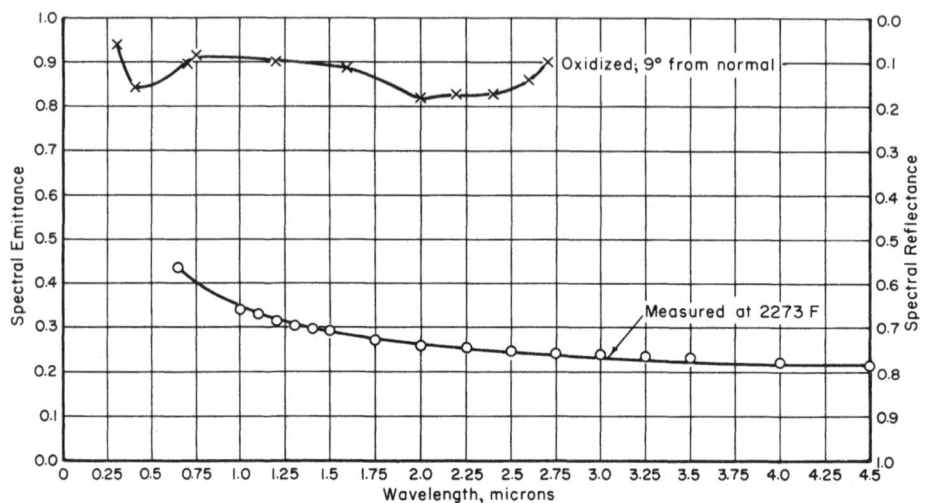

SPECTRAL EMITTANCE OF IRON

SPECTRAL EMITTANCE OF IRON--REFERENCE INFORMATION

Reference	Investigator	Symbol	Composition and Surface Condition	Test Method	Remarks
2	Price, D. J.	o	As rolled Purity - 99.96 per cent	Normal spectral emittance. Spectrometer-monochromator with thermopile detector. Modified hole-in-tube (open slit) specimen. Temperatures measured with optical pyrometer.	Measured in hydrogen atmosphere. Specimen at 2273 F. Data from tables.
18	Olson and Morris	×	Oxidized	Spectral reflectance at 9° from normal. Monochromator, integrating sphere, lead sulphide detector, and MgO standard. "Normal" illumination, hemispherical viewing.	Measured in air at room temperature. Data taken from reflectance curves.

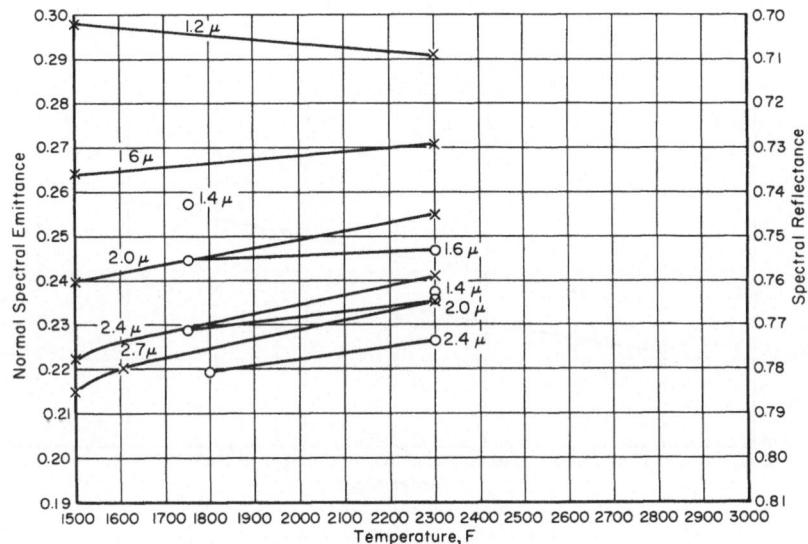

NORMAL SPECTRAL EMITTANCE OF IRON AT DIFFERENT WAVELENGTHS

NORMAL SPECTRAL EMITTANCE OF IRON AT DIFFERENT WAVELENGTHS--REFERENCE INFORMATION

Reference	Investigator	Symbol	Composition and Surface Condition	Test Method	Remarks
3	Lund and Ward	O	Polished with 000-grade emery paper and cleaned	Normal spectral emittance. Hole-in-tube method. Temperatures measured with thermocouples. Spectrometer-monochromator detector.	Measured in vacuum or hydrogen atmosphere. Data taken from curves.
4	Ward, L.	✕	Surface lapped with jewelers rouge	Normal spectral emittance. Modified hole-in-tube method. Temperatures measured with thermocouples. Spectrometer-monochromator detector.	Measured in hydrogen. Data taken from curves.

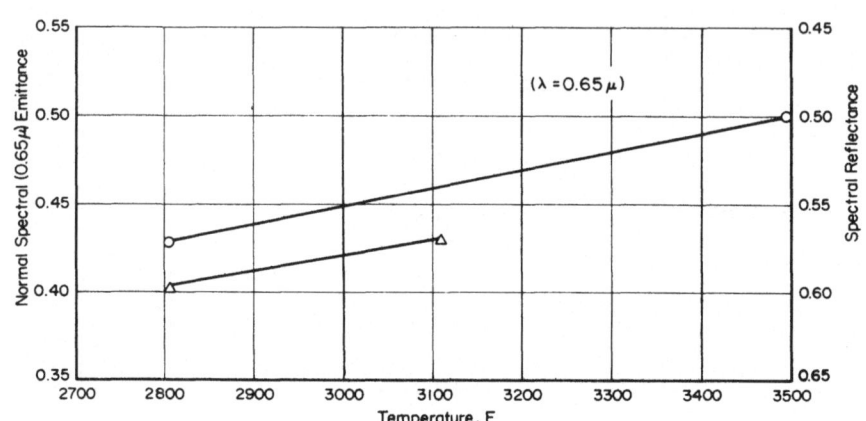

NORMAL SPECTRAL EMITTANCE OF MOLTEN IRON

NORMAL SPECTRAL EMITTANCE OF LIQUID IRON--REFERENCE INFORMATION

Reference	Investigator	Symbol	Composition and Surface Condition	Test Method	Remarks
5	Dastur and Gokcen	O	Melted in beryllia crucible; surface swept clean with preheated hydrogen	Normal spectral emittance. Temperatures measured with thermocouples and calibrated optical pyrometer. Blackbody hole in crucible bottom.	Measured in an argon-hydrogen atmosphere. (λ = 0.65μ)
6	Knowles and Sarjant	Δ	Clean	Normal spectral emittance. Immersion thermo-couples and optical pyrometer.	(λ = 0.65μ)

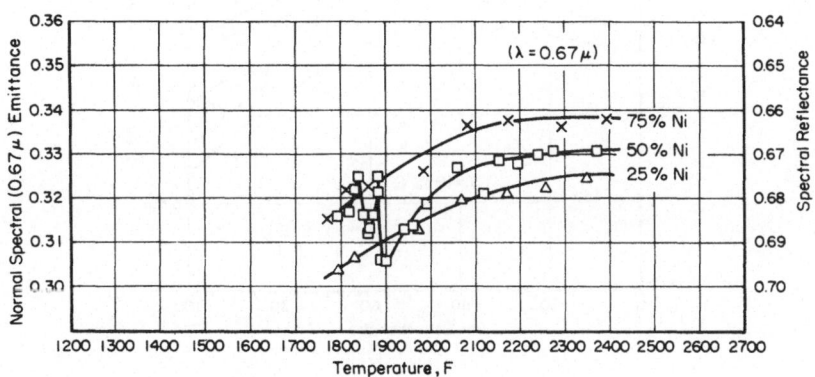

NORMAL SPECTRAL EMITTANCE OF IRON-NICKEL ALLOYS

NORMAL SPECTRAL EMITTANCE OF IRON-NICKEL ALLOYS--REFERENCE INFORMATION

Reference	Investigator	Symbol	Composition and Surface Condition	Test Method	Remarks
1	Wahlin, Zentner, and Martin		Mixed, sintered, and rolled from reagent quality powders	Normal spectral emittance. Hole-in-tube method.	Measured in vacuum or reducing atmospheres.
		✗	75 per cent nickel		$(\lambda = 0.67\mu)$
		□	50 per cent nickel		
		△	25 per cent nickel		

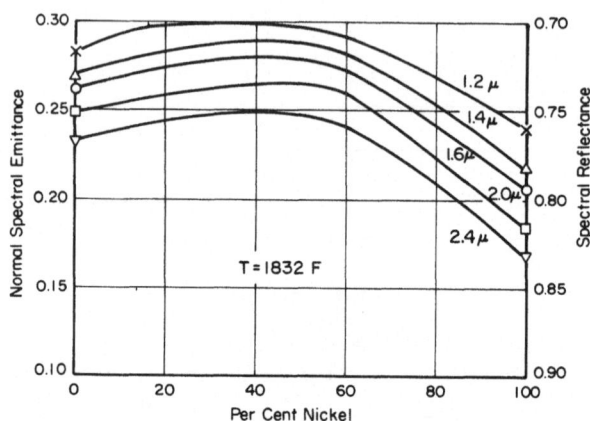

NORMAL SPECTRAL EMITTANCE OF IRON-NICKEL ALLOYS VERSUS COMPOSITION

NORMAL SPECTRAL EMITTANCE OF IRON-NICKEL ALLOYS VERSUS COMPOSITION--REFERENCE INFORMATION

Reference	Investigator	Symbol	Composition and Surface Condition	Test Method	Remarks
3	Lund and Ward		Polished with 000-grade emery paper, and cleaned	Normal spectral emittance. Hole-in-tube method. Temperatures measured with thermocouples. Spectrometer-mono-chromator detector.	Measured in vacuum or hydrogen atmos-phere. Data taken from curves.
		×	1.2 μ		
		△	1.4 μ		
		○	1.6 μ		
		□	2.0 μ		
		▽	2.4 μ		

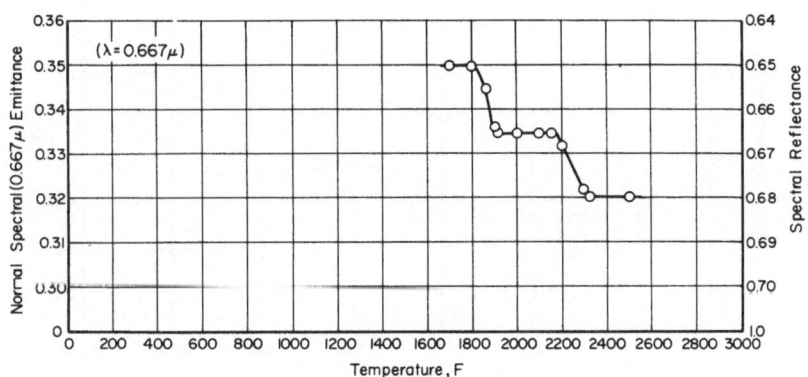

NORMAL SPECTRAL EMITTANCE OF IRON-18 PER CENT TUNGSTEN ALLOY

NORMAL SPECTRAL EMITTANCE OF IRON-18 PER CENT TUNGSTEN ALLOY--REFERENCE INFORMATION

Reference	Investigator	Symbol	Composition and Surface Condition	Test Method	Remarks
7	Knop, Jr., H. W.	O	As rolled	Normal spectral emittance. Hole-in-tube method. Calibrated optical pyrometer.	Measured in vacuum ($\lambda = 0.667\mu$). Data taken from curve.

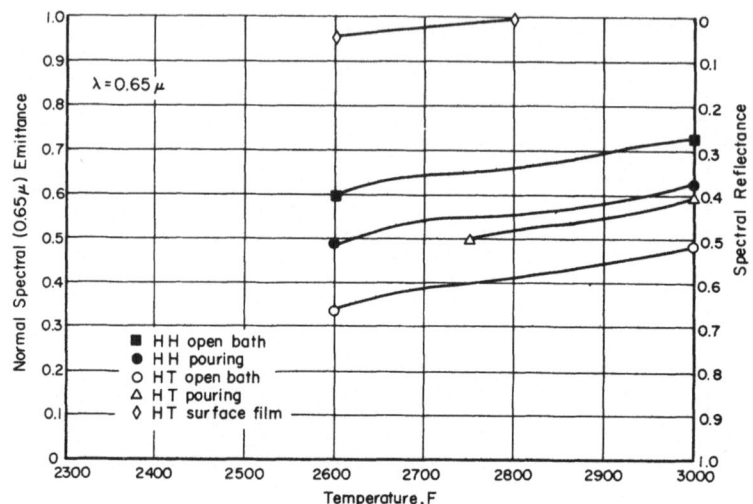

NORMAL SPECTRAL EMITTANCE OF MOLTEN TYPES HH AND HT IRON ALLOYS

NORMAL SPECTRAL EMITTANCE OF MOLTEN TYPES HH AND HT IRON ALLOYS--REFERENCE INFORMATION

Reference	Investigator	Symbol	Composition and Surface Condition	Test Method	Remarks
8	Gow, Brasunas, and Harder		Type HH Alloy (26 Cr, 12 Ni)	Normal spectral emittance.	Measured in air.
		■	Open bath condition	Surface temperatures measured with calibrated optical pyrometers.	($\lambda = 0.65\mu$)
		●	While pouring	Bath temperatures measured with immersion thermo-couples.	
			Type HT Alloy (36 Ni, 16 Cr)		
		O	Open bath condition		
		△	While pouring		
		◊	With surface film		

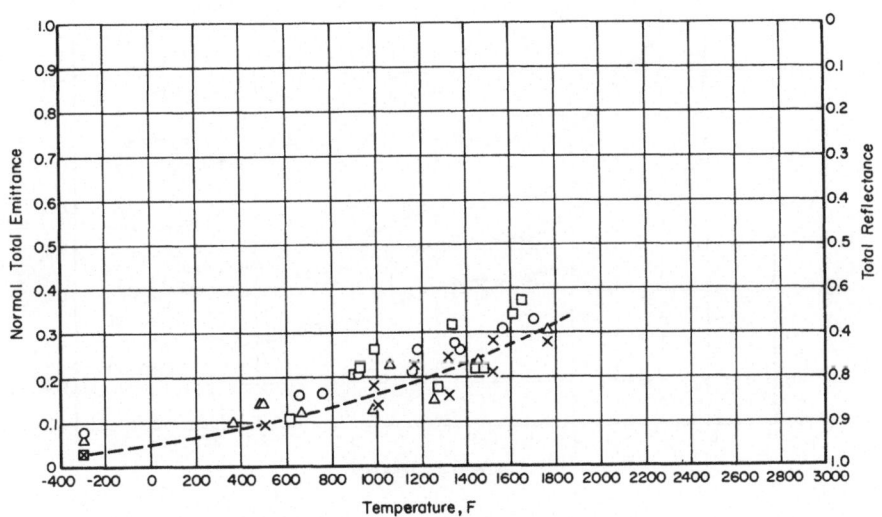

NORMAL TOTAL EMITTANCE OF ALLOY N-155 (MULTIMET)

NORMAL TOTAL EMITTANCE OF ALLOY N-155 (MULTIMET)--REFERENCE INFORMATION

Reference	Investigator	Symbol	Composition and Surface Condition	Test Method	Remarks
11	Betz, Olson, Schurin, and Morris	○ △ □ ✕	Oxidized As received Cleaned Polished	Normal total emittance. Resistance-heated specimen. Thermistor-bolometer detector. Comparison blackbody. Temperatures measured with thermocouples.	Measured in vacuum. Data taken from curves. (Dotted curve indicates probable polished condition)

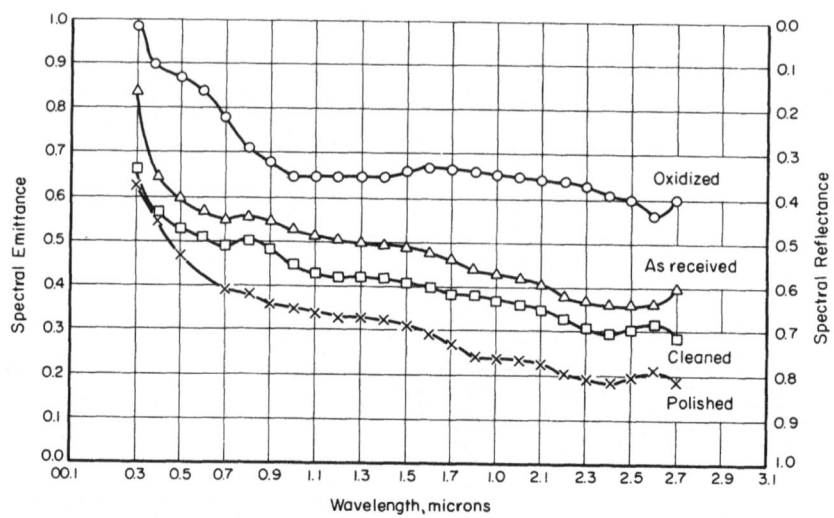

SPECTRAL EMITTANCE OF ALLOY N-155 (MULTIMET)

SPECTRAL EMITTANCE OF ALLOY N-155 (MULTIMET)--REFERENCE INFORMATION

Reference	Investigator	Symbol	Composition and Surface Condition	Test Method	Remarks
11	Betz, Olson, Schurin, and Morris	O Δ □ ✗	Oxidized As received Cleaned Polished	Spectral reflectance at 9° from normal. Integrating sphere. Monochromator with lead sulphide detector. "Normal" illumination, hemispherical viewing.	Measured in air at room temperature.

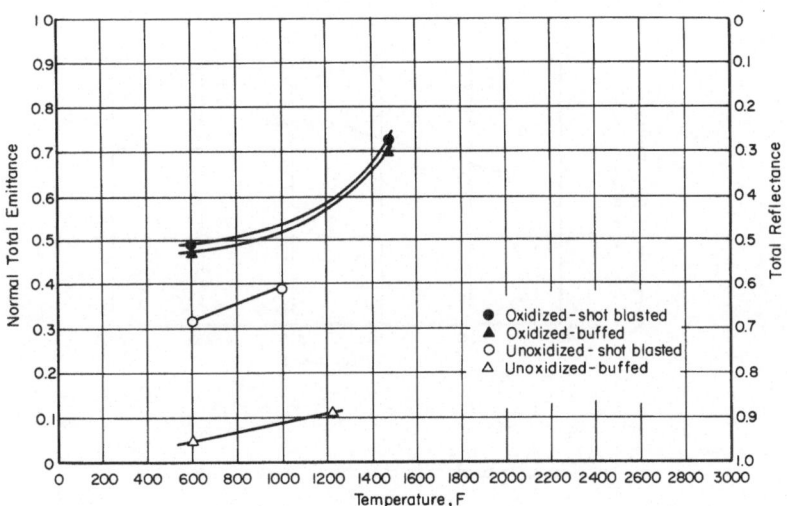

NORMAL TOTAL EMITTANCE OF NICKEL

NORMAL TOTAL EMITTANCE OF NICKEL--REFERENCE INFORMATION

Reference	Investigator	Symbol	Composition and Surface Condition	Test Method	Remarks
9	Sully, Brandes, and Waterhouse	●	Oxidized at 1650 F, shot blasted	Normal total emittance. Thermopile detector. Comparison blackbody. Temperatures measured with thermocouples. Self-resistance-heated specimen.	Measured in air. Data taken from curves.
		▲	Oxidized, buffed		
		○	Unoxidized, shot blasted		
		△	Unoxidized, buffed		

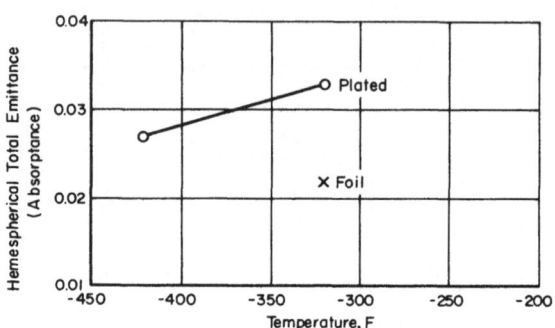

HEMISPHERICAL TOTAL EMITTANCE OF NICKEL AT CRYOGENIC TEMPERATURES

HEMISPHERICAL TOTAL EMITTANCE OF NICKEL AT CRYOGENIC TEMPERATURES--REFERENCE INFORMATION

Reference	Investigator	Symbol	Composition and Surface Condition	Test Method	Remarks
10	Fulk and Reynolds		Commercially pure, cleaned, as-received condition	Hemispherical total absorptance. Electrically cali-brated calorimeter. Absorbed heat measured by boil-off rate of liquid nitrogen.	Measured in vacuum. Results for room-temperature radiation.
		o	Nickel plated on copper		
		x	0.004-inch-thick foil.		

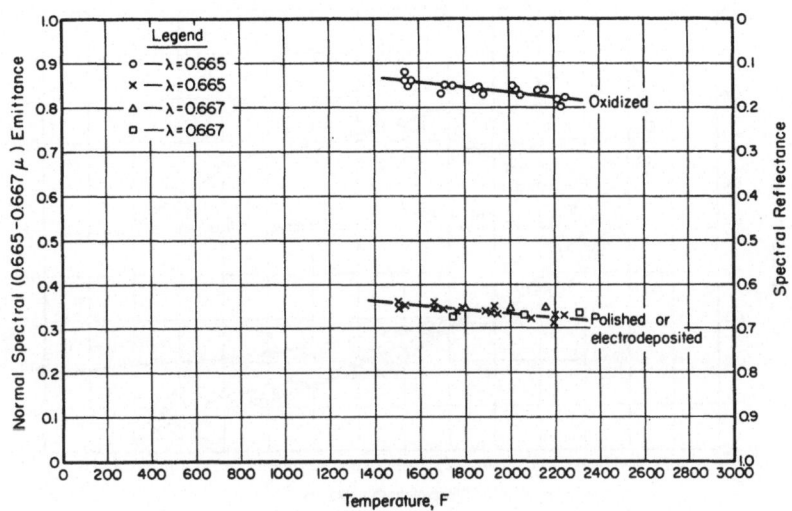

NORMAL SPECTRAL EMITTANCE OF NICKEL

NORMAL SPECTRAL EMITTANCE OF NICKEL—REFERENCE INFORMATION

Reference	Investigator	Symbol	Composition and Surface Condition	Test Method	Remarks
11	Betz, Olson, Schurin, and Morris	o ✕	Oxidized Polished	Normal spectral emittance. Self-resistance heating. Modified hole-in-tube method.	Measured in vacuum ($\lambda = 0.665\mu$).
12	Wahlin and Wright	Δ	Electrolytically deposited on stainless steel	Normal spectral emittance. Hole-in-tube method. Temperatures checked with thermocouples.	Measured in reducing atmosphere ($\lambda = 0.667\mu$). Data taken from curves.
13	Wahlin and Knop	□	Electrolytically deposited on stainless steel	Normal spectral emittance. Hole-in-tube method. Temperatures checked with thermocouples.	Measured in reducing atmosphere ($\lambda = 0.667\mu$). Data taken from curves.

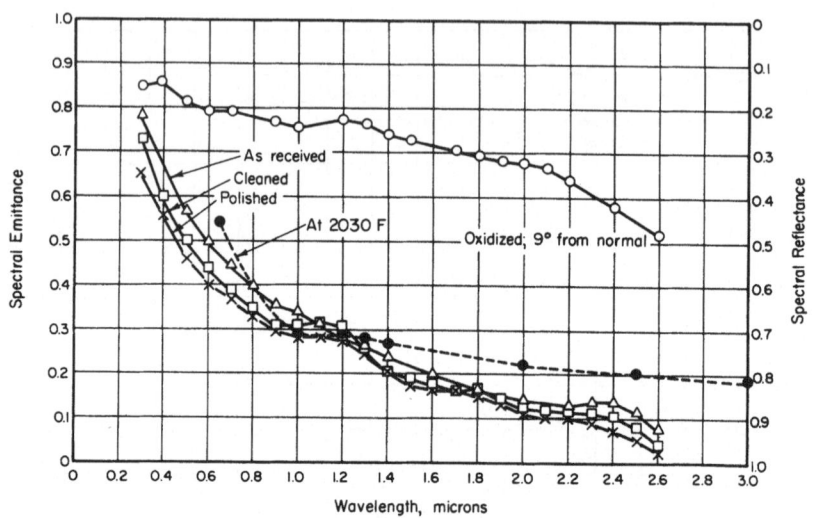

SPECTRAL EMITTANCE OF NICKEL

SPECTRAL EMITTANCE OF NICKEL--REFERENCE INFORMATION

Reference	Investigator	Symbol	Composition and Surface Condition	Test Method	Remarks
11	Betz, Olson, Schurin, and Morris	O △ □ ×	Oxidized As received Cleaned Polished	Spectral reflectance at 9° from normal. Integrating sphere. Monochromator with lead sulphide detector. "Normal" illumination, hemispherical viewing.	Measured in air at room temperature. Data taken from curves.
2	Price, D. J.	●	As rolled Purity – 99.97 per cent	Normal spectral emittance. Spectrometer-monochromator with thermopile detector. Modified hole-in-tube (open slit) specimen. Temperatures measured with optical pyrometer.	Measured in hydrogen atmosphere. Specimen at 2030 F. Data taken from tables.

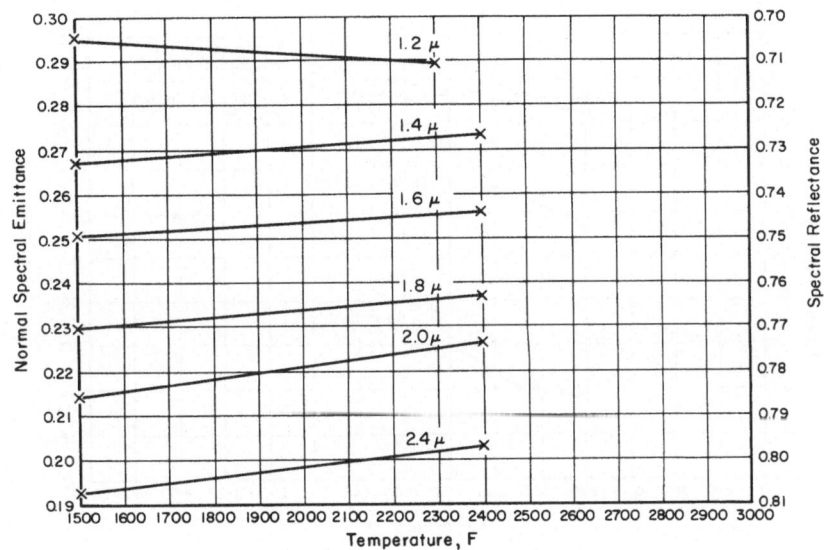

NORMAL SPECTRAL EMITTANCE OF NICKEL AT DIFFERENT WAVELENGTHS

NORMAL SPECTRAL EMITTANCE OF NICKEL AT DIFFERENT WAVELENGTHS--REFERENCE INFORMATION

Reference	Investigator	Symbol	Composition and Surface Condition	Test Method	Remarks
4	Ward, L.	✕	Surface lapped with jewelers rouge.	Normal spectral emittance. Modified hole-in-tube method. Temperatures measured with thermocouples. Spectrometer-mono-chromator detector.	Measured in hydrogen. Data taken from curves.

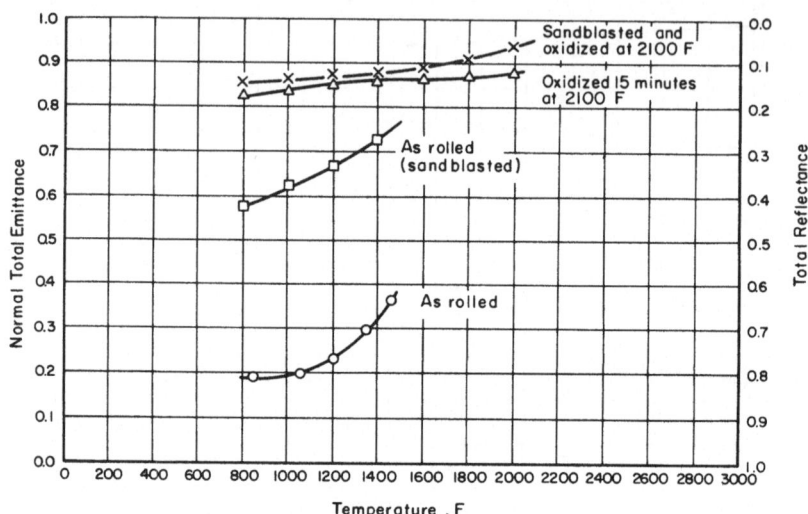

NORMAL TOTAL EMITTANCE OF NICHROME V

NORMAL TOTAL EMITTANCE OF NICHROME V--REFERENCE INFORMATION

Reference	Investigator	Symbol	Composition and Surface Condition	Test Method	Remarks
15	DeCorso and Coit	×	Sandblasted and oxidized 15 minutes at 2100 F	Normal total emittance. Thermopile detector. Comparison blackbody. Temperatures measured with thermocouples. Resistance-heated specimens.	Measured in air. Data taken from curves.
		△	Oxidized 15 minutes at 2100 F		
		□	Sandblasted; as rolled		
		○	As rolled (Nominal composition: 80 Ni, 20 Cr)		

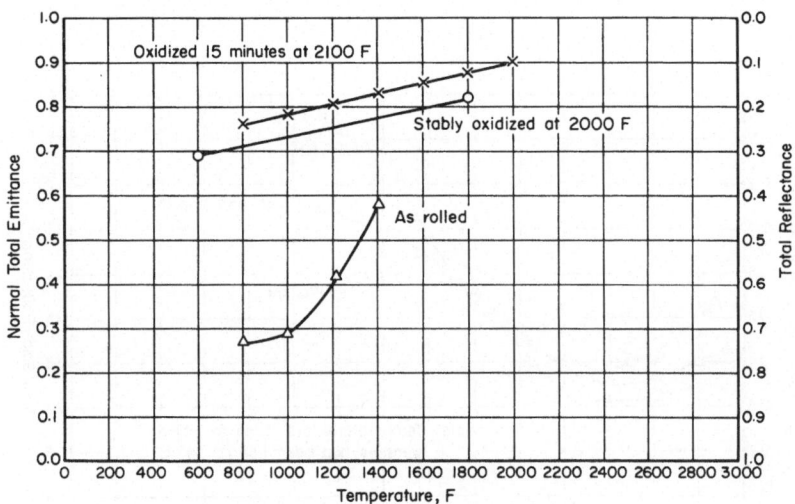

NORMAL TOTAL EMITTANCE OF INCONEL

NORMAL TOTAL EMITTANCE OF INCONEL--REFERENCE INFORMATION

Reference	Investigator	Symbol	Composition and Surface Condition	Test Method	Remarks
14	O'Sullivan and Wade	○	Stably oxidized at 2000 F	Normal total emittance. Thermopile detector. Comparison blackbody. Temperatures measured with thermocouples. Resistance-heated specimens.	Measured in air. Data taken from curves.
15	DeCorso and Coit	△ ✕	As rolled Heated in air for 15 minutes at 2100 F	Normal total emittance. Thermopile detector. Comparison blackbody. Temperatures measured with thermocouples. Resistance-heated specimens.	Measured in air. Data taken from curves.

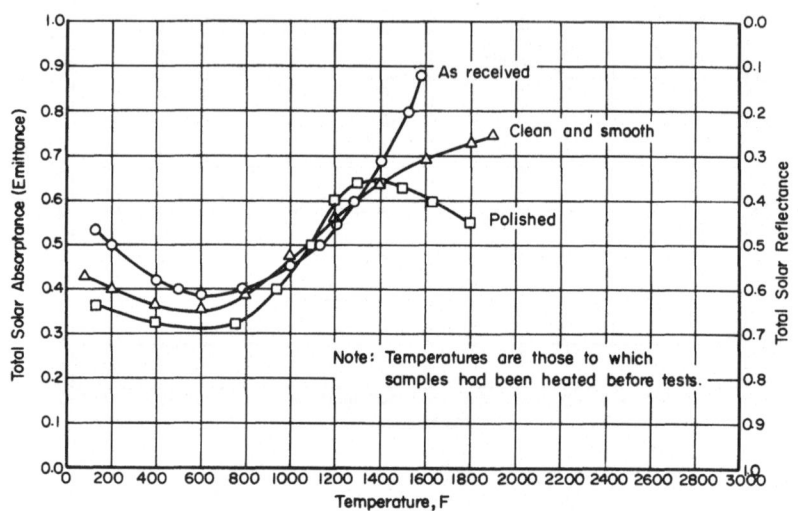

TOTAL SOLAR ABSORPTANCE OF INCONEL B AT 100 F

TOTAL SOLAR ABSORPTANCE OF INCONEL B--REFERENCE INFORMATION

Reference	Investigator	Symbol	Composition and Surface Condition	Test Method	Remarks
16	Wilkes, G. B.	○ △ □	As received Clean and smooth Polished	Total solar absorptance. Comparison standards. Comparison pyroheliometer. Output measured with thermocouples.	Measured in air at 100 F. Temperatures are those to which samples had been heated previously.

117

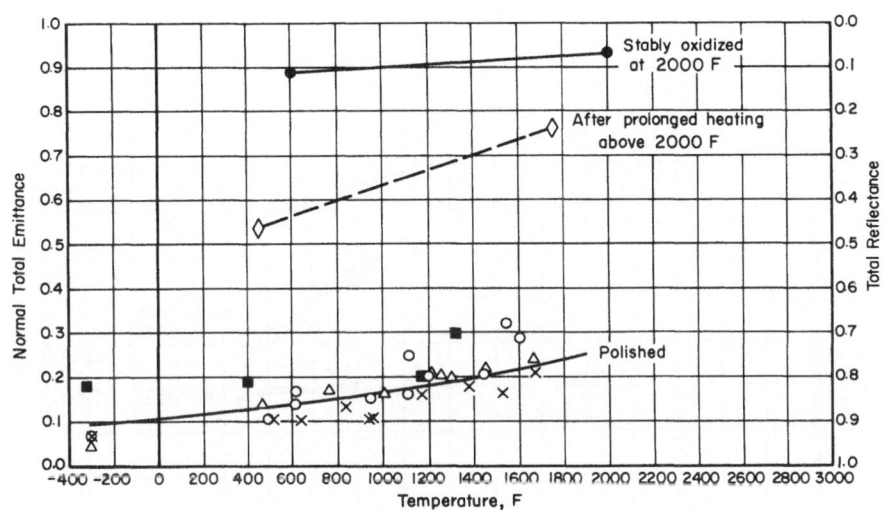

NORMAL TOTAL EMITTANCE OF INCONEL X

NORMAL TOTAL EMITTANCE OF INCONEL X--REFERENCE INFORMATION

Reference	Investigator	Symbol	Composition and Surface Condition	Test Method	Remarks
11	Betz, Olson, Shurin, and Morris	O Δ ×	Oxidized As received or wiped clean Polished	Normal total emittance. Resistance-heated specimens. Thermistor-bolometer detector. Comparison blackbody. Temperatures measured with thermocouples.	Measured in vacuum. Data taken from curves.
14	O'Sullivan and Wade	●	Stably oxidized at 2000 F	Normal total emittance. Thermopile detector. Comparison blackbody. Temperatures measured with thermocouples.	Measured in air. Data taken from curves.
16	Wilkes, G. B.	■ ◊	Polished After prolonged heating and cycling above 2000 F (some oxide indicated)	Normal total emittance. Total-radiation detector. Comparison blackbody. Temperatures measured with thermocouples.	Measured in a 10-micron pressure of helium.

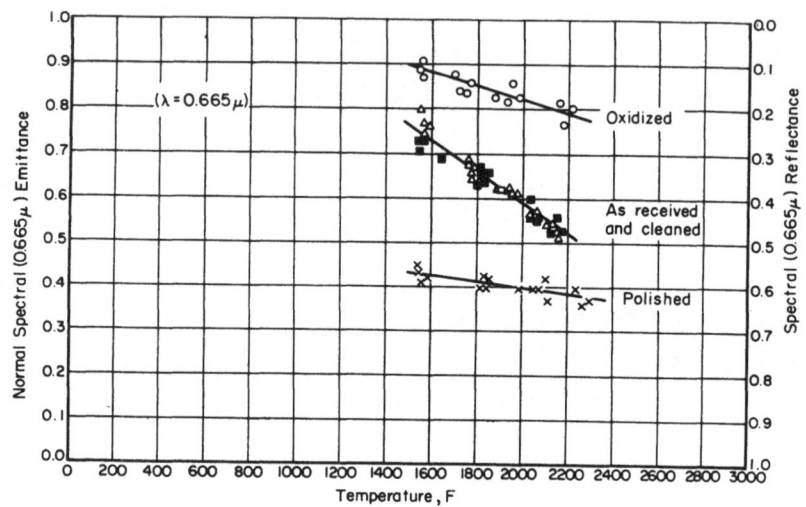

NORMAL SPECTRAL EMITTANCE OF INCONEL X

NORMAL SPECTRAL EMITTANCE OF INCONEL X--REFERENCE INFORMATION

Reference	Investigator	Symbol	Composition and Surface Condition	Test Method	Remarks
11	Betz, Olson, Schurin, and Morris	o △ ■ ×	Oxidized As received Cleaned Polished	Normal spectral emittance. Modified hole-in-tube method. Drilled blackbody hole. Temperatures measured with thermocouples.	Measured in vacuum ($\lambda = 0.665\mu$). Data taken from curves.

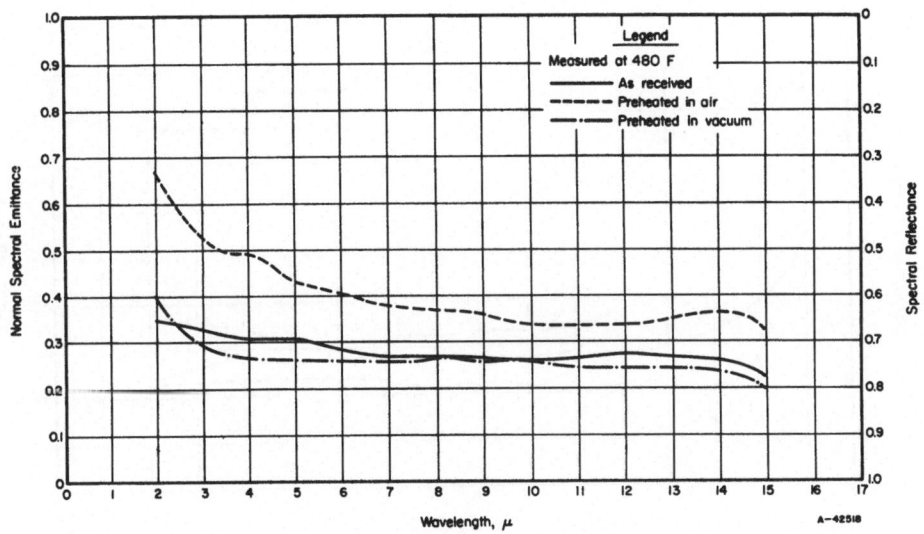

NORMAL SPECTRAL EMITTANCE OF INCONEL X AT 480 F

NORMAL SPECTRAL EMITTANCE OF INCONEL X AT 480 F—REFERENCE INFORMATION

Reference	Investigator	Symbol	Composition and Surface Condition	Test Method	Remarks
19	Adams, J. G.		As received Heated 30 minutes in air at 1500 C Heated 30 minutes in 6.8 x 10^{-5} mm Hg pressure at 1500 C	Normal spectral emittance. Furnace-heated disk specimen. Comparison blackbody (Hohlraun). Spectrometer-monochromator with photomultiplier, lead sulphide, and thermocouple detectors. Temperatures measured with thermocouples.	Measured in air.

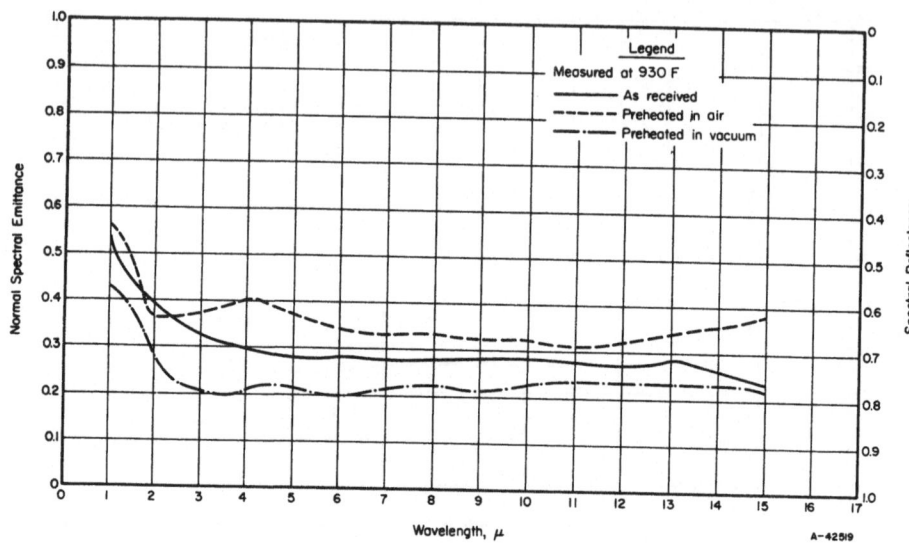

NORMAL SPECTRAL EMITTANCE OF INCONEL X AT 930 F

NORMAL SPECTRAL EMITTANCE OF INCONEL X AT 930 F--REFERENCE INFORMATION

Reference	Investigator	Symbol	Composition and Surface Condition	Test Method	Remarks
19	Adams, J. G.		As received Heated 30 minutes in air at 1500 C Heated 30 minutes in 6.8 x 10⁻⁵ mm Hg pressure at 1500 C	Normal spectral emittance. Furnace-heated disk specimen. Comparison blackbody (Hohlraun). Spectrometer-monochromator with photomultiplier, lead sulphide, and thermocouple detectors. Temperatures measured with thermocouples.	Measured in air.

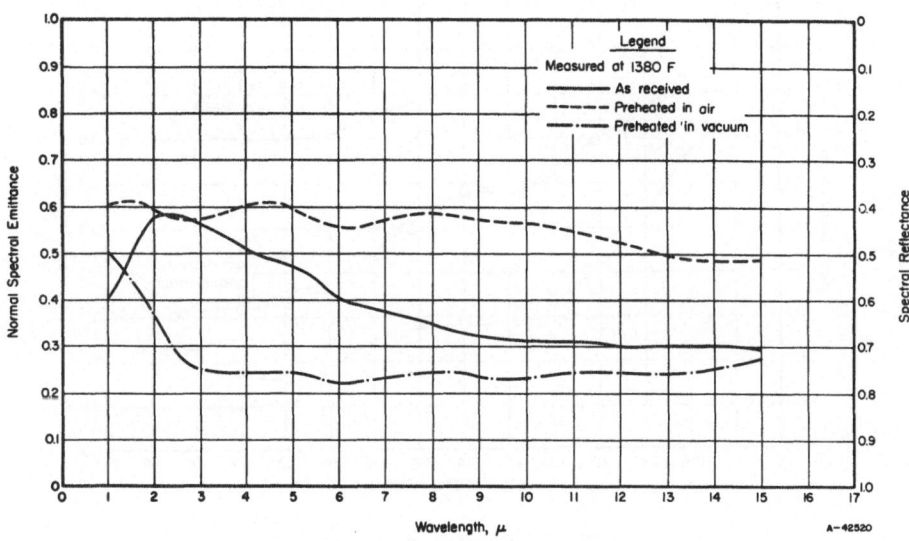

NORMAL SPECTRAL EMITTANCE OF INCONEL X AT 1380 F

NORMAL SPECTRAL EMITTANCE OF INCONEL X AT 1380 F--REFERENCE INFORMATION

Reference	Investigator	Symbol	Composition and Surface Condition	Test Method	Remarks
19	Adams, J. G.		As received Heated 30 minutes in air at 1500 C Heated 30 minutes in 6.8 x 10⁻⁵ mm Hg pressure at 1500 C	Normal spectral emittance. Furnace-heated disk specimen. Comparison blackbody (Hohlraun). Spectrometer-monochromator with photomultiplier, lead sulphide, and thermocouple detectors. Temperatures measured with thermocouples.	Measured in air.

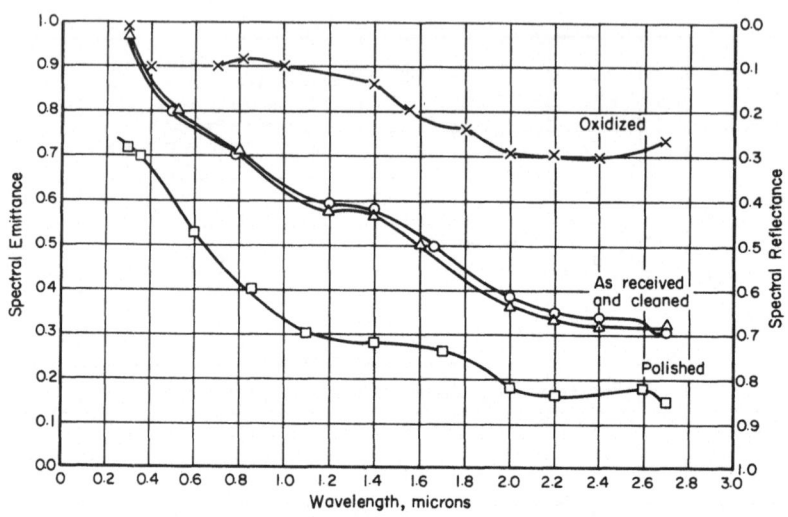

SPECTRAL EMITTANCE OF INCONEL X

SPECTRAL EMITTANCE OF INCONEL X--REFERENCE INFORMATION

Reference	Investigator	Symbol	Composition and Surface Condition	Test Method	Remarks
18	Olson and Morris	O Δ □ ×	As received Cleaned Polished Oxidized	Spectral reflectance at 9° from normal. Monochromator, integrating sphere, lead sulphide detector, and MgO standard. "Normal" (9°) illumination, hemispherical viewing.	Measured in air at room temperature. Data taken from reflectance curves.

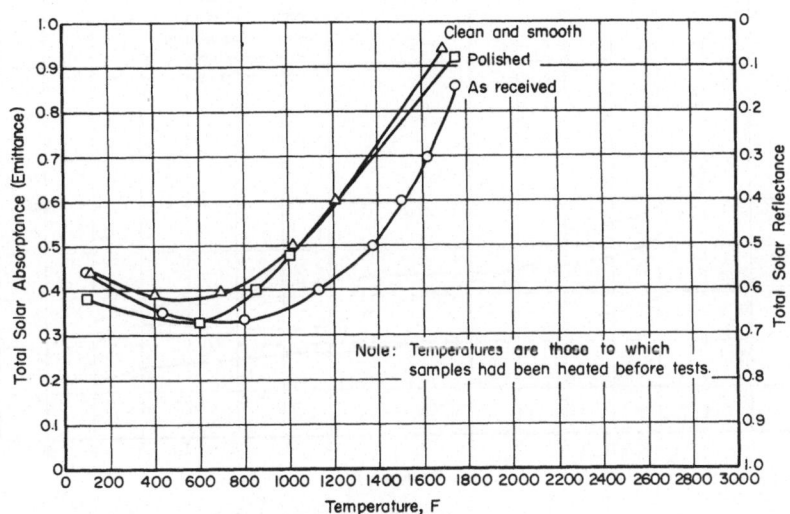

TOTAL SOLAR ABSORPTANCE OF INCONEL X AT 100 F

TOTAL SOLAR ABSORPTANCE OF INCONEL X--REFERENCE INFORMATION

Reference	Investigator	Symbol	Composition and Surface Condition	Test Method	Remarks
16	Wilkes, G. B.	o △ □	As received Clean and smooth Polished	Total solar absorptance. Comparison standards. Comparison pyro- heliometer. Output measured with thermocouples.	Measured in air at 100 F. Temperatures are those to which samples had been heated previously.

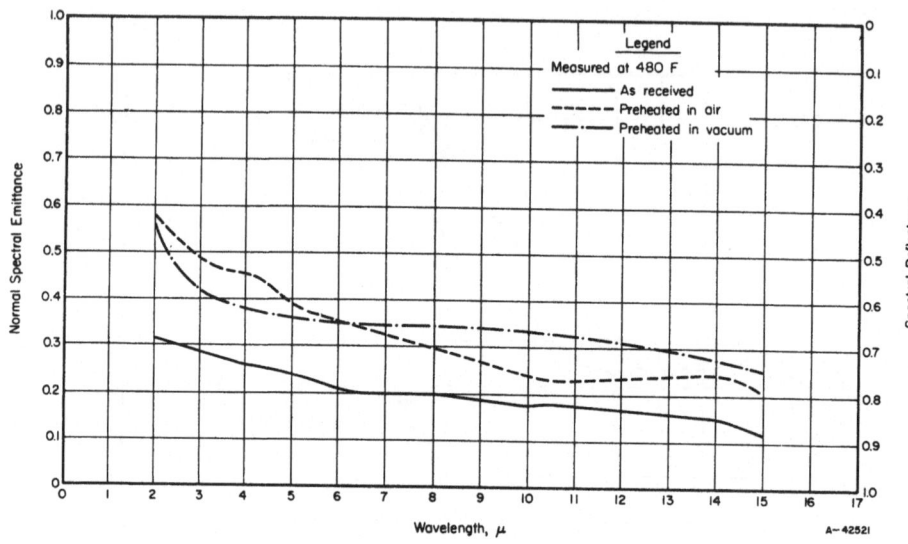

NORMAL SPECTRAL EMITTANCE OF INCONEL 702 AT 480 F

NORMAL SPECTRAL EMITTANCE OF INCONEL 702 AT 480 F--REFERENCE INFORMATION

Reference	Investigator	Symbol	Composition and Surface Condition	Test Method	Remarks
19	Adams, J. G.		As received Heated 30 minutes in air at 1800 F Heated 30 minutes in 7.6 x 10^{-5} mm Hg pressure at 1800 F	Normal spectral emittance. Furnace-heated disk specimen. Comparison blackbody (Hohlraun). Spectrometer-monochromator with photomultiplier, lead sulphide, and thermocouple detectors. Temperatures measured with thermocouples.	Measured in air.

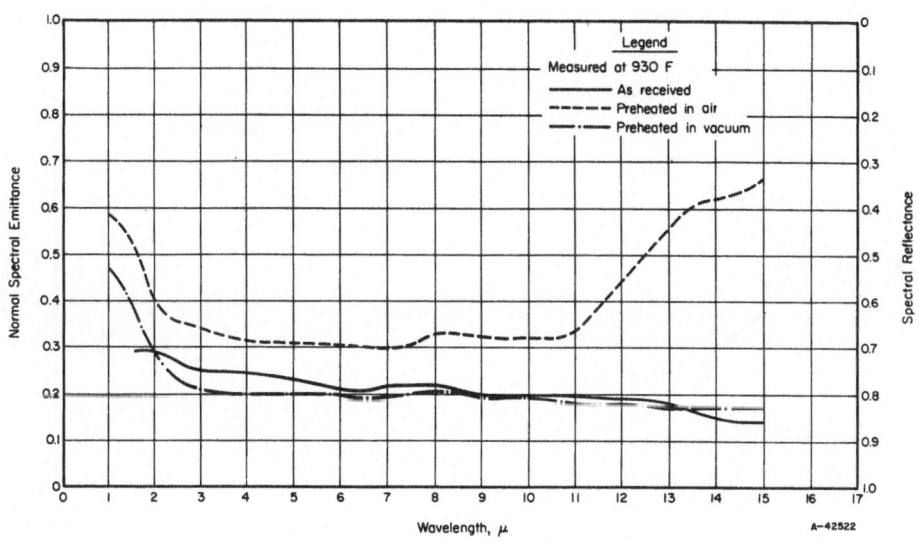

NORMAL SPECTRAL EMITTANCE OF INCONEL 702 AT 930 F

NORMAL SPECTRAL EMITTANCE OF INCONEL 702 AT 930 F—-REFERENCE INFORMATION

Reference	Investigator	Symbol	Composition and Surface Condition	Test Method	Remarks
19	Adams, J. G.		As received Heated 30 minutes in air at 1800 F Heated 30 minutes in 7.6 x 10^{-5} mm Hg pressure at 1800 F	Normal spectral emittance. Furnace-heated disk specimen. Comparison blackbody (Hohlraun). Spectrometer-monochromator with photomultiplier, lead sulphide, and thermocouple detectors. Temperatures measured with thermocouples.	Measured in air.

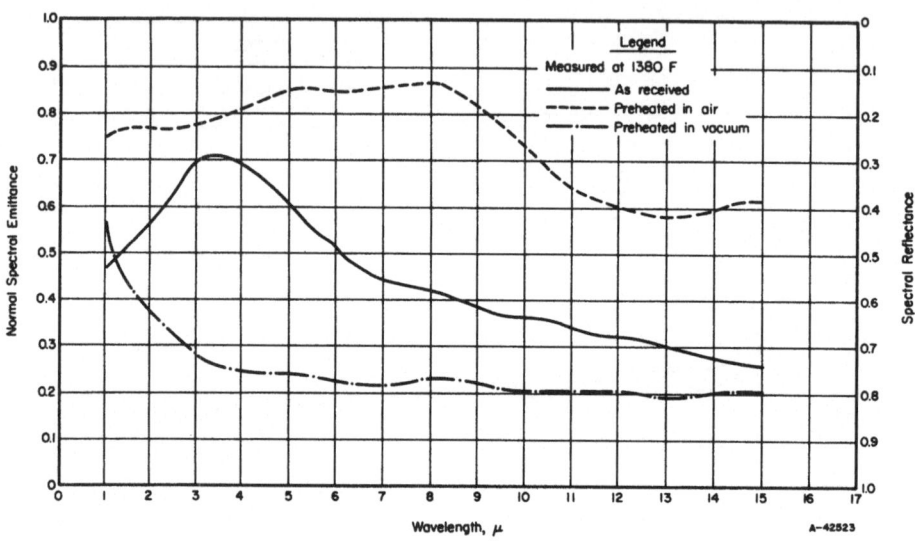

NORMAL SPECTRAL EMITTANCE OF INCONEL 702 AT 1380 F

NORMAL SPECTRAL EMITTANCE OF INCONEL 702 AT 1380 F--REFERENCE INFORMATION

Reference	Investigator	Symbol	Composition and Surface Condition	Test Method	Remarks
19	Adams, J. G.		As received Heated 30 minutes in air at 1800 F Heated 30 minutes in 7.6 x 10^{-5} mm Hg pressure at 1800 F	Normal spectral emittance. Furnace-heated disk specimen. Comparison blackbody (Hohlraun). Spectrometer-monochromator with photomultiplier, lead sulphide, and thermocouple detectors. Temperatures measured with thermocouples.	Measured in air.

127

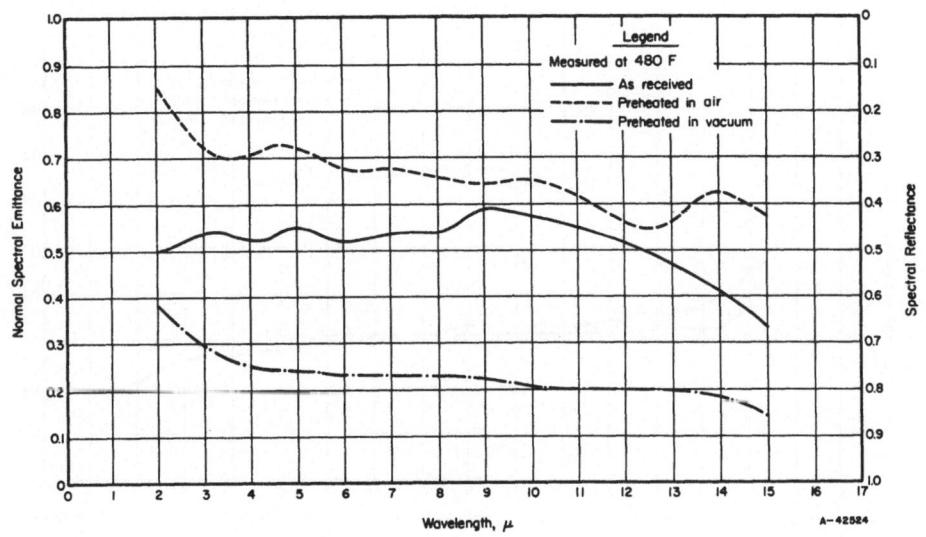

NORMAL SPECTRAL EMITTANCE OF UDIMET 500 AT 480 F

NORMAL SPECTRAL EMITTANCE OF UDIMET 500 AT 480 F--REFERENCE INFORMATION

Reference	Investigator	Symbol	Composition and Surface Condition	Test Method	Remarks
19	Adams, J. G.		As received Heated 30 minutes in air at 1800 F Heated 30 minutes in 7.6 x 10⁻⁵ mm Hg pressure at 1800 F	Normal spectral emittance. Furnace-heated disk speci-men. Comparison blackbody (Hohlraun). Spectrometer-mono-chromator with photo-multiplier, lead sulphide, and thermo-couple detectors. Temperatures measured with thermocouples.	Measured in air.

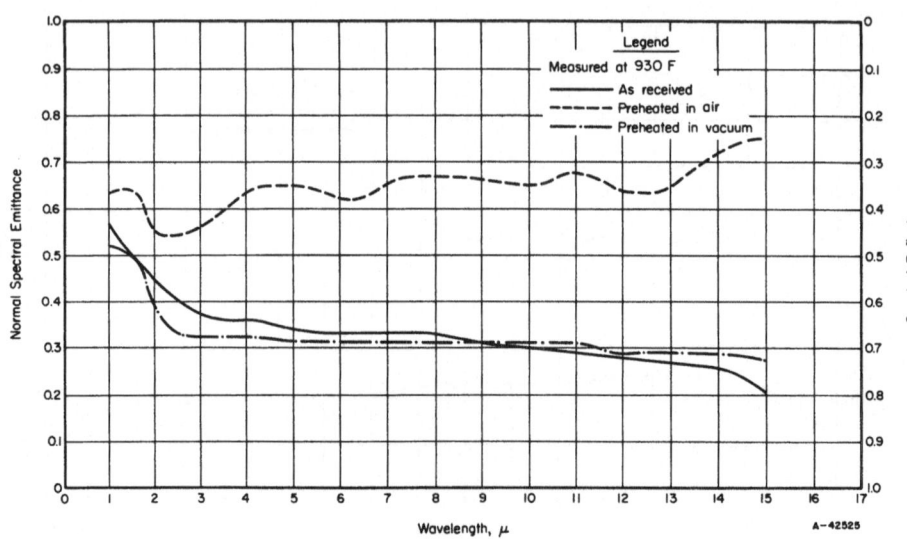

NORMAL SPECTRAL EMITTANCE OF UDIMET 500 AT 930 F

NORMAL SPECTRAL EMITTANCE OF UDIMET 500 AT 930 F—REFERENCE INFORMATION

Reference	Investigator	Symbol	Composition and Surface Condition	Test Method	Remarks
19	Adams, J. G.		As received Heated 30 minutes in air at 1800 F Heated 30 minutes in 7.6 x 10^{-5} mm Hg pressure at 1800 F	Normal spectral emittance. Furnace-heated disk specimen. Comparison blackbody (Hohlraun). Spectrometer-monochromator with photomultiplier, lead sulphide, and thermocouple detectors. Temperatures measured with thermocouples.	Measured in air.

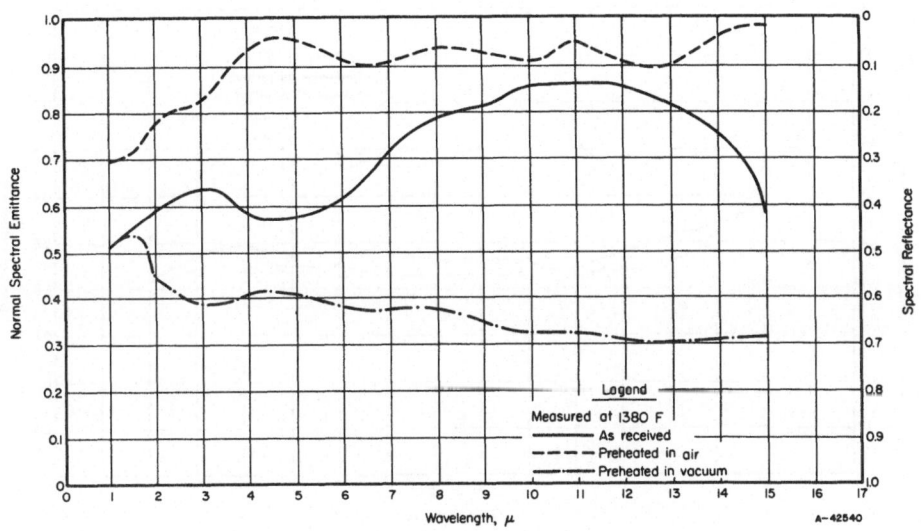

NORMAL SPECTRAL EMITTANCE OF UDIMET 500 AT 1380 F

NORMAL SPECTRAL EMITTANCE OF UDIMET 500 AT 1380 F--REFERENCE INFORMATION

Reference	Investigator	Symbol	Composition and Surface Condition	Test Method	Remarks
19	Adams, J. G.		As received Heated 30 minutes in air at 1800 F Heated 30 minutes in 7.6 x 10^{-5} mm Hg pressure at 1800 F	Normal spectral emittance. Furnace-heated disk specimen. Comparison blackbody (Hohlraun). Spectrometer-monochromator with photomultiplier, lead sulphide, and thermocouple detectors. Temperatures measured with thermocouples.	Measured in air.

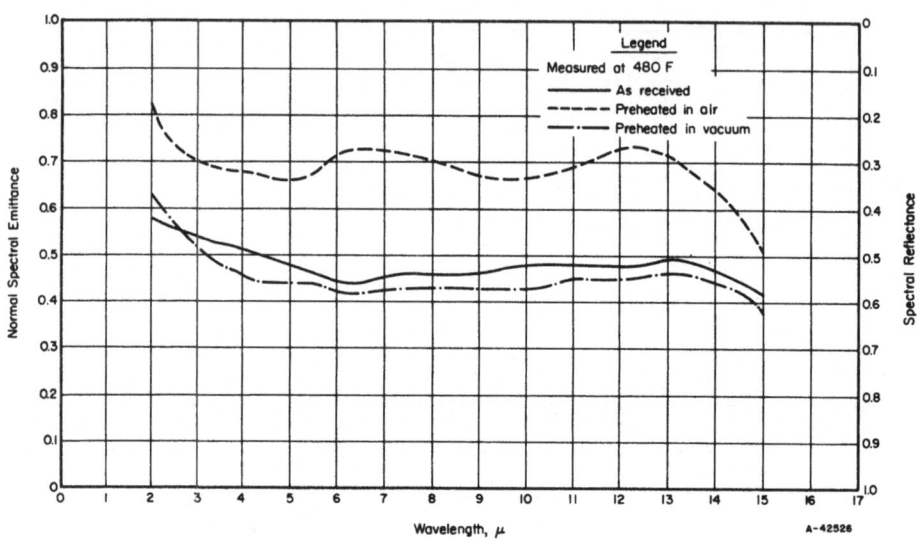

NORMAL SPECTRAL EMITTANCE OF RENÉ 41 AT 480 F

NORMAL SPECTRAL EMITTANCE OF RENE 41 AT 480 F--REFERENCE INFORMATION

Reference	Investigator	Symbol	Composition and Surface Condition	Test Method	Remarks
19	Adams, J. G.		As received Heated 30 minutes in air at 1800 F Heated 30 minutes in 7.6 x 10^{-5} mm Hg pressure at 1800 F	Normal spectral emittance. Furnace-heated disk specimen. Comparison blackbody (Hohlraun). Spectrometer-monochromator with photomultiplier, lead sulphide, and thermocouple detectors. Temperatures measured with thermocouples.	Measured in air.

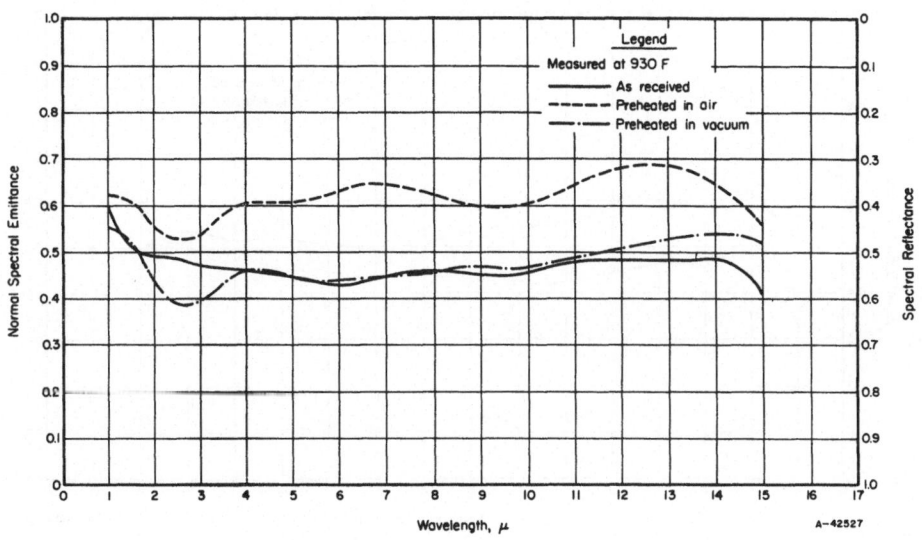

NORMAL SPECTRAL EMITTANCE OF RENÉ 41 AT 930 F

NORMAL SPECTRAL EMITTANCE OF RENE 41 AT 930 F--REFERENCE INFORMATION

Reference	Investigator	Symbol	Composition and Surface Condition	Test Method	Remarks
19	Adams, J. G.		As received Heated 30 minutes in air at 1800 F Heated 30 minutes in 7.6 x 10^{-5} mm Hg pressure at 1800 F	Normal spectral emittance. Furnace-heated disk specimen. Comparison blackbody (Hohlraun). Spectrometer-monochromator with photomultiplier, lead sulphide, and thermocouple detectors. Temperatures measured with thermocouples.	Measured in air.

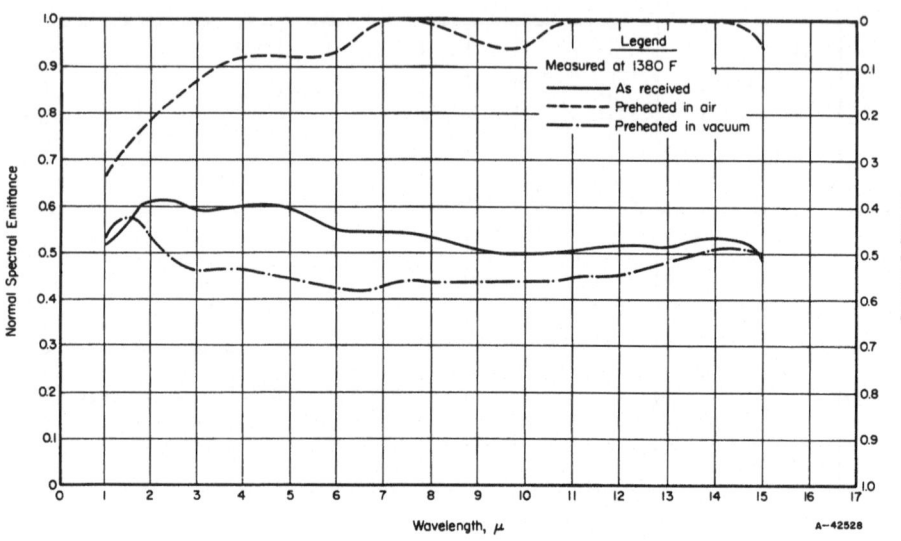

NORMAL SPECTRAL EMITTANCE OF RENÉ 41 AT 1380 F

NORMAL SPECTRAL EMITTANCE OF RENE 41 AT 1380 F--REFERENCE INFORMATION

Reference	Investigator	Symbol	Composition and Surface Condition	Test Method	Remarks
19	Adams, J. G.		As received Heated 30 minutes in air at 1800 F Heated 30 minutes in 7.6 x 10^{-5} mm Hg pressure at 1800 F	Normal spectral emittance. Furnace-heated disk specimen. Comparison blackbody (Hohlraun). Spectrometer-monochromator with photomultiplier, lead sulphide, and thermocouple detectors. Temperatures measured with thermocouples.	Measured in air.

133

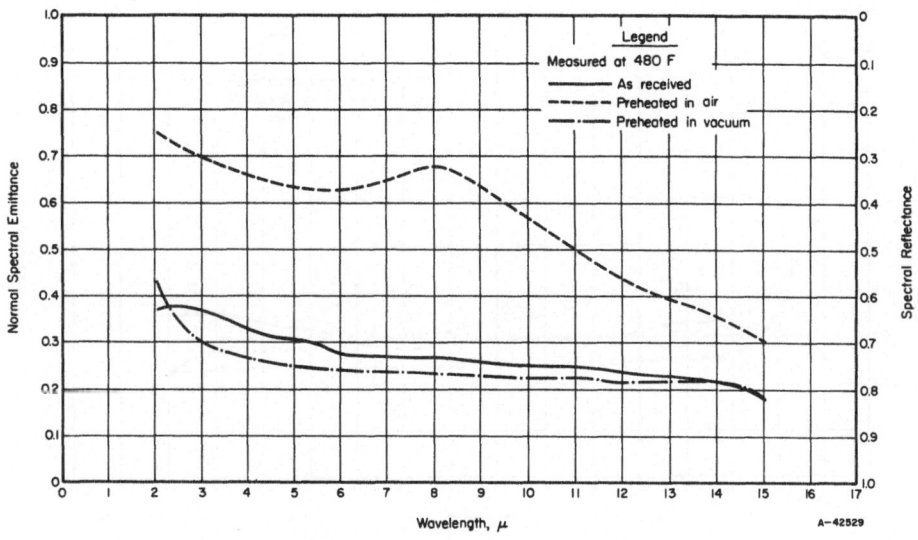

NORMAL SPECTRAL EMITTANCE OF ASTROLOY AT 480 F

NORMAL SPECTRAL EMITTANCE OF ASTROLOY AT 480 F--REFERENCE INFORMATION

Reference	Investigator	Symbol	Composition and Surface Condition	Test Method	Remarks
19	Adams, J. G.		As received Heated 30 minutes in air at 1800 F Heated 30 minutes in 7.6 x 10^{-5} mm Hg pressure at 1800 F	Normal spectral emittance. Furnace-heated disk specimen. Comparison blackbody (Hohlraun). Spectrometer-monochromator with photomultiplier, lead sulphide, and thermocouple detectors. Temperatures measured with thermocouples.	Measured in air.

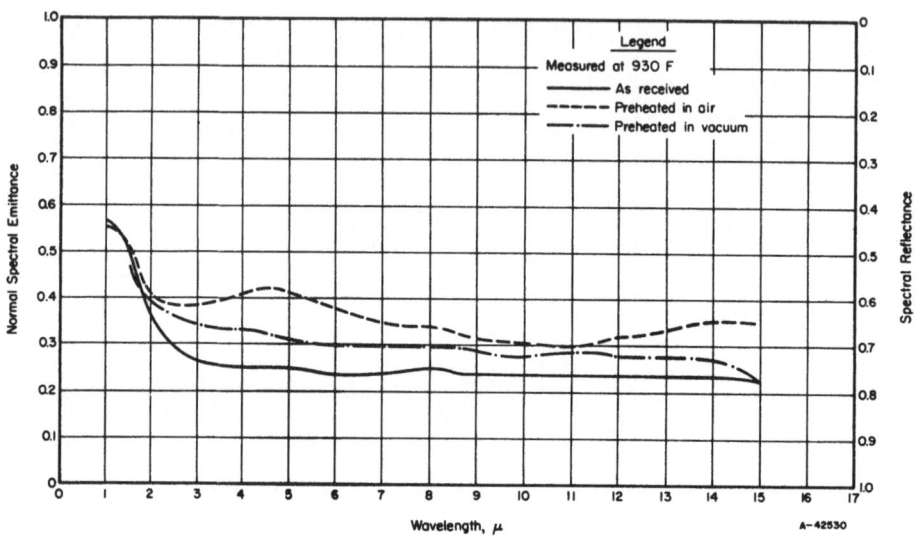

NORMAL SPECTRAL EMITTANCE OF ASTROLOY AT 930 F

NORMAL SPECTRAL EMITTANCE OF ASTROLOY AT 930 F--REFERENCE INFORMATION

Reference	Investigator	Symbol	Composition and Surface Condition	Test Method	Remarks
19	Adams, J. G.		As received Heated 30 minutes in air at 1800 F Heated 30 minutes in 7.6 x 10^{-5} mm Hg pressure at 1800 F	Normal spectral emittance. Furnace-heated disk specimen. Comparison blackbody (Hohlraun). Spectrometer-monochromator with photomultiplier, lead sulphide, and thermocouple detectors. Temperatures measured with thermocouples.	Measured in air.

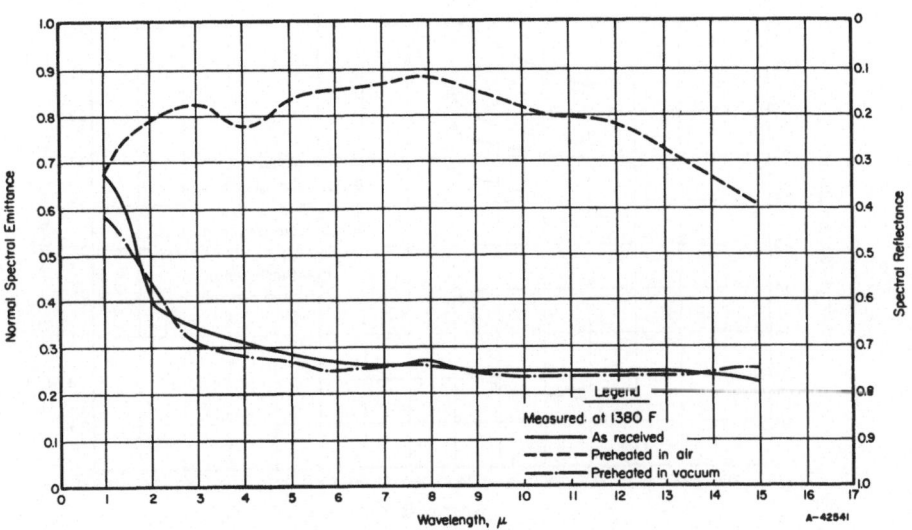

NORMAL SPECTRAL EMITTANCE OF ASTROLOY AT 1380 F

NORMAL SPECTRAL EMITTANCE OF ASTROLOY AT 1380 F--REFERENCE INFORMATION

Reference	Investigator	Symbol	Composition and Surface Condition	Test Method	Remarks
19	Adams, J. G.		As received Heated 30 minutes in air at 1800 F Heated 30 minutes in 7.6 x 10^{-5} mm Hg pressure at 1800 F	Normal spectral emittance. Furnace-heated disk specimen. Comparison blackbody (Hohlraun). Spectrometer-monochromator with photomultiplier, lead sulphide, and thermocouple detectors. Temperatures measured with thermocouples.	Measured in air.

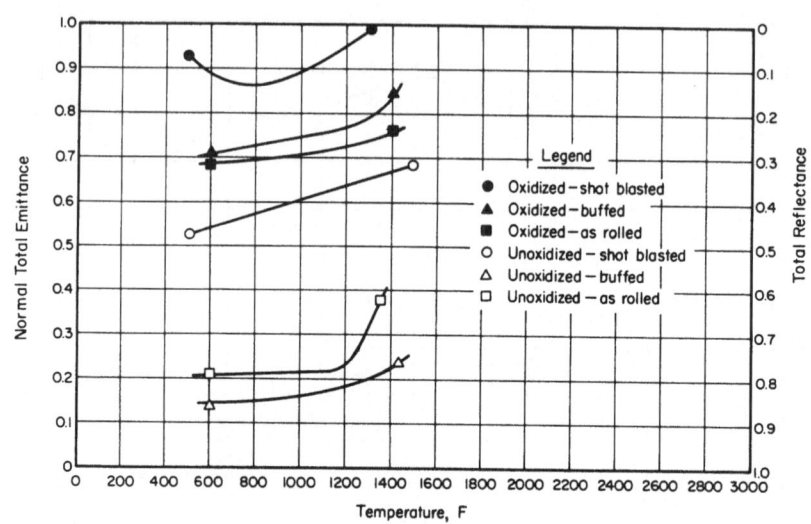

NORMAL TOTAL EMITTANCE OF NIMONIC 75

NORMAL TOTAL EMITTANCE OF NIMONIC 75--REFERENCE INFORMATION

Reference	Investigator	Symbol	Composition and Surface Condition	Test Method	Remarks
9	Sully, Brandes, and Waterhouse	● ▲ ■ ○ △ □	Oxidized, shot blasted Oxidized, buffed Oxidized, as rolled Unoxidized, shot blasted Unoxidized, buffed Unoxidized, as rolled (Oxidized at 2200 F) composition not given	Normal total emittance. Thermopile detector. Comparison blackbody. Temperatures measured with thermocouples. Self-resistance- heated specimen.	Measured in air. Data taken from curves.

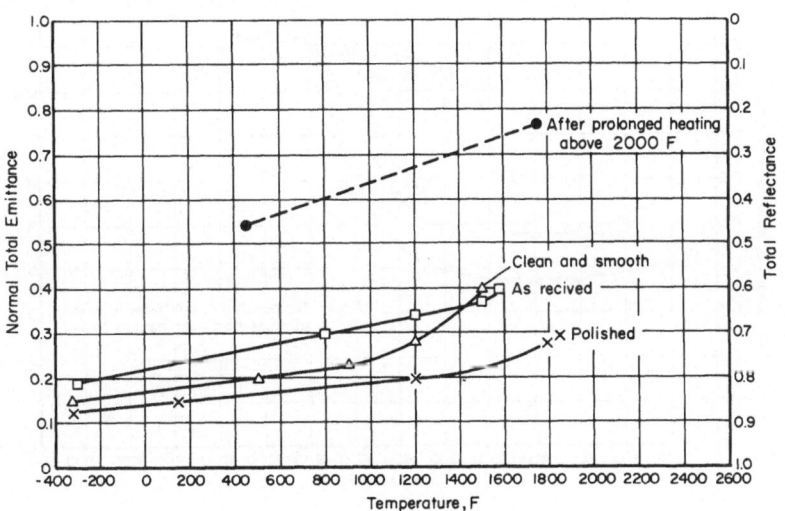

NORMAL TOTAL EMITTANCE OF K-MONEL 5700

NORMAL TOTAL EMITTANCE OF K-MONEL 5700—REFERENCE INFORMATION

Reference	Investigator	Symbol	Composition and Surface Condition	Test Method	Remarks
16	Wilkes, G. B.	✕ △ □ ●	Polished Clean and smooth As received After prolonged heating and cycling above 2000 F (some oxide indicated)	Normal total emittance. Total-radiation detector. Comparison blackbody. Temperatures measured with thermocouples.	Measured in a 10-micron pressure of helium.

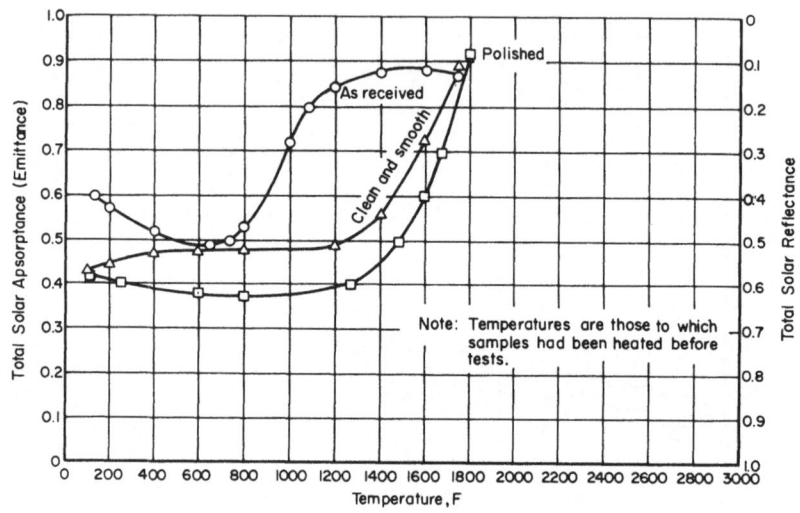

TOTAL SOLAR ABSORPTANCE OF K-MONEL 5700 AT 100 F

TOTAL SOLAR ABSORPTANCE OF K-MONEL 5700--REFERENCE INFORMATION

Reference	Investigator	Symbol	Composition and Surface Condition	Test Method	Remarks
16	Wilkes, G. B.	□ △ ○	Polished Clean and smooth As received	Total solar absorptance. Comparison standards. Comparison pyro- heliometer. Output measured with thermocouples.	Measured in air at 100 F. Temperatures are those to which samples had been heated previously.

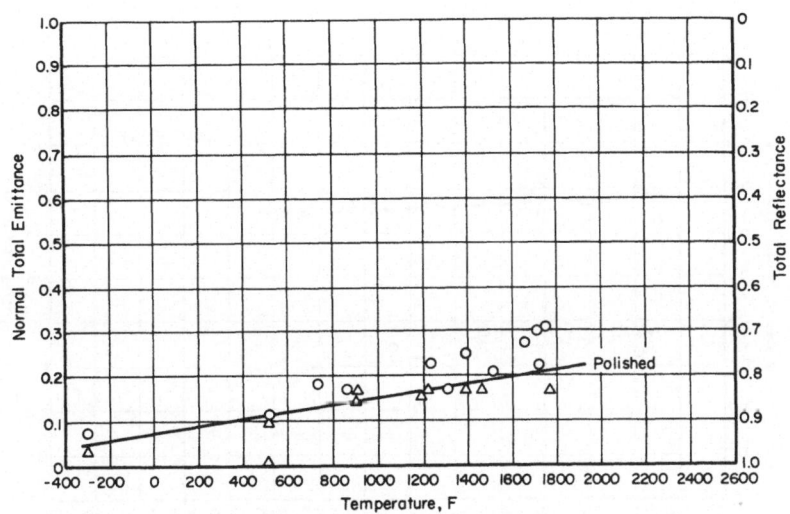

NORMAL TOTAL EMITTANCE OF HAYNES ALLOY B

NORMAL TOTAL EMITTANCE OF HAYNES ALLOY B--REFERENCE INFORMATION

Reference	Investigator	Symbol	Composition and Surface Condition	Test Method	Remarks
11	Betz, Olson, Schurin, and Morris	△	Polished: 15 microinches rms	Normal total emittance. Resistance-heated specimen. Thermistor-bolometer detector. Comparison blackbody. Temperatures measured with thermocouples.	Measured in vacuum.
		○	Polished: 2 microinches rms		

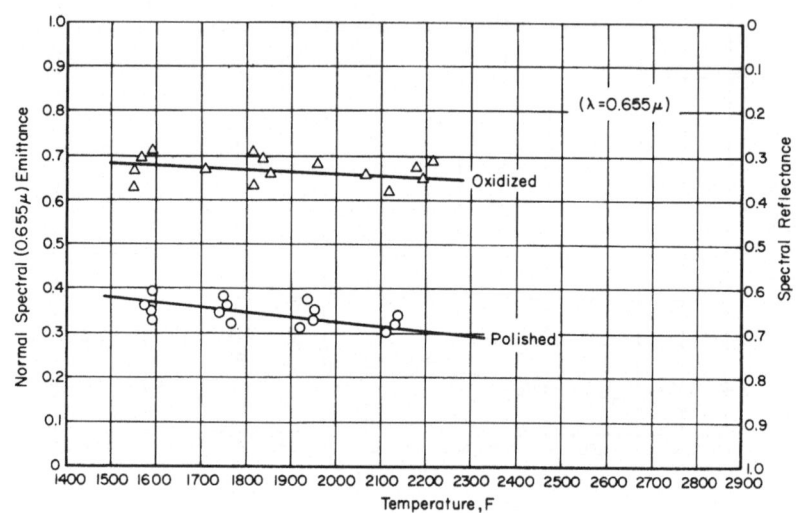

NORMAL SPECTRAL EMITTANCE OF HAYNES ALLOY B

NORMAL SPECTRAL EMITTANCE OF HAYNES ALLOY B--REFERENCE INFORMATION

Reference	Investigator	Symbol	Composition and Surface Condition	Test Method	Remarks
11	Betz, Olson, Schurin, and Morris	△	Polished: 2 microinches rms	Normal spectral emittance.	Measured in vacuum ($\lambda = 0.655\mu$)
		○	2 microinches rms - oxidized	Modified hole-in-tube method. Drilled blackbody hole. Temperatures measured with thermocouples.	Data taken from curves.

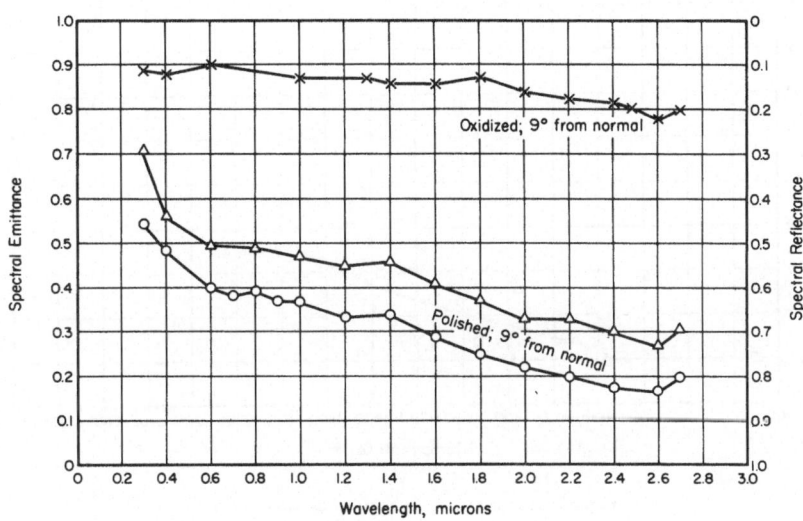

SPECTRAL EMITTANCE OF HAYNES ALLOY B

SPECTRAL EMITTANCE OF HAYNES ALLOY B--REFERENCE INFORMATION

Reference	Investigator	Symbol	Composition and Surface Condition	Test Method	Remarks
18	Olson and Morris	×	Oxidized	Spectral reflectance at 9° from normal. Monochromator, integrating sphere, lead sulphide detector, and MgO standard. "Normal" illumination, hemispherical viewing.	Measured in air at room temperature. Data taken from reflectance curve.
11	Betz, Olson, Schurin, and Morris	○	Polished: 2 microinches rms	Spectral reflectance at 9° from normal. Monochromator, integrating sphere reflectometer, lead sulphide detector, and MgO standard. "Normal" illumination, hemispherical viewing.	Measured in air at room temperature. Data taken from reflectance curves.
		△	Polished: 15 microinches rms		

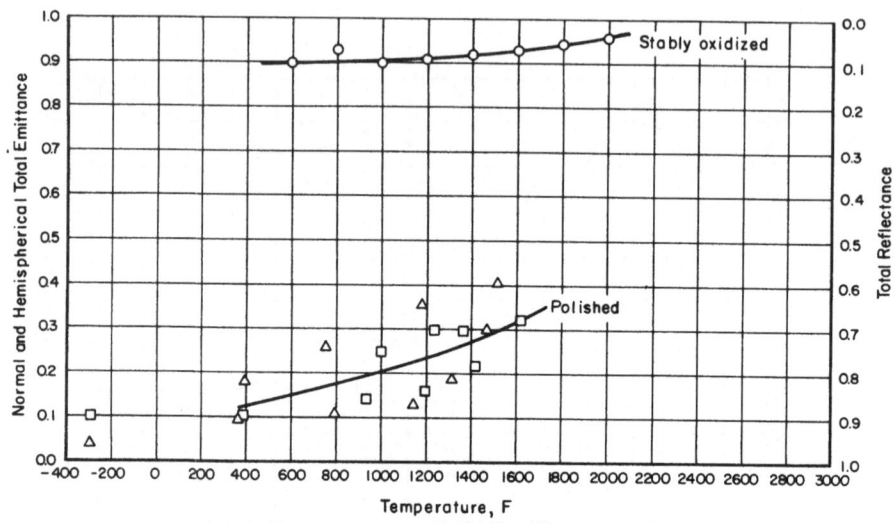

NORMAL AND HEMISPHERICAL TOTAL EMITTANCE OF HAYNES ALLOY C

HEMISPHERICAL TOTAL EMITTANCE OF HAYNES ALLOY C--REFERENCE INFORMATION

Reference	Investigator	Symbol	Composition and Surface Condition	Test Method	Remarks
17	Wade, W. R.	o	Stably oxidized at 2000 F (nominal composition: 52-60 Ni 16-18 Mo 15.5-17.5 Cr 4.5-7 Fe)	Hemispherical total emittance. Total radiation pyrometer. Comparison blackbody. Temperatures measured with thermocouples.	Measured in air. Normal total emittance equals hemispherical total emittance for this specimen. Data taken from curve.
11	Betz, Olson, Schurin, and Morris	△ □	Polished: 2 microinches rms Polished: 15 microinches rms	Normal total emittance. Resistance-heated specimens. Thermistor-bolometer detector. Comparison blackbody. Temperatures measured with thermocouples.	Measured in vacuum. Data taken from curves.

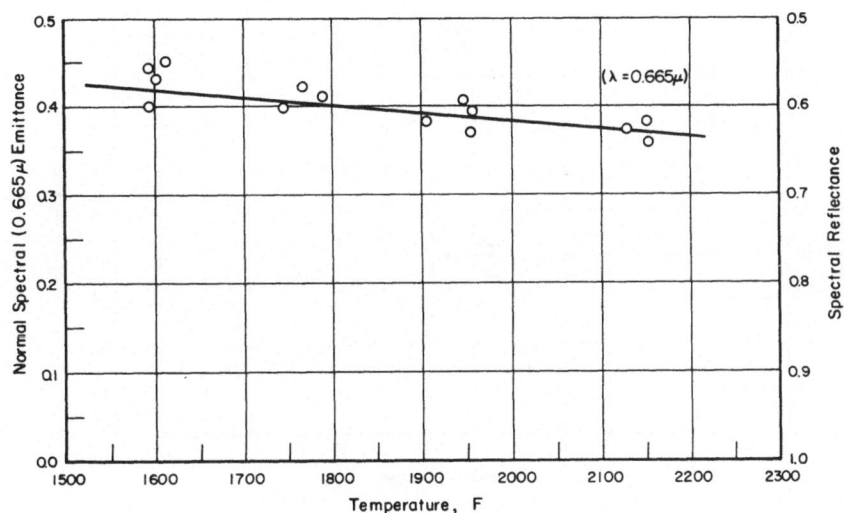

NORMAL SPECTRAL EMITTANCE OF HAYNES ALLOY C

NORMAL SPECTRAL EMITTANCE OF HAYNES ALLOY C--REFERENCE INFORMATION

Reference	Investigator	Symbol	Composition and Surface Condition	Test Method	Remarks
11	Betz, Olson, Schurin, and Morris	o	Polished: 2 and 15 microinches rms	Normal spectral emittance. Modified hole-in-tube method. Drilled blackbody hole. Temperatures measured with thermocouples.	Measured in vacuum. (λ = 0.655μ). Data taken from curves.

144

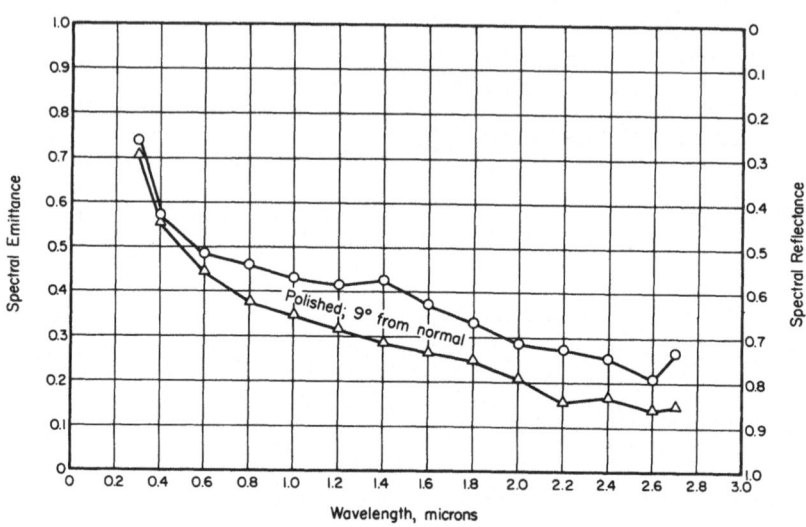

SPECTRAL EMITTANCE OF HAYNES ALLOY C

SPECTRAL EMITTANCE OF HAYNES ALLOY C--REFERENCE INFORMATION

Reference	Investigator	Symbol	Composition and Surface Condition	Test Method	Remarks
11	Betz, Olson, Schurin, and Morris	△	Polished: 2 microinches rms	Spectral reflectance at 9° from normal. Monochromator, inte- grating sphere re- flectometer, lead sulphide detector, and MgO standard. "Normal" illumi- nation, hemispherical viewing.	Measured in air at room temperature. Data taken from reflectance curves.
		○	Polished: 15 microinches rms		

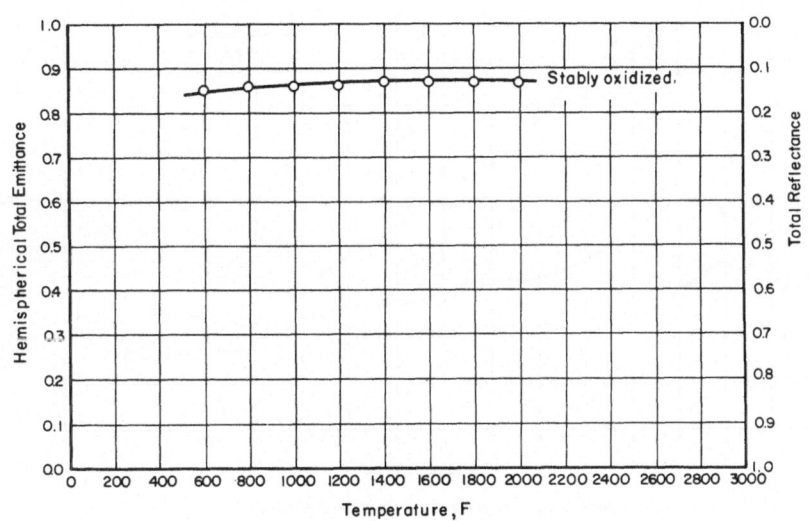

HEMISPHERICAL TOTAL EMITTANCE OF HAYNES ALLOY X

HEMISPHERICAL TOTAL EMITTANCE OF HAYNES ALLOY X--REFERENCE INFORMATION

Reference	Investigator	Symbol	Composition and Surface Condition	Test Method	Remarks
17	•Wade, W. R.	o	Stably oxidized at 2000 F (nominal composition: 42-52 Ni 20-23 Cr 8-10 Mo)	Hemispherical total emittance. Total-radiation pyro-meter. Comparison blackbody. Temperatures measured with thermocouples.	Measured in air. Normal total emittance equals hemispherical total emittance for this specimen. Data taken from curve.

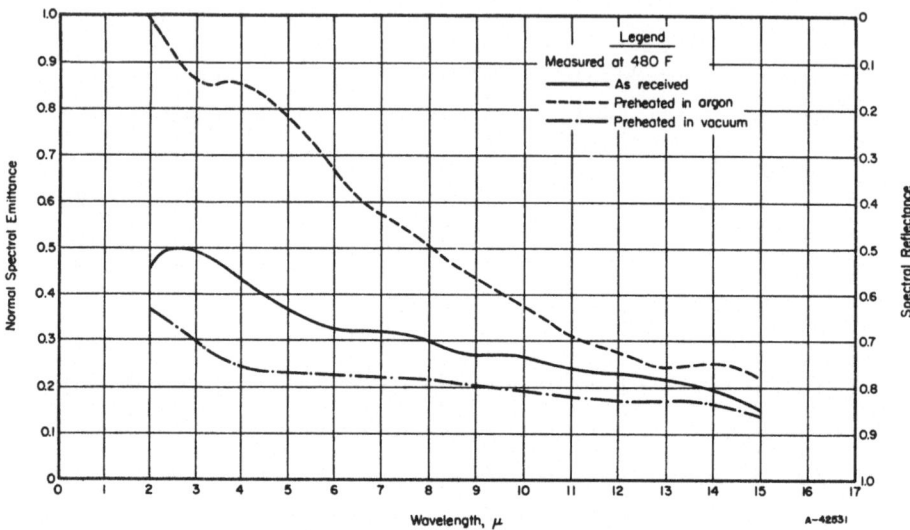

NORMAL SPECTRAL EMITTANCE OF HAYNES ALLOY X AT 480 F

NORMAL SPECTRAL EMITTANCE OF HAYNES ALLOY X AT 480 F—REFERENCE INFORMATION

Reference	Investigator	Symbol	Composition and Surface Condition	Test Method	Remarks
19	Adams, J. G.		As received Heated 30 minutes in argon at 2000 F Heated 30 minutes in 2.5 x 10⁻⁵ mm Hg pressure at 2000 F	Normal spectral emittance. Furnace-heated disk specimen. Comparison blackbody (Hohlraun). Spectrometer-monochromator with photomultiplier, lead sulphide, and thermocouple detectors. Temperatures measured with thermocouples.	Measured in argoh.

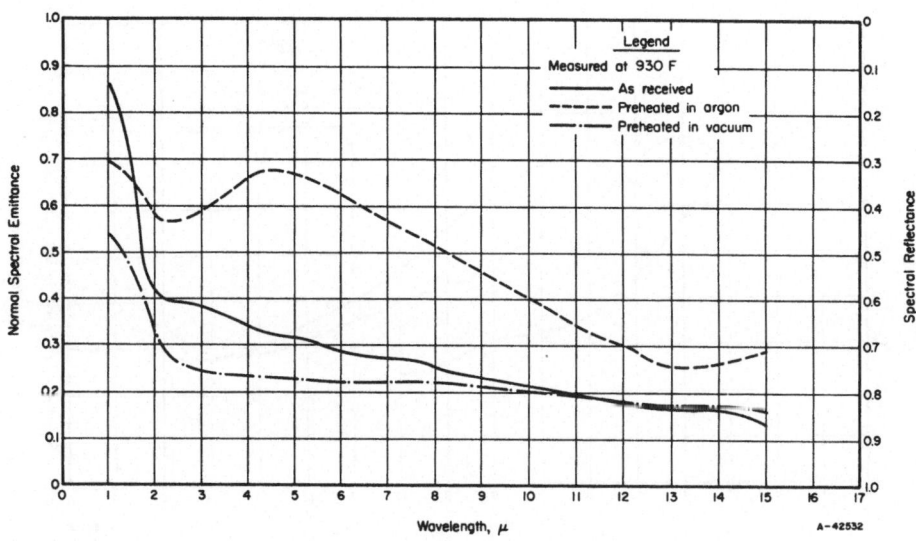

NORMAL SPECTRAL EMITTANCE OF HAYNES ALLOY X AT 930 F

NORMAL SPECTRAL EMITTANCE OF HAYNES ALLOY X AT 930 F--REFERENCE INFORMATION

Reference	Investigator	Symbol	Composition and Surface Condition	Test Method	Remarks
19	Adams, J. G.		As received Heated 30 minutes in argon at 2000 F Heated 30 minutes in 2.5 x 10⁻⁵ mm Hg pressure at 2000 F	Normal spectral emittance. Furnace-heated disk specimen. Comparison blackbody (Hohlraun). Spectrometer-monochromator with photomultiplier, lead sulphide, and thermocouple detectors. Temperatures measured with thermocouples.	Measured in argon.

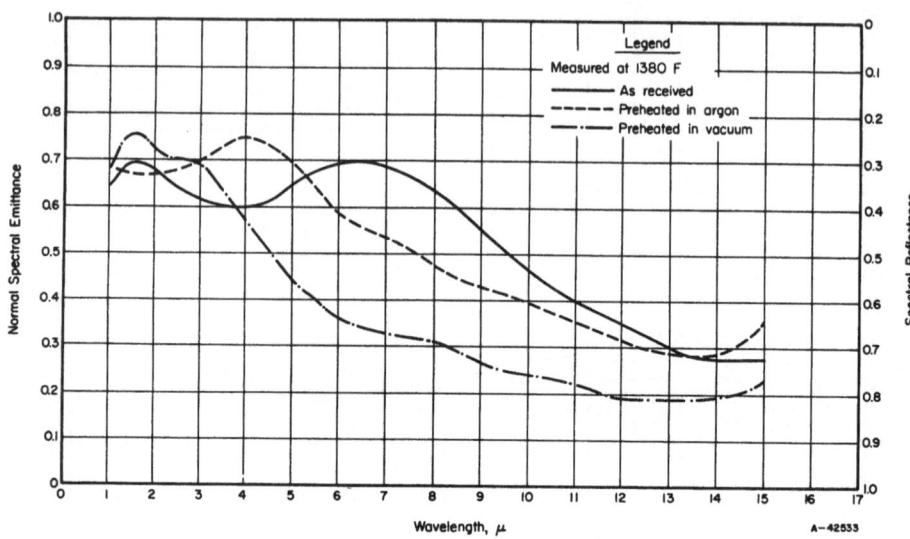

NORMAL SPECTRAL EMITTANCE OF HAYNES ALLOY X AT 1380 F

NORMAL SPECTRAL EMITTANCE OF HAYNES ALLOY X AT 1380 F--REFERENCE INFORMATION

Reference	Investigator	Symbol	Composition and Surface Condition	Test Method	Remarks
19	Adams, J. G.		As received Heated 30 minutes in argon at 2000 F Heated 30 minutes in 2.5 x 10^{-5} mm Hg pressure at 2000 F	Normal spectral emittance. Furnace-heated disk specimen. Comparison blackbody (Hohlraun). Spectrometer-monochromator with photomultiplier, lead sulphide, and thermocouple detectors. Temperatures measured with thermocouples.	Measured in argon.

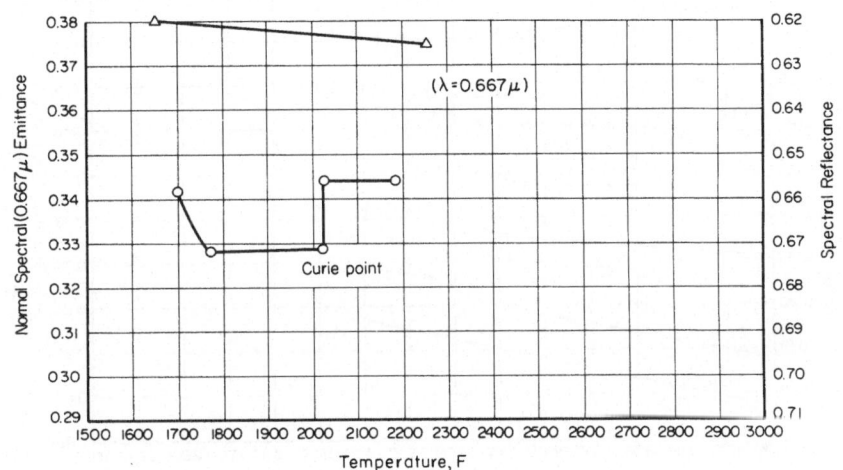

NORMAL SPECTRAL EMITTANCE OF COBALT

NORMAL SPECTRAL EMITTANCE OF COBALT--REFERENCE INFORMATION

Reference	Investigator	Symbol	Composition and Surface Condition	Test Method	Remarks
13	Wahlin and Knop	○	Electrolytically deposited on stainless steel	Normal spectral emittance. Hole-in-tube method.	Measured in vacuum. Data taken from curve. ($\lambda = 0.667\mu$)
12	Wahlin and Wright	△	Electrolytically deposited on stainless steel	Normal spectral emittance. Hole-in-tube method.	Measured in vacuum. Data taken from curve. ($\lambda = 0.667\mu$)

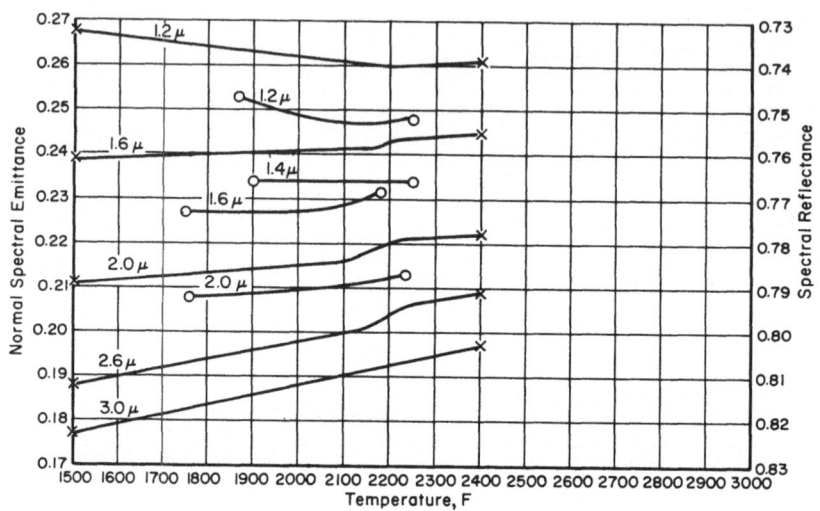

NORMAL SPECTRAL EMITTANCE OF COBALT AT DIFFERENT WAVELENGTHS

NORMAL SPECTRAL EMITTANCE OF COBALT AT DIFFERENT WAVELENGTHS--REFERENCE INFORMATION

Reference	Investigator	Symbol	Composition and Surface Condition	Test Method	Remarks
3	Lund and Ward	O	Polished with 000-grade emery paper, and cleaned	Normal spectral emittance. Hole-in-tube method. Temperatures measured with thermocouples. Spectrometer-mono-chromator detector.	Measured in vacuum, or hydrogen atmos-phere. Data taken from curves.
4	Ward, L.	x	Surface lapped with jewelers rouge	Normal spectral emittance. Hole-in-tube method. Temperatures measured with thermocouples. Spectrometer-mono-chromator detector.	Measured in hydrogen. Data taken from curves.

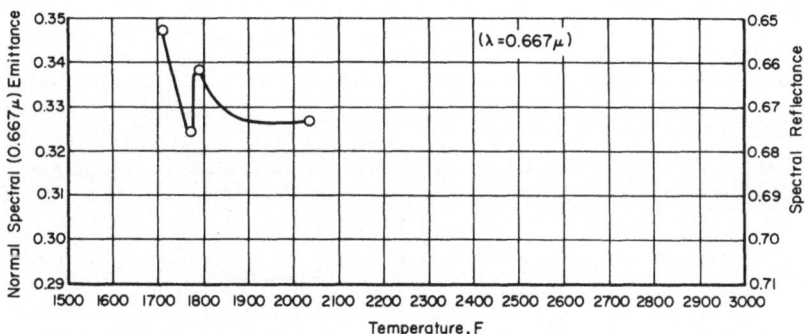

NORMAL SPECTRAL EMITTANCE OF COBALT-35 PER CENT NICKEL ALLOY

NORMAL SPECTRAL EMITTANCE OF COBALT-35 PER CENT NICKEL ALLOY--REFERENCE INFORMATION

Reference	Investigator	Symbol	Composition and Surface Condition	Test Method	Remarks
13	Wahlin and Knop	○	As received	Normal spectral emittance. Hole-in-tube method.	Measured in vacuum. Data taken from curve.

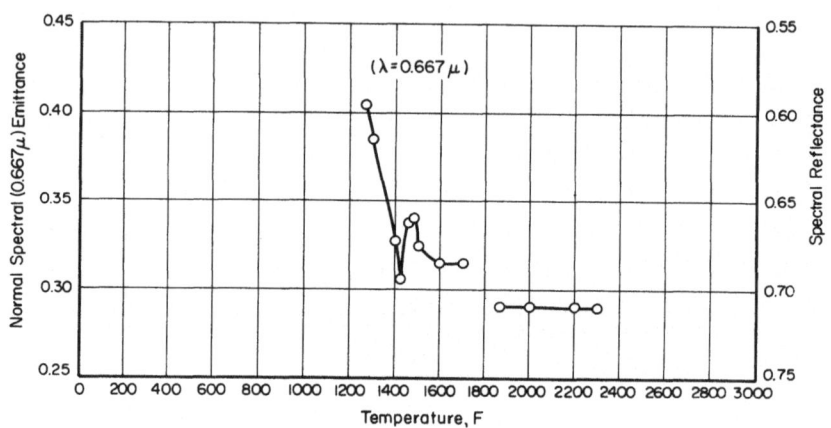

NORMAL SPECTRAL EMITTANCE OF COBALT-40 PER CENT IRON ALLOY

NORMAL SPECTRAL EMITTANCE OF COBALT-40 PER CENT IRON ALLOY--REFERENCE INFORMATION

Reference	Investigator	Symbol	Composition and Surface Condition	Test Method	Remarks
7	Knop, Jr., H. W.	o	As rolled	Normal spectral emittance. Hole-in-tube method. Calibrated optical pyrometer.	Measured in vacuum. ($\lambda = 0.667\mu$) Data taken from curve.

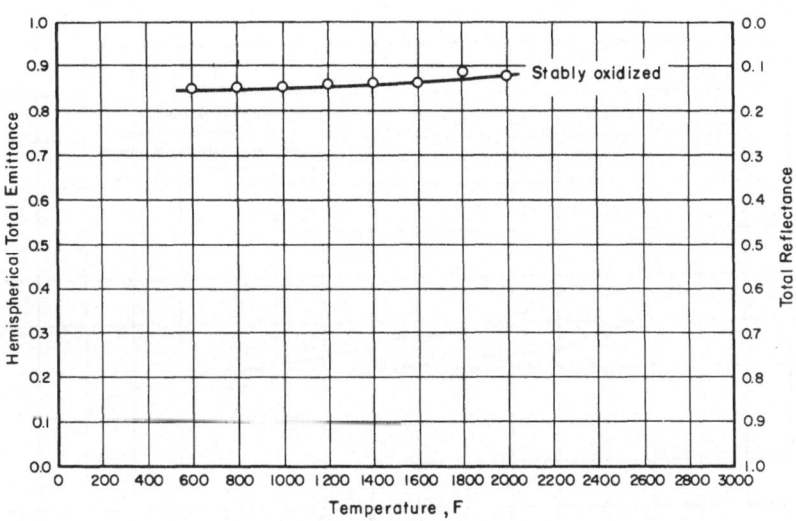

HEMISPHERICAL TOTAL EMITTANCE OF HAYNES ALLOY 25

HEMISPHERICAL TOTAL EMITTANCE OF HAYNES ALLOY 25--REFERENCE INFORMATION

Reference	Investigator	Symbol	Composition and Surface Condition	Test Method	Remarks
17	Wade, W. R.	○	Stably oxidized at 2000 F (nominal composition: 46–53 Co 19–21 Cr 9–11 Ni 14–16 W)	Hemispherical total emittance. Total-radiation pyrometer. Comparison blackbody. Temperatures measured with thermocouples.	Measured in air. Normal total emittance equals hemispherical total emittance for this specimen. Data taken from curve.

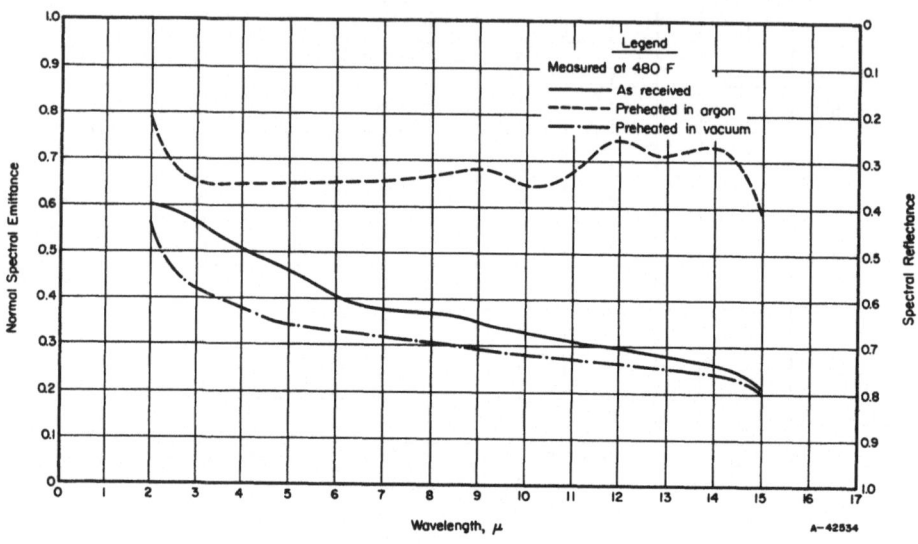

NORMAL SPECTRAL EMITTANCE OF HAYNES ALLOY ·25 AT 480 F

NORMAL SPECTRAL EMITTANCE OF HAYNES ALLOY 25 AT 480 F—REFERENCE INFORMATION

Reference	Investigator	Symbol	Composition and Surface Condition	Test Method	Remarks
19	Adams, J. G.		As received Heated 30 minutes in air at 1800 F Heated 30 minutes in 7.6 x 10^{-5} mm Hg pressure at 1800 F	Normal spectral emittance. Furnace-heated disk specimen. Comparison blackbody (Hohlraun). Spectrometer-monochromator with photomultiplier, lead sulphide, and thermocouple detectors. Temperatures measured with thermocouples.	Measured in air.

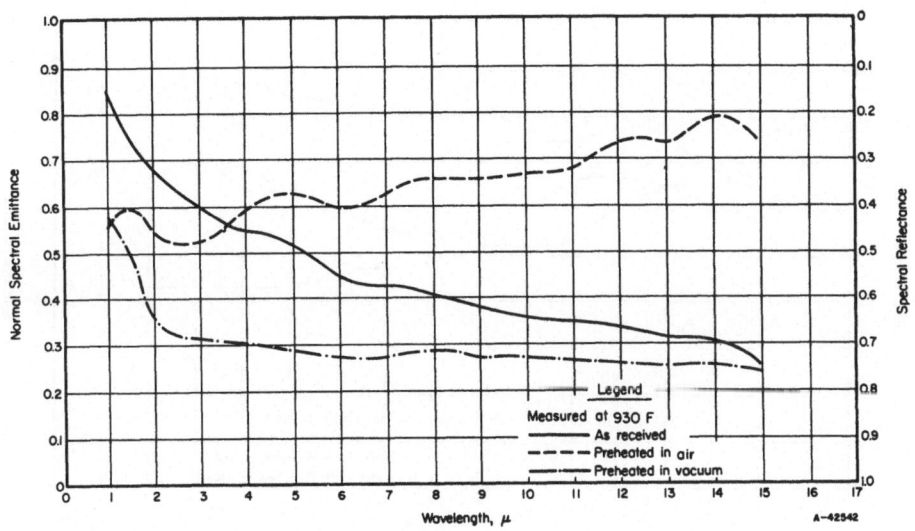

NORMAL SPECTRAL EMITTANCE OF HAYNES ALLOY 25 AT 930 F

NORMAL SPECTRAL EMITTANCE OF HAYNES ALLOY 25 AT 930 F--REFERENCE INFORMATION

Reference	Investigator	Symbol	Composition and Surface Condition	Test Method	Remarks
19	Adams, J. G.		As received Heated 30 minutes in air at 1800 F Heated 30 minutes in 7.6 x 10^{-5} mm Hg pressure at 1800 F	Normal spectral emittance. Furnace-heated disk specimen. Comparison blackbody (Hohlraun). Spectrometer-mono-chromator with photomultiplier, lead sulphide, and thermocouple detectors. Temperatures measured with thermocouples.	Measured in air.

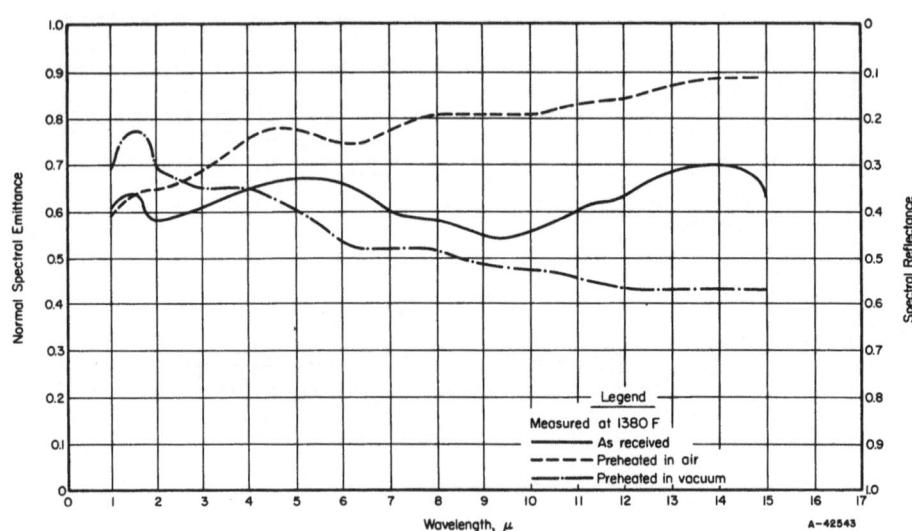

NORMAL SPECTRAL EMITTANCE OF HAYNES ALLOY 25 AT 1380 F

NORMAL SPECTRAL EMITTANCE OF HAYNES ALLOY 25 AT 1380 F--REFERENCE INFORMATION

Reference	Investigator	Symbol	Composition and Surface Condition	Test Method	Remarks
19	Adams, J. G.		As received Heated 30 minutes in air at 1800 F Heated 30 minutes in 7.6 x 10^{-5} mm Hg pressure at 1800 F	Normal spectral emittance. Furnace heated disk specimen. Comparison blackbody (Hohlraun). Spectrometer-monochromator with photomultiplier, lead sulphide, and thermocouple detectors. Temperatures measured with thermocouples.	Measured in air.

REFERENCES

(1) Wahlin, H. B. , Zentner, R. , and Martin, J. , "The Spectral Emissivity of Iron-Nickel Alloys", J. Appl. Phys. , 23 (1), 107-108 (January, 1952).

(2) Price, D. J. , "The Emissivity of Hot Metals in the Infra-red", Proc. Phys. Soc. (London), 59, 118-131 (1947).

(3) Lund, H. , and Ward, L. , "The Spectral Emissivities of Iron, Nickel, and Cobalt", Proc. Phys. Soc. , B, 65, 535-540 (1952).

(4) Ward, L. , "The Variation With Temperature of the Spectral Emissivities of Iron, Nickel, and Cobalt", Proc. Phys. Soc. B, 69, 339-343 (1956).

(5) Dastur, M. N. , and Gokcen, N. A. , "Optical Temperature Scale and Emissivity of Liquid Iron", Metals Trans. , 185, 665-667 (October, 1949).

(6) Knowles, D. , and Sarjant, R. J. , "Emissivity of Molten Iron and Steel", Iron and Steel Inst. J. , 155, 577-592 (1947).

(7) Knop, H. W. , Jr. , "The Emissivity of Iron-Tungsten and Iron-Cobalt Alloys", Phys. Rev. , 74 (10), 1413-1416 (November 15, 1948).

(8) Gow, J. T. , Brasunas, A. DeS. , and Harder, O. E. , "The Liquidus-Solidus Temperatures and Emissivities of Some Commercial Heat-Resistant Alloys", Trans. AIME, 162, 156-174 (1945).

(9) Sully, A. H. , Brandes, E. A. , and Waterhouse, R. B. , "Some Measurements of the Total Emissivity of Metals and Pure Refractory Oxides and the Variation of Emissivity with Temperature", Brit. J. Appl. Phys. , 3, 97-101 (March, 1952).

(10) Fulk, M. M. , and Reynolds, M. M. , "Emissivities of Metallic Surfaces at 76°K", J. Appl. Phys. , 28 (12), 1464-1467 (December, 1957).

(11) Betz, H. T. , Olson, H. O. , Schurin, B. D. , and Morris, J. C. , "Determination of Emissivity and Reflectivity Data on Aircraft Structural Materials", WADC TR 56-222, Part II, Contract No. AF 33(616)-3002.

(12) Wahlin, H. B. , and Wright, R. , "Emissivities and Temperature Scales of the Iron Group", J. Appl. Phys. , 13, 40-42 (January, 1942).

(13) Wahlin, H. B. , and Knop, H. W. , Jr. , "The Spectral Emissivity of Iron and Cobalt", Phys. Rev. , 74 (6), 687-689 (September 15, 1948).

(14) O'Sullivan, W. J. , Jr. , and Wade, W. R. , "Theory and Apparatus for Measurement of Emissivity for Radiative Cooling of Hypersonic Aircraft With Data for Inconel and Inconel X", NACA TN 4121 (1957).

(15) DeCorso, S. M. , and Coit, R. L. , "Measurement of Total Emissivities of Gas-Turbine Combustor Materials", Trans. ASME, 77, 1189-1197 (November, 1955).

(16) Wilkes, G. B. , "Total Normal Emissivities and Solar Absorptivities of Materials", WADC TR 54-42 (March, 1954).

(17) Wade, W. R. , "Measurements of Total Hemispherical Emissivity of Several Stably Oxidized Metals and Some Refractory Oxide Coatings", NASA Memorandum No. 1-20-59L (January, 1959).

(18) Olson, H. O. , and Morris, J. C. , "Determination of Emissivity and Reflectivity Data on Aircraft Structural Materials", WADC TR 56-222, Part II, Suppl. 1 (October, 1958).

(19) Adams, J. G. , "The Determination of Spectral Emissivities, Reflectivities, and Absorptivities of Materials and Coatings", Northrup Corporation Report No. NOR-61-189 (August 3, 1961).

RADIATIVE PROPERTY DATA

Chromium, Columbium, Molybdenum, Tantalum, and
Tungsten and Their Alloys

TABLE OF CONTENTS

TABLE OF CONTENTS
(Continued)

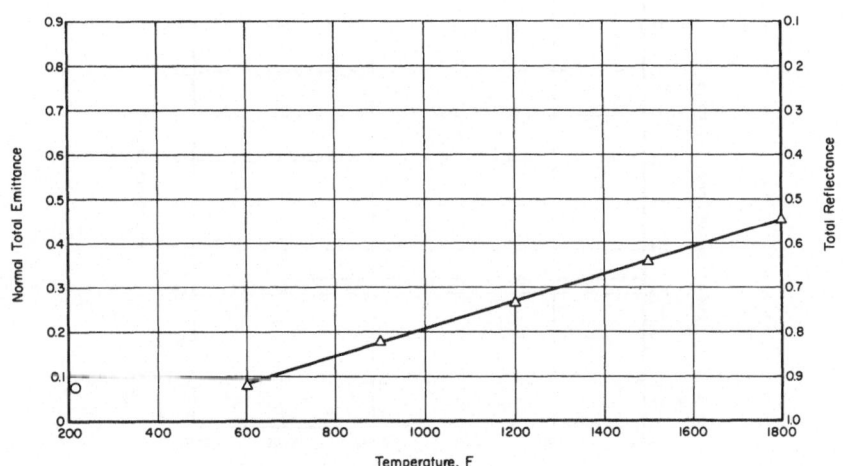

NORMAL TOTAL EMITTANCE OF CHROMIUM

NORMAL TOTAL EMITTANCE OF CHROMIUM--REFERENCE INFORMATION

Reference	Investigator	Symbol	Composition and Surface Condition	Test Method	Remarks
1	Barnes, Forsythe, and Adams	O	Polished.	Normal total emittance. Disk specimen heated by contact with heat-diffusion disk. Thermopile detector. Temperatures measured with thermocouples. Comparison blackbody.	Measured in air. Data taken from table.
2	Betz, Olson, Schurin, and Morris	△	As received, cleaned, polished, and oxidized.	Normal total emittance. Electrically heated strip specimen. Thermistor detector. Comparison blackbody. Temperatures measured with thermocouples.	Measured in vacuum. Data taken from table.

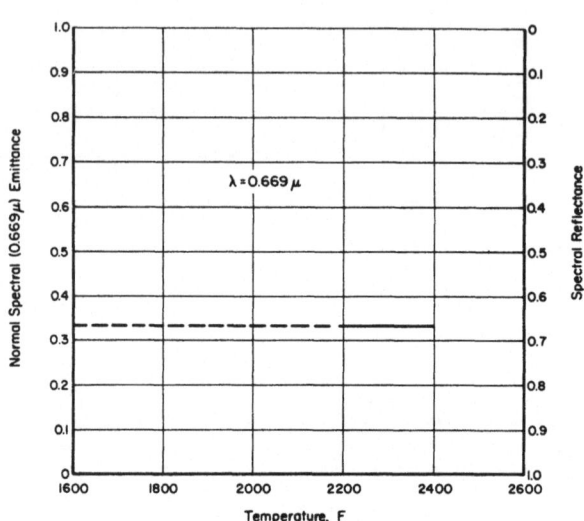

$\lambda = 0.669\,\mu$

NORMAL SPECTRAL EMITTANCE OF CHROMIUM

NORMAL SPECTRAL EMITTANCE OF CHROMIUM--REFERENCE INFORMATION

Reference	Investigator	Symbol	Composition and Surface Condition	Test Method	Remarks
3	Wahlin, H. B.		Chromium electrodeposited on brass tube; brass then removed with nitric acid. Heat treated (reduced) in hydrogen for 1 week at 2200 F.	Normal spectral emittance. Hole-in-tube method. Temperatures measured with optical pyrometer.	Measured in vacuum or hydrogen. Data taken from discussion. (Lower temperature not defined, top temperature only is given and emittance independent of temperature.) ($\lambda = 0.669\,\mu$)

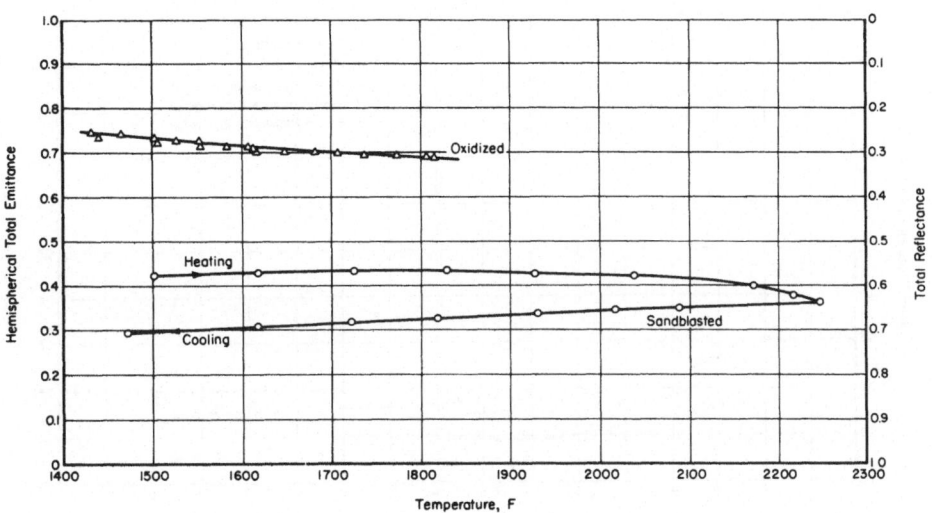

HEMISPHERICAL TOTAL EMITTANCE OF COLUMBIUM

HEMISPHERICAL TOTAL EMITTANCE OF COLUMBIUM--REFERENCE INFORMATION

Reference	Investigator	Symbol	Composition and Surface Condition	Test Method	Remarks
4	Pratt & Whitney Aircraft	O	Sand blasted with No. G-25 grit.	Hemispherical total emittance. Hole-in-tube method. Temperatures measured with thermocouples. Measured power input to center section.	Measured in vacuum. Data taken from tables.
		Δ	Oxidized in air 5 minutes at 1200 F. [Powder oxide (Cb_2O_5) removed by brushing to leave black CbO.]		

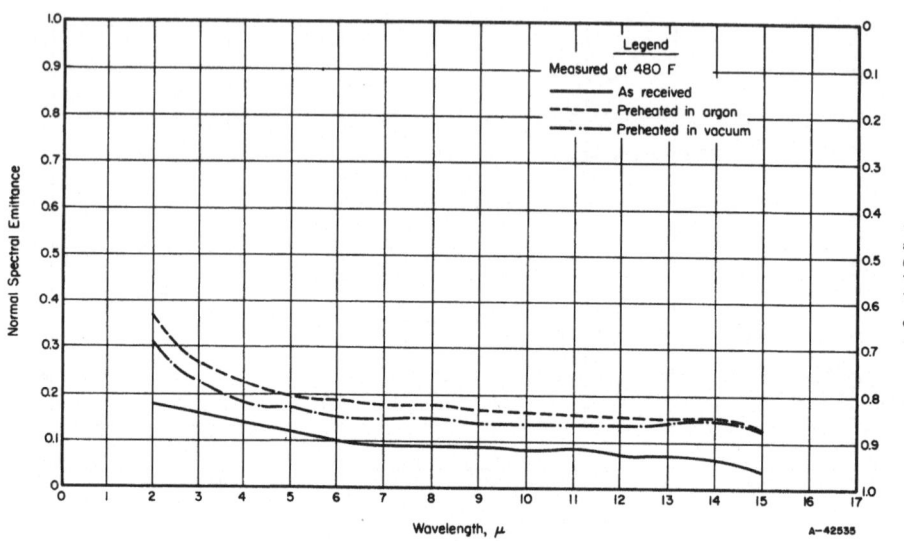

NORMAL SPECTRAL EMITTANCE OF COLUMBIUM AT 480 F

NORMAL SPECTRAL EMITTANCE OF COLUMBIUM AT 480 F--REFERENCE INFORMATION

Reference	Investigator	Symbol	Composition and Surface Condition	Test Method	Remarks
25	Adams, J. G.		As received Heated 30 minutes in argon at 2000 F Heated 30 minutes in 2.2 x 10^{-5} mm Hg pressure at 2000 F	Normal spectral emittance. Furnace-heated disk specimen. Comparison blackbody (Hohlraun). Spectrometer-monochromator with photomultiplier, lead sulphide, and thermocouple detectors. Temperatures measured with thermocouples.	Measured in argon.

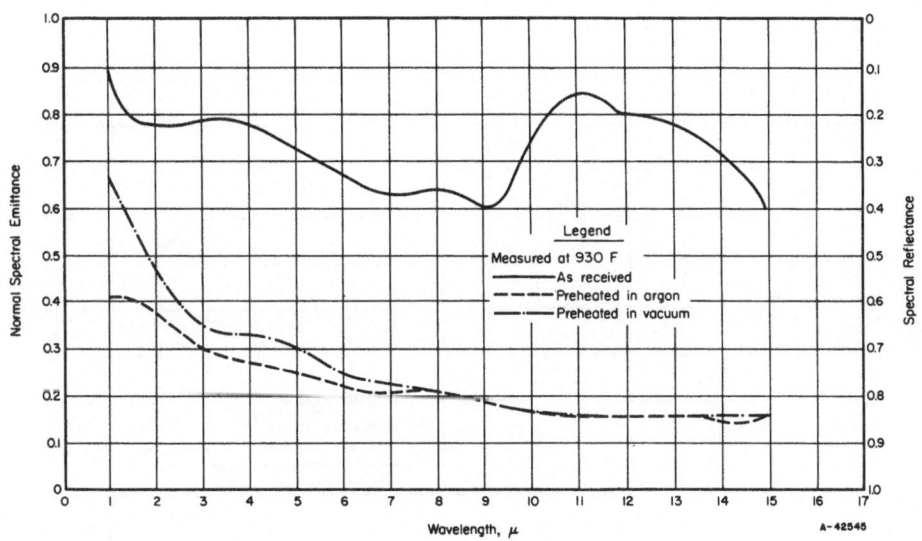

NORMAL SPECTRAL EMITTANCE OF COLUMBIUM AT 930 F

NORMAL SPECTRAL EMITTANCE OF COLUMBIUM AT 930 F--REFERENCE INFORMATION

Reference	Investigator	Symbol	Composition and Surface Condition	Test Method	Remarks
25	Adams, J. G.		As received Heated 30 minutes in argon at 2000 F Heated 30 minutes in 2.2 x 10^{-5} mm Hg pressure at 2000 F	Normal spectral emittance. Furnace-heated disk specimen. Comparison blackbody (Hohlraun). Spectrometer-monochromator with photomultiplier, lead sulphide, and thermocouple detectors. Temperatures measured with thermocouples.	Measured in argon.

166

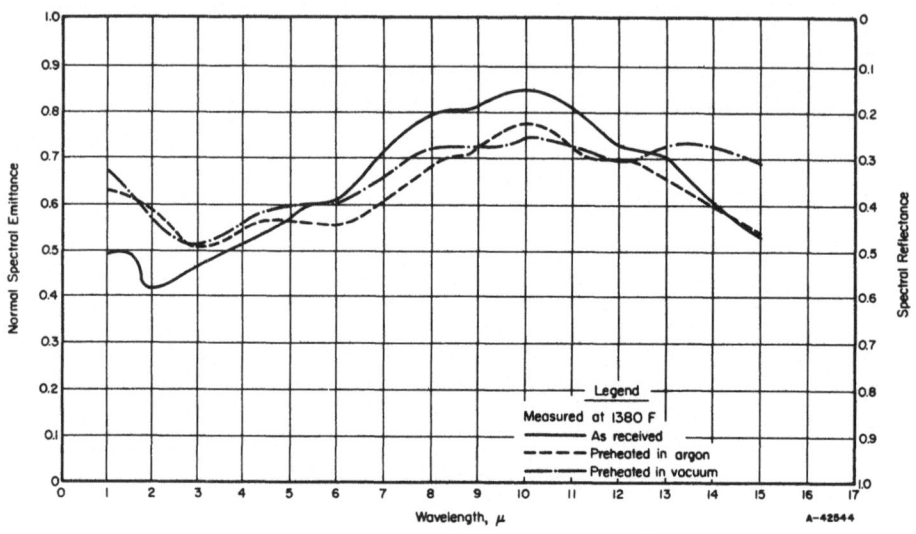

NORMAL SPECTRAL EMITTANCE OF COLUMBIUM AT 1380 F

NORMAL SPECTRAL EMITTANCE OF COLUMBIUM AT 1380 F—REFERENCE INFORMATION

Reference	Investigator	Symbol	Composition and Surface Condition	Test Method	Remarks
25	Adams, J. G.		As received Heated 30 minutes in argon at 2000 F Heated 30 minutes in 2.2 x 10^{-5} mm Hg pressure at 2000 F	Normal spectral emittance. Furnace-heated disk specimen. Comparison blackbody (Hohlraun). Spectrometer-monochromator with photomultiplier, lead sulphide, and thermocouple detectors. Temperatures measured with thermocouples.	Measured in argon.

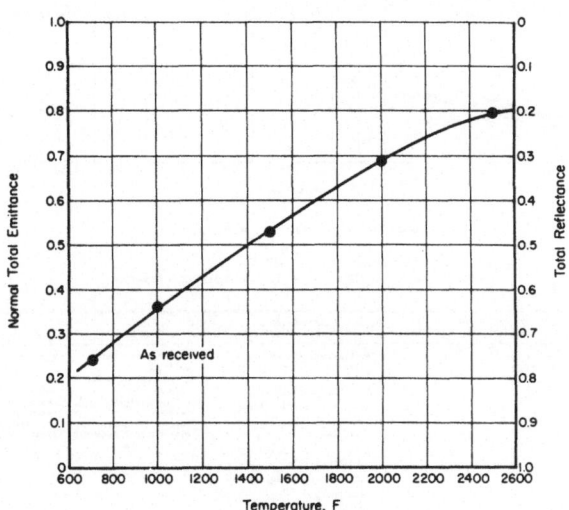

NORMAL TOTAL EMITTANCE OF 10Ti-10Mo COLUMBIUM ALLOY

NORMAL TOTAL EMITTANCE OF 10Ti-10Mo COLUMBIUM ALLOY--REFERENCE INFORMATION

Reference	Investigator	Symbol	Composition and Surface Condition	Test Method	Remarks
5	Anthony and Pearl		Cb-10Ti-10Mo as received. Bare Oxidation occurred during test.	Normal total emittance. Induction heated specimen. Thermopile detector. Comparison blackbody. Temperatures measured with thermocouples and optical pyrometer.	Measured in continuous purge of helium gas.

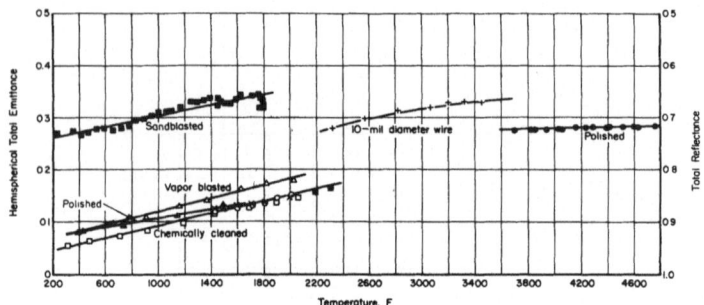

HEMISPHERICAL TOTAL EMITTANCE OF MOLYBDENUM

HEMISPHERICAL TOTAL EMITTANCE OF MOLYBDENUM--REFERENCE INFORMATION

Reference	Investigator	Symbol	Composition and Surface Condition	Test Method	Remarks
4	Pratt & Whitney Aircraft		As received and chemically cleaned. Purity not given. As-received specimen Hole-in-tube measurements:	Hemispherical total emittance. Hole-in-tube and re-sistance-heated-strip-specimen methods.	Measured in vacuum. Data taken from curves. Numerous temperature cycles were made on each specimen.
		O	Temperature measured with pyrometer.	Temperatures measured with optical pyrome-ter or thermocouple.	All specimens were from the same stock of material.
		X	Temperature measured with thermocouples.		
			Heated-strip measurements.		
		☐	Run No. 1		
		△	Vapor blasted First heating. Fourth heating.		
4	Pratt & Whitney Aircraft		Blasted with No. 90 (PWC3043A) aluminum oxide.	Hemispherical total emittance. Resistance-heated strip specimen. Temperatures measured with thermocouples.	
		■	Held at temperature 17 hours at 892 F and again at 1475 F.		
7	Glasier, Allen, and Saldinger	●	Polished with 0000 abrasive paper.	Hemispherical total emittance. Power dissipated from electrically heated rod specimen. Brightness temper-atures measured with optical pyrometer - converted to true temperatures using values obtained at the melting point.	Measured in flow of argon gas. Data taken from curves.
8	Rudkin, R. L.	+	10-mil-diameter wire. Surface changes noted after heating.	Hemispherical total emittance. Power dissipated per unit length of electrically heated wire. Temperatures measured with a two-color photoelectric pyrometer, an optical pyrometer and known resistivity versus temperature data.	Measured in vacuum. Data taken from curve.
6	Butler, Jenkins, Rudkin, and Laughridge	▲	Highly polished. Vacuum arc cast, machined, extruded, recrystallized, and rolled.	Hemispherical total emittance. Disk specimen. Temperatures measured with thermocouples. Emittance calculated from the mass, specific heat, and rate of change of temperature of the specimen.	Measured in vacuum. Data taken from curves.

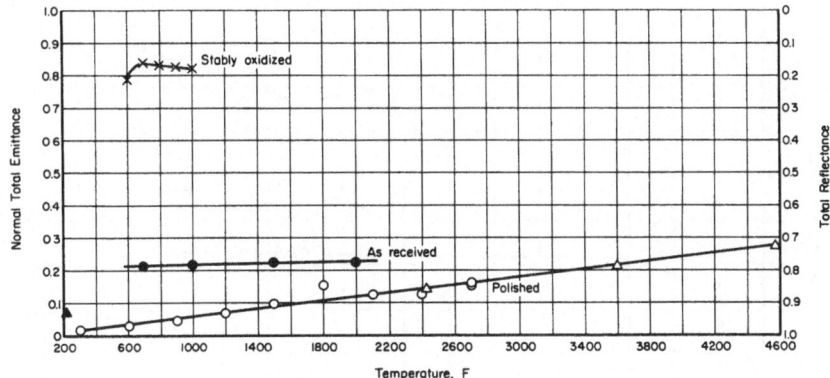

NORMAL TOTAL EMITTANCE OF MOLYBDENUM

NORMAL TOTAL EMITTANCE OF MOLYBDENUM--REFERENCE INFORMATION

Reference	Investigator	Symbol	Composition and Surface Condition	Test Method	Remarks
11	Wade, W. R.	X	Stably oxidized for 75 minutes at 1000 F.	Normal total emittance. Resistance-heated strip specimens. Commercial thermopile-radiation detector. Comparison blackbody. Temperatures measured with thermocouples.	Measured in air. Hemispherical total emittance calculated from data taken at 0, 30, 45, and 60 degrees from the normal. Data taken from curves.
2	Betz, Olson, Schurin, and Morris	O	Polished (buffed).	Normal total emittance. Resistance-heated strip specimens. Comparison blackbody. Temperatures measured with thermocouples. Thermistor detector.	Measured in vacuum. Data taken from tables.
5	Anthony and Pearl	●	As received.	Normal total emittance. Induction-heated specimen. Thermopile detector. Comparison blackbody. Temperatures measured with thermocouples.	Measured in purge of helium gas. Data taken from table.
1	Barnes, Forsythe, and Adams	▲	Polished. Composition not given.	Normal total emittance. Disk specimen. Thermopile detector. Comparison blackbody. Temperatures measured with thermocouples. Specimen heated by contact with copper heat-diffusion plate.	Measured in dry hydrogen. Data taken from table.
10	Coffman, Coulson, and Kibler	△	Highly polished. Composition not given.	Normal total emittance. Induction heated specimen. Blackbody hole in specimen. Total detector. Temperatures measured with optical pyrometer.	Measured in positive pressure of argon. Data taken from curves.

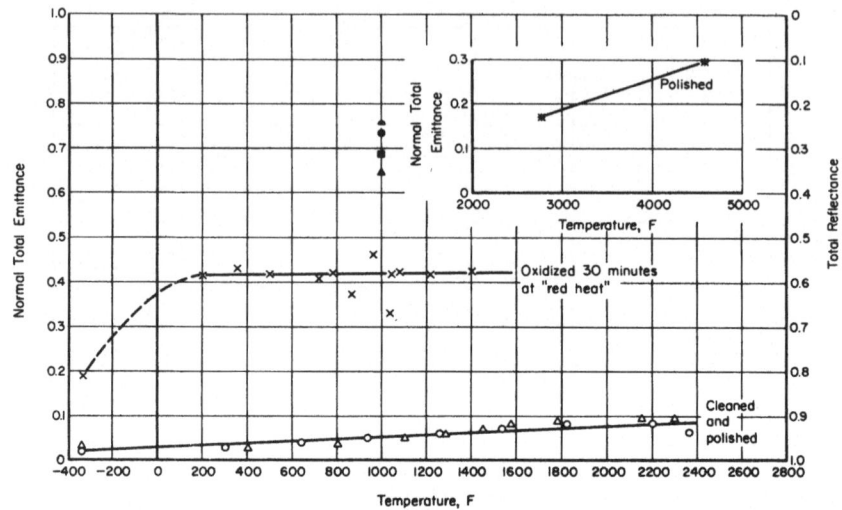

NORMAL TOTAL EMITTANCE OF TANTALUM

NORMAL TOTAL EMITTANCE OF TANTALUM--REFERENCE INFORMATION

Reference	Investigator	Symbol	Composition and Surface Condition	Test Method	Remarks
2	Betz, Olson, Schurin, and Morris	O	Composition "pure". Cleaned.	Normal total emittance. Resistance heated specimen.	Measured in vacuum. Data taken from curves.
		△	Polished.	Thermistor-bolometer detector.	
		✕	Oxidized 30 minutes at "red heat".	Comparison blackbody. Temperatures measured with thermocouples.	
24	Wade, W. R.	▲	Composition not given. Oxidized 50 minutes at 1000 F.	Total normal emittance. Thermopile detector. Comparison blackbody. Temperatures measured with thermocouples	Measured in air. Data taken from curves.
		■	Oxidized 60 minutes at 1000 F.		
		●	Oxidized 80 minutes at 1000 F.		
		◓	Oxidized 110 minutes at 1000 F.		
			NOTE: The oxide formed was flaky, porous, and unstable.		
10	Coffman, Coulson, and Kibler	✳	Highly polished. Composition not given.	Normal total emittance. Induction-heated specimen. Comparison blackbody hole in specimen. Total detector. Temperatures measured with optical pyrometer.	Measured in positive pressure of argon. Data taken from curves.

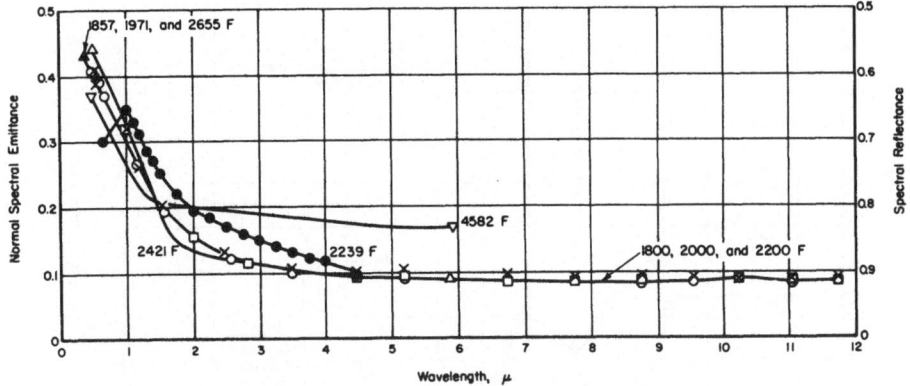

NORMAL SPECTRAL EMITTANCE OF MOLYBDENUM

NORMAL SPECTRAL EMITTANCE OF MOLYBDENUM--REFERENCE INFORMATION

Reference	Investigator	Symbol	Composition and Surface Condition	Test Method	Remarks
12	Pratt & Whitney Aircraft		Purity and surface condition not given.	Normal spectral emittance. Modified hole-in-tube method.	Measured in vacuum at 1800, 2000,
		O	1800 F	Temperatures measured	and 2200 F.
		□	2000 F	with calibrated optical	Data taken
		X	2200 F	pyrometer. Double-beam, ratio recording spectrophotometer. Thermocouple, photomultiplier, or lead-sulphide detectors.	from curve.
14	Price, D. J.	●	Commercially pure. Polished.	Normal spectral emittance. Hole-in-tube method. Spectrometer and thermopile detector. Temperatures measured with optical pyrometer.	Measured in static atmosphere of hydrogen. Data taken from table. Measured at 2239 F.
13	Taylor, J. E.	▲	Purity 99.9 per cent. Cold rolled and highly polished.	Normal spectral emittance. Hole-in-tube method. Monochromator and photomultiplier detector. Temperatures measured with optical pyrometer.	Measured in vacuum of 10^{-7} mm Hg or better. Data taken as average of curves at temperatures of 2655, 1971, and 1857 F.
10	Coffman, Coulson, and Kibler		Highly polished. Composition not given.	Normal spectral emittance. Induction-heated specimen. Monochromator and detector.	Measured in positive pressure of argon.
		△	2421 F	Comparison blackbody hole in specimen.	Data taken from curves.
		▽	4582 F	Temperature measured with optical pyrometer.	

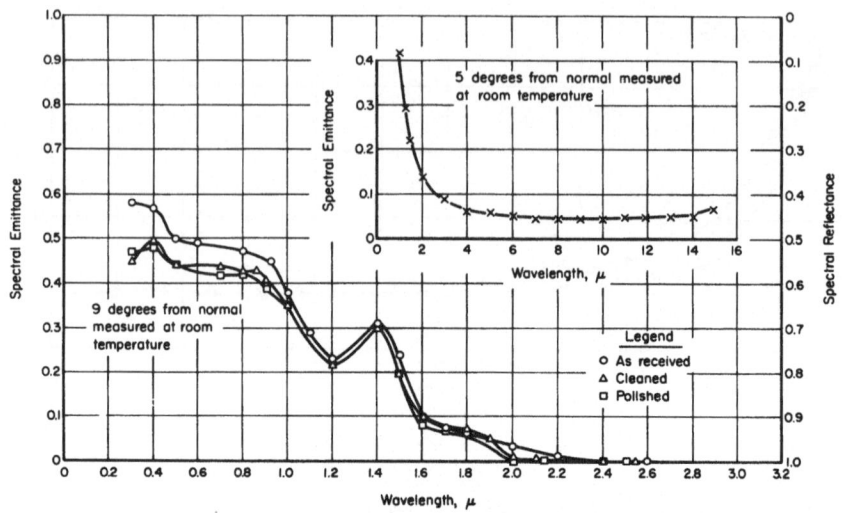

SPECTRAL EMITTANCE OF MOLYBDENUM

SPECTRAL EMITTANCE OF MOLYBDENUM--REFERENCE INFORMATION

Reference	Investigator	Symbol	Composition and Surface Condition	Test Method	Remarks
16	Betz, Olson, Schurin, and Morris	O △ □	As received. Cleaned. Polished Composition not given.	Spectral reflectance. Incident radiation 9 degrees to normal. Spectrophotometer with integrating sphere and lead sulphide detector. Normal (9 degrees) illumination. Diffuse (hemispherical) viewing.	Measured in air. Data taken from curves. Measured at room temperature.
15	Gier, Dunkle, and Bevans	X	Commercially pure. Surface condition not given.	Spectral reflectance at 5 degrees to normal. Gier-Dunkle blackbody reflectometer. Monochromator detector and amplifier. Temperatures measured with thermocouples. Diffuse illumination. Normal (5 degrees) viewing.	Measured in air at room temperature. Data taken from curve.

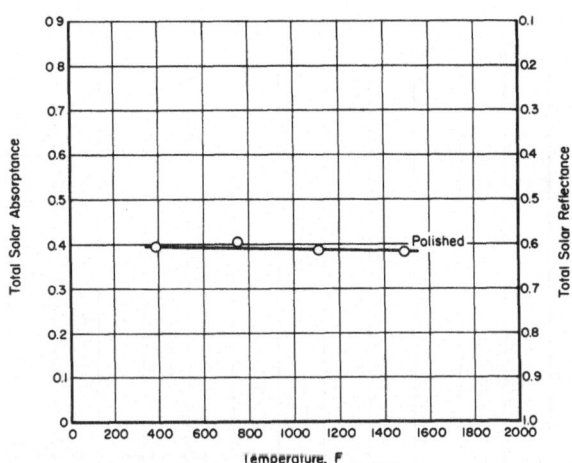

TOTAL SOLAR ABSORPTANCE OF MOLYBDENUM

TOTAL SOLAR ABSORPTANCE OF MOLYBDENUM--REFERENCE INFORMATION

Reference	Investigator	Symbol	Composition and Surface Condition	Test Method	Remarks
6	Butler, Jenkins, Rudkin, and Laughridge	O	Highly polished. Vacuum arc cast, machined, extruded, recrystallized, and rolled.	Total solar absorptance. Carbon-arc-image furnace. Disk specimen. Temperatures measured with thermocouples. Absorptance calculated from mass, specific heat, rate of change of temperature, and known irradiance of the surface. (Solar spectrum simulated by carbon arc.)	Measured in vacuum. Data taken from curves.

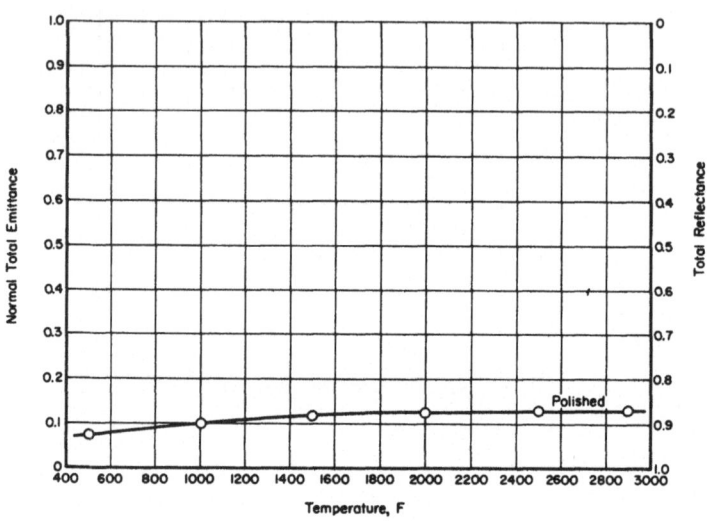

NORMAL TOTAL EMITTANCE OF MOLYBDENUM-0.5 PER CENT TITANIUM ALLOY

NORMAL TOTAL EMITTANCE OF MOLYBDENUM-0.5 PER CENT TITANIUM ALLOY—REFERENCE INFORMATION

Reference	Investigator	Symbol	Composition and Surface Condition	Test Method	Remarks
17	Fieldhouse and Long	O	Nominal composition. Surface roughness 32 micro-inches rms.	Normal total emittance. Induction-heated specimen. Monochromator, thermo-couple detector. Comparison blackbody. Temperatures measured with micro-optical pyrometer. Facilities provided for angular measurements.	Measured in vacuum. Data taken from curve. Normal total emittance equals hemispherical total emittance for this sample.

175

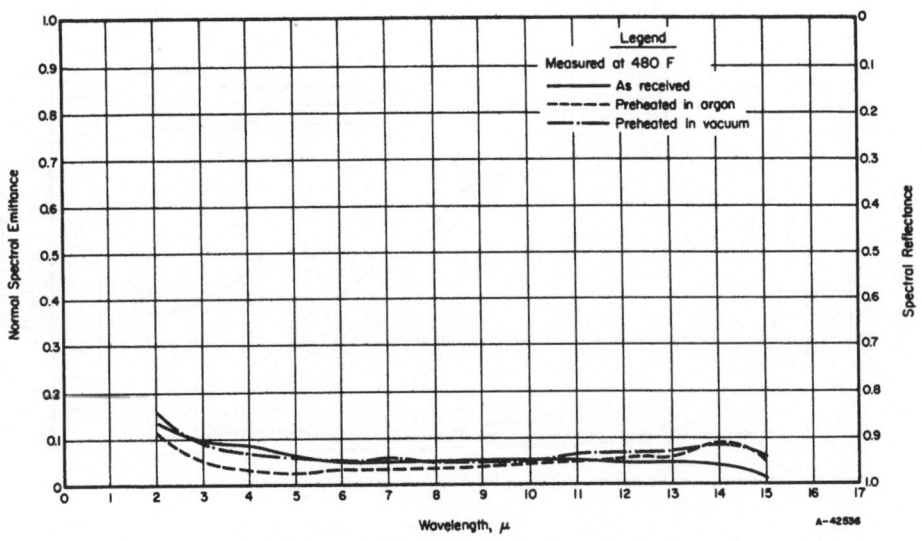

NORMAL SPECTRAL EMITTANCE OF MOLYBDENUM-0.5 TITANIUM AT 480 F

NORMAL SPECTRAL EMITTANCE OF MOLYBDENUM + 0.5 TITANIUM AT 480 F--REFERENCE INFORMATION

Reference	Investigator	Symbol	Composition and Surface Condition	Test Method	Remarks
25	Adams, J. G.		As received Heated 30 minutes in argon at 2000 F Heated 30 minutes in 2.2 x 10^-5 mm Hg pressure at 2000 F	Normal spectral emittance. Furnace-heated disk specimen. Comparison blackbody (Hohlraun). Spectrometer-monochromator with photomultiplier, lead sulphide, and thermocouple detectors. Temperatures measured with thermocouples.	Measured in argon.

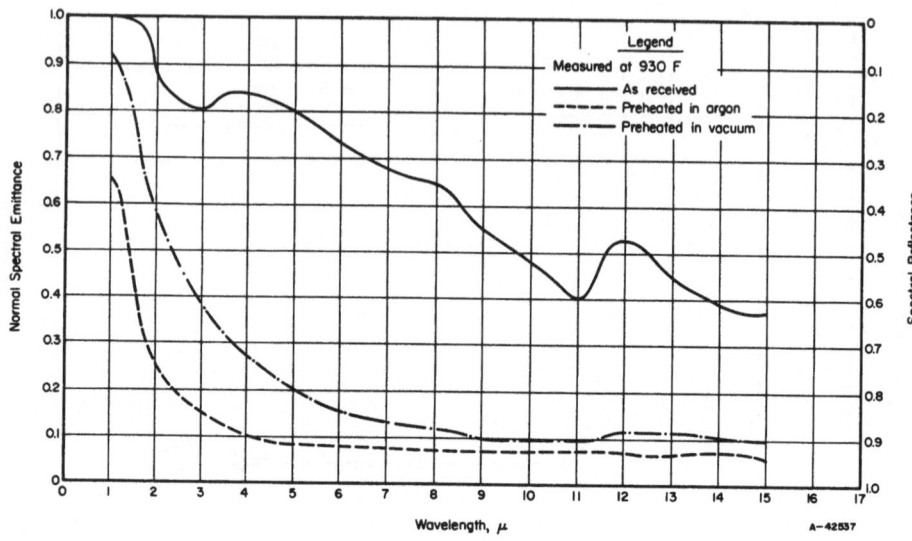

NORMAL SPECTRAL EMITTANCE OF MOLYBDENUM - 0.5 TITANIUM AT 930 F

NORMAL SPECTRAL EMITTANCE OF MOLYBDENUM + 0.5 TITANIUM AT 930 F--REFERENCE INFORMATION

Reference	Investigator	Symbol	Composition and Surface Condition	Test Method	Remarks
25	Adams, J. G.		As received Heated 30 minutes in argon at 2000 F Heated 30 minutes in 2.2 x 10^{-5} mm Hg pressure at 2000 F	Normal spectral emittance. Furnace-heated disk specimen. Comparison blackbody (Hohlraun). Spectrometer-monochromator with photomultiplier, lead sulphide, and thermocouple detectors. Temperatures measured with thermocouples.	Measured in argon.

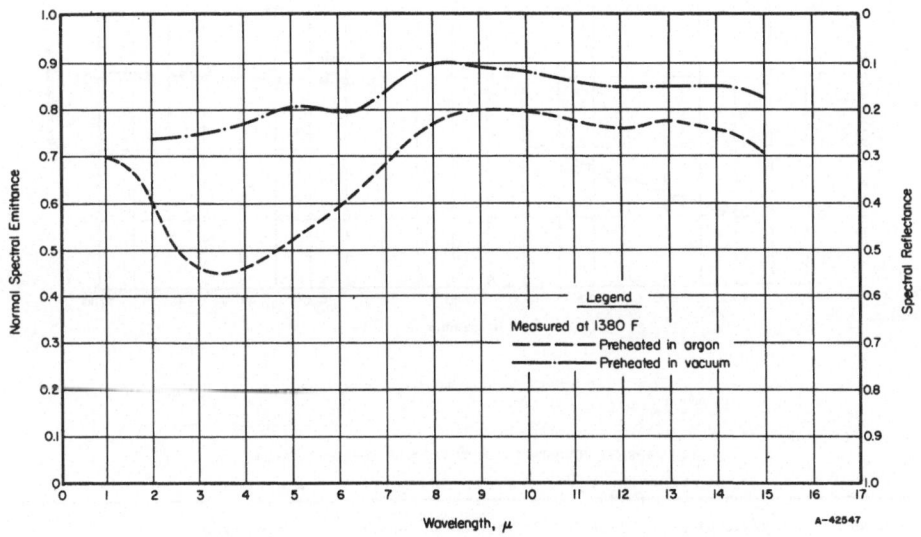

NORMAL SPECTRAL EMITTANCE OF MOLYBDENUM – 0.5 TITANIUM AT 1380 F

NORMAL SPECTRAL EMITTANCE OF MOLYBDENUM + 0.5 TITANIUM AT 1380 F--REFERENCE INFORMATION

Reference	Investigator	Symbol	Composition and Surface Condition	Test Method	Remarks
25	Adams, J. G.		Heated 30 minutes in argon at 2000 F Heated 30 minutes in 2.2 x 10⁻⁵ mm Hg pressure at 2000 F	Normal spectral emittance. Furnace-heated disk specimen. Comparison blackbody (Hohlraun). Spectrometer-monochromator with photomultiplier, lead sulphide, and thermocouple detectors. Temperatures measured with thermocouples.	Measured in argon.

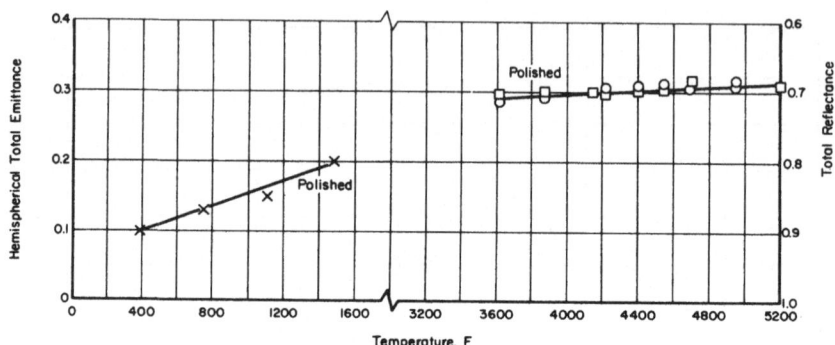

HEMISPHERICAL TOTAL EMITTANCE OF TANTALUM

HEMISPHERICAL TOTAL EMITTANCE OF TANTALUM--REFERENCE INFORMATION

Reference	Investigator	Symbol	Composition and Surface Condition	Test Method	Remarks
7	Glasier, Allen, and Saldinger		Polished with 0000 abrasive paper.	Hemispherical total emittance.	Measured in flow of argon gas.
		O	Fansteel material.	Power dissipated from electrically heated rod specimen.	Data taken from curves.
		☐	National Research material.	Brightness temperatures measured with optical pyrometer – converted to true temperatures using values obtained at the melting point.	
6	Butler, Jenkins, Rudkin, and Laughridge	X	Highly polished. Commercially pure, arc cast, or sintered.	Hemispherical total emittance. Carbon-arc-image furnace. Disk specimen. Temperatures measured with thermocouples. Emittance calculated from mass, specific heat, and rate of change of temperature of specimen.	Measured in vacuum. Data taken from curves.

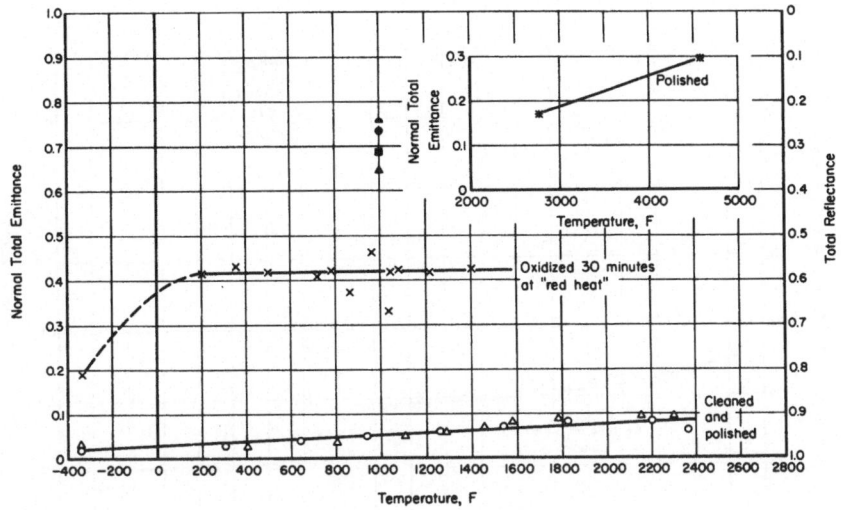

NORMAL TOTAL EMITTANCE OF TANTALUM

NORMAL TOTAL EMITTANCE OF TANTALUM--REFERENCE INFORMATION

Reference	Investigator	Symbol	Composition and Surface Condition	Test Method	Remarks
2	Betz, Olson, Schurin, and Morris	O	Composition "pure". Cleaned.	Normal total emittance. Resistance heated specimen.	Measured in vacuum. Data taken from curves.
		△	Polished.	Thermistor-bolometer detector.	
		✕	Oxidized 30 minutes at "red heat".	Comparison blackbody. Temperatures measured with thermocouples.	
24	Wade, W. R.	▲	Composition not given. Oxidized 50 minutes at 1000 F.	Total normal emittance. Thermopile detector. Comparison blackbody. Temperatures measured with thermocouples	Measured in air. Data taken from curves.
		■	Oxidized 60 minutes at 1000 F.		
		●	Oxidized 80 minutes at 1000 F.		
		⬤	Oxidized 110 minutes at 1000 F.		
			NOTE: The oxide formed was flaky, porous, and unstable.		
10	Coffman, Coulson, and Kibler	✶	Highly polished. Composition not given.	Normal total emittance. Induction-heated specimen. Comparison blackbody hole in specimen. Total detector. Temperatures measured with optical pyrometer.	Measured in positive pressure of argon. Data taken from curves.

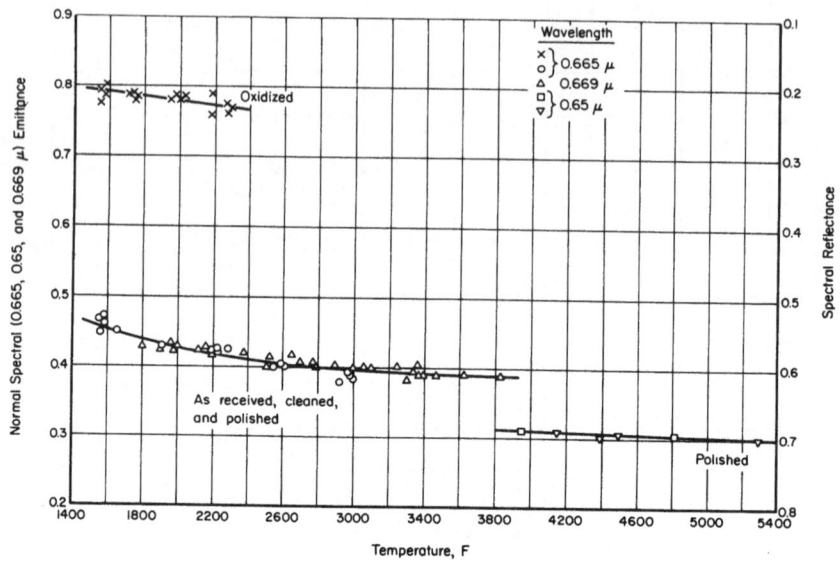

NORMAL SPECTRAL EMITTANCE OF TANTALUM

NORMAL SPECTRAL EMITTANCE OF TANTALUM--REFERENCE INFORMATION

Reference	Investigator	Symbol	Composition and Surface Condition	Test Method	Remarks
2	Betz, Olson, Schurin, and Morris	O	As received, cleaned, and polished.	Normal spectral emittance.	Measured in vacuum.
		✕	Oxidized for 30 minutes at red heat.	Modified hole-in-tube method. Optical pyrometer.	$(\lambda = 0.665\mu)$ Data taken from curves.
18	Fiske, M. D.	△	As received. Purity: 99.9 per cent.	Normal spectral emittance. Hole-in-tube method. Calibrated optical pyrometer.	Measured in vacuum. $(\lambda = 0.669\mu)$ Data taken from curve.
7	Glasier, Allen, and Saldinger	▽	Polished with 0000 abrasive papers.	Normal spectral emittance. Electrically heated rod specimen. Brightness temperatures measured with optical pyrometer. True temperatures obtained using values at melting point and heat-flow rates.	Measured in flow of argon gas. Data taken from curves.
		□	Fansteel material.		$(\lambda = 0.65\mu)$
			National Research material.		

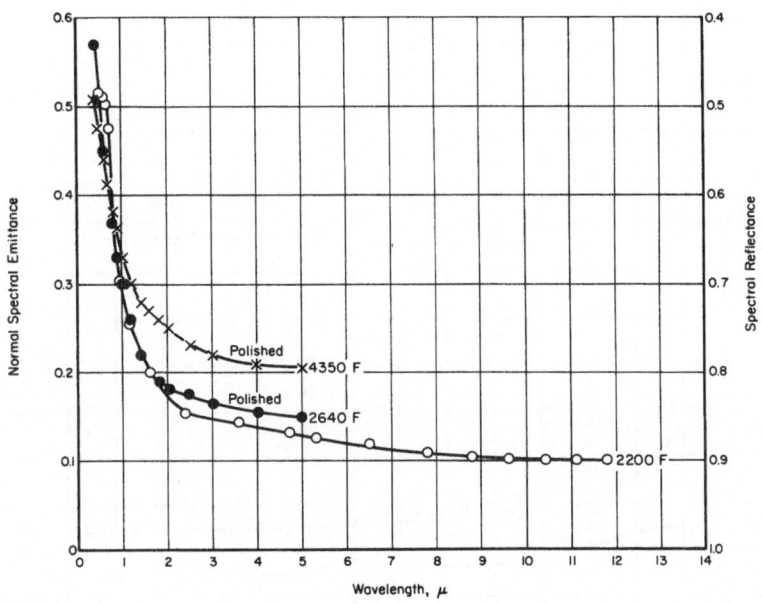

NORMAL SPECTRAL EMITTANCE OF TANTALUM

NORMAL SPECTRAL EMITTANCE OF TANTALUM--REFERENCE INFORMATION

Reference	Investigator	Symbol	Composition and Surface Condition	Test Method	Remarks
4	Pratt & Whitney Aircraft	O	Purity and surface condition not given.	Normal spectral emittance. Modified hole-in-tube method. Temperature measured with calibrated optical pyrometer. Double beam, ratio recording spectrophotometer. Thermocouple, photomultiplier, or lead-sulphide detectors.	Measured in vacuum. Data taken from curve. Measured at 2200 F.
19	Riethof, T. R.		Highly polished. Composition not given.	Normal spectral emittance. Induction heated specimen. Comparison blackbody hole in specimen. Monochromator and photomultiplier or thermocouple detectors. Temperature measured with optical pyrometer.	Measured in purified argon or vacuum. Data taken from curves.
		X	4350 F		
		●	2640 F		

182

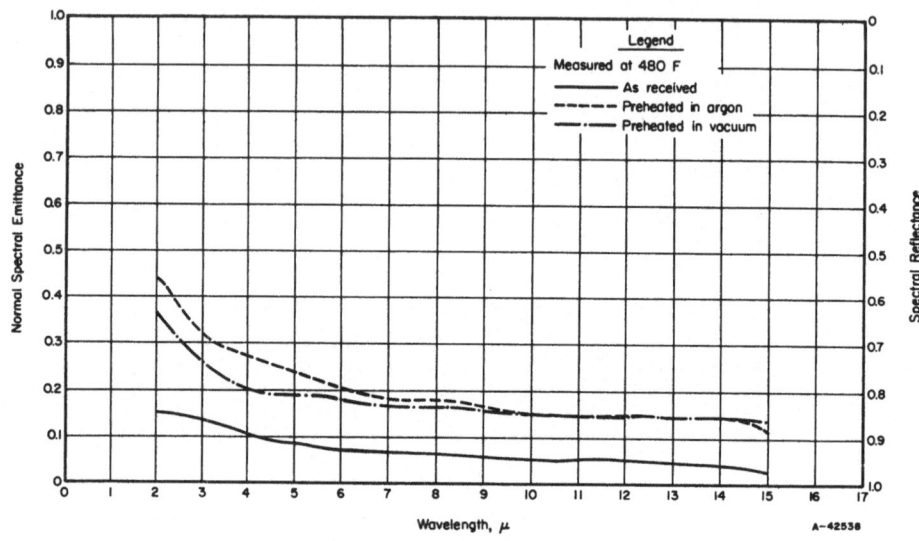

NORMAL SPECTRAL EMITTANCE OF TANTALUM AT 480 F

NORMAL SPECTRAL EMITTANCE OF TANTALUM AT 480 F—REFERENCE INFORMATION

Reference	Investigator	Symbol	Composition and Surface Condition	Test Method	Remarks
25	Adams, J. G.		As received Heated 30 minutes in argon at 2000 F Heated 30 minutes in 2.2 x 10⁻⁵ mm Hg pressure at 2000 F	Normal spectral emittance. Furnace-heated disk specimen. Comparison blackbody (Hohlraun). Spectrometer-monochromator with photomultiplier, lead sulphide, and thermocouple detectors. Temperatures measured with thermocouples.	Measured in argon.

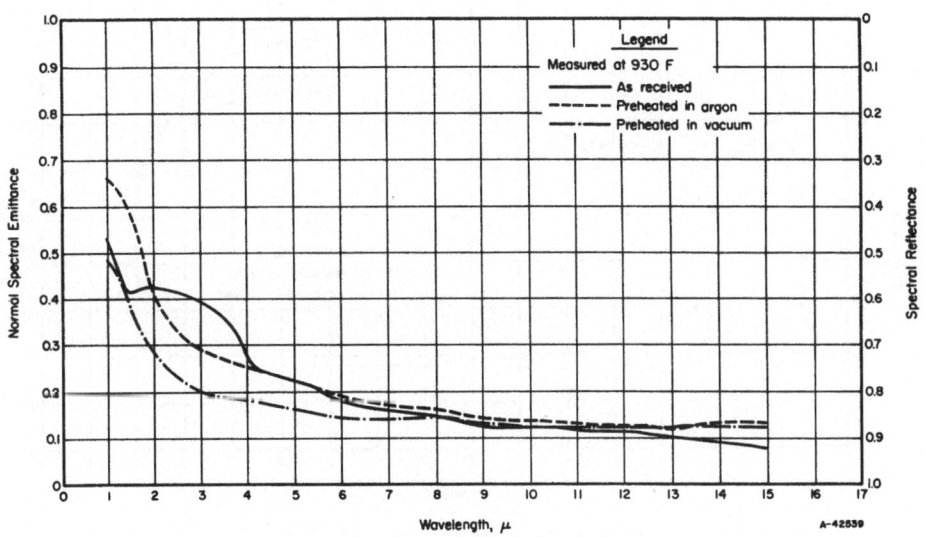

NORMAL SPECTRAL EMITTANCE OF TANTALUM AT 930 F

NORMAL SPECTRAL EMITTANCE OF TANTALUM AT 930 F—REFERENCE INFORMATION

Reference	Investigator	Symbol	Composition and Surface Condition	Test Method	Remarks
25	Adams, J. G.		As received Heated 30 minutes in argon at 2000 F Heated 30 minutes in 2.2 x 10^{-5} mm Hg pressure at 2000 F	Normal spectral emittance. Furnace-heated disk specimen. Comparison blackbody (Hohlraun). Spectrometer-monochromator with photomultiplier, lead sulphide, and thermocouple detectors. Temperatures measured with thermocouples.	Measured in argon.

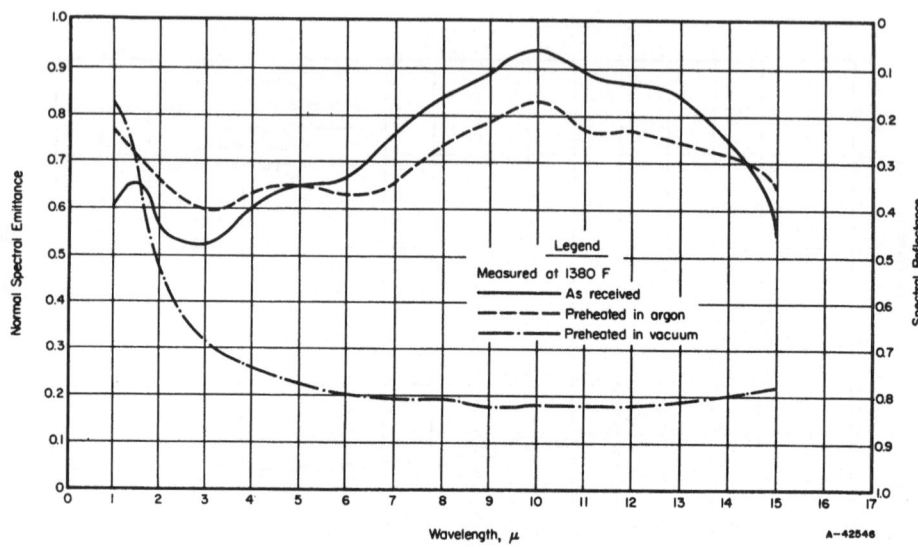

NORMAL SPECTRAL EMITTANCE OF TANTALUM AT 1380 F

NORMAL SPECTRAL EMITTANCE OF TANTALUM AT 1380 F--REFERENCE INFORMATION

Reference	Investigator	Symbol	Composition and Surface Condition	Test Method	Remarks
25	Adams, J. G.		As received Heated 30 minutes in argon at 2000 F Heated 30 minutes in 2.2 x 10⁻⁵ mm Hg pressure at 2000 F	Normal spectral emittance. Furnace-heated disk specimen. Comparison blackbody (Hohlraun). Spectrometer-monochromator with photomultiplier, lead sulphide, and thermocouple detectors. Temperatures measured with thermocouples.	Measured in argon.

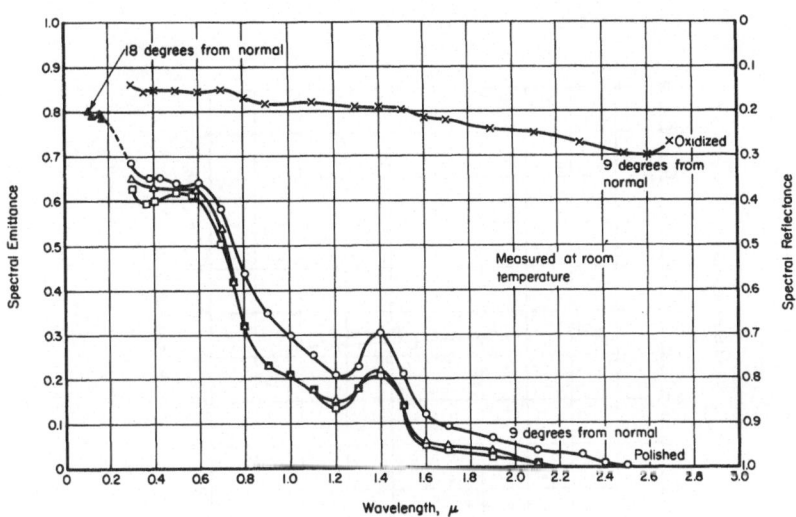

SPECTRAL EMITTANCE OF TANTALUM

SPECTRAL EMITTANCE OF TANTALUM--REFERENCE INFORMATION

Reference	Investigator	Symbol	Composition and Surface Condition	Test Method	Remarks
2	Betz, Olson, Schurin, and Morris	△ □ ○ ✕	As received. Cleaned. Polished. Oxidized.	Spectral reflectance at 9 degrees to normal incidence. Integrating sphere, spectrophotometer, mono-chromator, lead sulphide detector. Normal (9 degrees) illumination and diffuse viewing.	Measured in air. Data taken from curves.
20	Fabre and Romand	▲	Evaporated film. Polished appearance Thickness: 1 per cent or less transmission in the visible region.	Spectral reflectance at 18 degrees from the normal. Monochromator, photo-multiplier detector.	Measured in vacuum. Data taken from table.

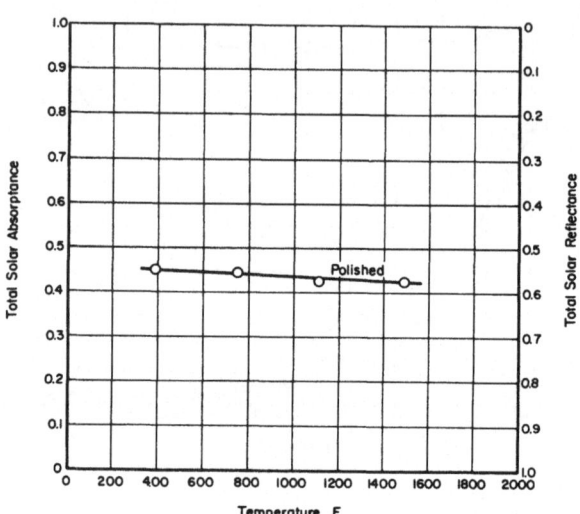

TOTAL SOLAR ABSORPTANCE OF TANTALUM

TOTAL SOLAR ABSORPTANCE OF TANTALUM--REFERENCE INFORMATION

Reference	Investigator	Symbol	Composition and Surface Condition	Test Method	Remarks
6	Butler, Jenkins, Rudkin, and Laughridge	O	Highly polished. Commercially pure, arc cast or sintered.	Total solar absorptance. Carbon-arc-image furnace. Disk specimen. Temperatures measured with thermocouples. Absorptance calculated from mass, specific heat, rate of change of temperature, and known irradiance of surface. (Solar spectral distribution simulated by carbon arc.)	Measured in vacuum. Data taken from curves.

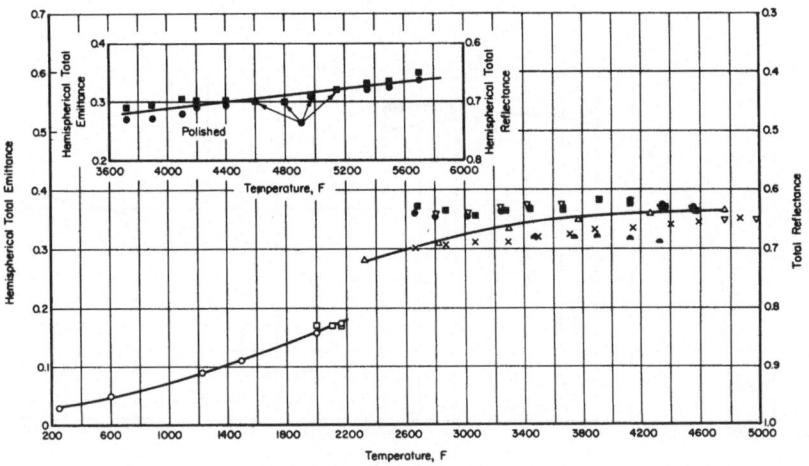

HEMISPHERICAL TOTAL EMITTANCE OF TUNGSTEN

HEMISPHERICAL TOTAL EMITTANCE OF TUNGSTEN--REFERENCE INFORMATION

Reference	Investigator	Symbol	Composition and Surface Condition	Test Method	Remarks
7	Glasier, Allen, and Saldinger		Polished with 0000 abrasive papers.	Hemispherical total emittance.	Measured in flow of argon gas.
		●	Fansteel material.	Power dissipated from electrically heated rod specimen.	Data taken from curves.
		■	Wah Chang material.	Brightness temperatures measured with optical pyrometer - converted to true temperature using values obtained at melting point.	
9	Rudkin, Parker, and Westover	△	Spectrographically pure straight wire (0.010-inch diameter).	Hemispherical total emittance. Electrically heated wire. Measured power input to constant temperature zone. Temperatures measured with two-color photoelectric pyrometer.	Measured in vacuum. Data taken from curve. Investigators estimated accuracy ± 10 per cent.
4	Pratt & Whitney Aircraft		Purity and surface condition not given.	Hemispherical total emittance. Power dissipated from resistance-heated strip specimen. Temperatures measured with thermocouples. (Optical pyrometer in spectral-hole-in-tube method.)	Measured in vacuum of 10^{-5} to 10^{-9} mm of Hg. Data taken from table -and curve.
		○	Total emittance equipment.		
		□	Spectral emittance equipment.		
21	Allen, R. D.		Porous tungsten.	Hemispherical total emittance. Heat radiated from solid rod of resistance-heated material to cold walls. Temperature calculated from brightness temperature and spectral emittance data.	Measured in positive pressure of argon. Data taken from curves.
			Per cent of theoretical density:		
		×	90		
		⊠	90		
		▽	70		
		●	70		
		◓	70		

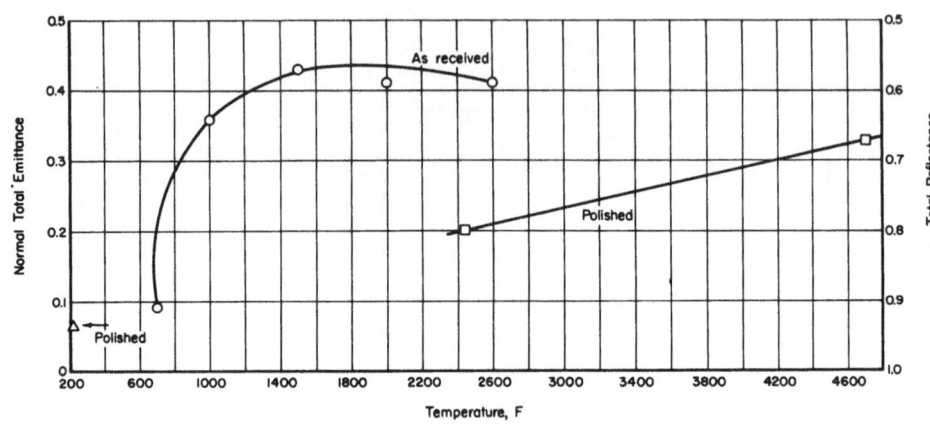

NORMAL TOTAL EMITTANCE OF TUNGSTEN

NORMAL TOTAL EMITTANCE OF TUNGSTEN--REFERENCE INFORMATION

Reference	Investigator	Symbol	Composition and Surface Condition	Test Method	Remarks
10	Coffman, Coulson, and Kibler	□	Highly polished. Composition not given.	Normal total emittance. Induction-heated specimen. Comparison blackbody hole in specimen. Total detector. Temperatures measured with optical pyrometer.	Measured in positive pressure of argon. Data taken from curve. Individual data points not given.
5	Anthony and Pearl	O	As received. Composition not given.	Normal total emittance. Induction-heated specimen. Thermopile detector. Comparison blackbody. Temperatures measured with thermocouples and optical pyrometer.	Measured in continous purge of helium gas. Data taken from curve.
1	Barnes, Forsythe, and Adams	Δ	Polished. Composition not given.	Normal total emittance. Disk specimen. Thermopile detector. Comparison blackbody. Temperatures measured with thermocouples. Specimen heated by contact with copper heat-diffusion plate.	Measured in dry hydrogen. Data taken from table.

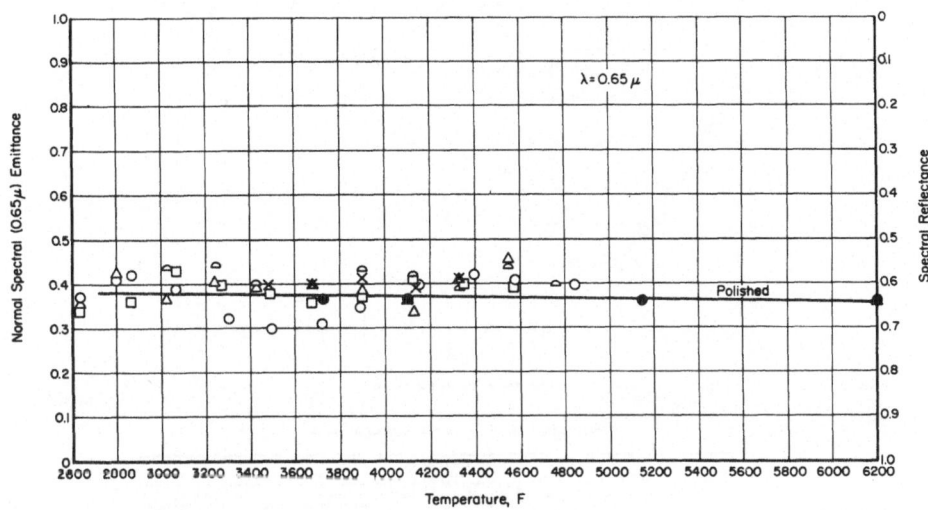

NORMAL TOTAL EMITTANCE OF TUNGSTEN

NORMAL SPECTRAL EMITTANCE OF TUNGSTEN--REFERENCE INFORMATION

Reference	Investigator	Symbol	Composition and Surface Condition	Test Method	Remarks
21	Allen, R. D.		Porous tungsten.	Normal spectral emittance. Determined using brightness temperature, heat-flow rate and the derivative of brightness temperature with respect to heat-flow rate. Resistance-heated rod specimen.	Measured in positive pressure of argon. Data taken from curves. ($\lambda = 0.65 \mu$)
			Per cent of theoretical density:		
		○	90		
		□	90		
		×	70		
		△	70		
		◠	70		
	Glasier, Allen, and Saldinger		Polished with 0000 abrasive papers.	Normal spectral emittance. Electrically heated rod specimen. Brightness temperature measured with optical pyrometer. True temperature obtained using values at melting point and heat-flow rate.	Measured in flow of argon gas. Data taken from curves. ($\lambda = 0.65 \mu$)
		●	Fansteel material.		
		▲	Wah Chang material.		

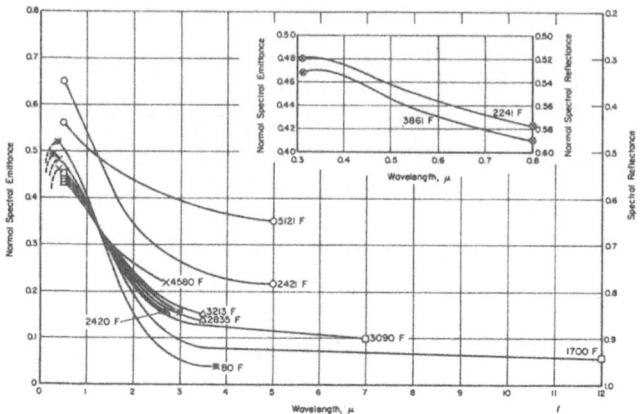

NORMAL SPECTRAL EMITTANCE OF TUNGSTEN

NORMAL SPECTRAL EMITTANCE OF TUNGSTEN—REFERENCE INFORMATION

Reference	Investigator	Symbol	Composition and Surface Condition	Test Method	Remarks
4	Pratt & Whitney Aircraft	X	Purity and surface condition not given.	Normal spectral emittance. Modified hole-in-tube method. Temperatures measured with calibrated optical pyrometer. Double-beam, ratio recording spectrophotometer. Photomultiplier, thermocouple, or lead-sulphide detectors.	Measured in vacuum of 10^{-8} mm Hg or better at 1700, 2110, and 3090 F. Data taken from curves.
10	Coffman, Coulson, and Kibler	O	Highly polished. Composition not given.	Normal spectral emittance. Induction-heated specimen. Monochromator and detector. Blackbody hole in specimen. Temperature measured with optical pyrometer.	Measured in positive pressure of argon (1.3 atm). Data taken from curves. Only those curves for the highest and lowest temperatures are shown. Others fell between these two.
19	Riethof, T. R.	△	Highly polished. Composition not given.	Normal spectral emittance. Induction-heated specimen. Comparison blackbody hole in specimen. Monochromator and photomultiplier and thermocouple detectors. Temperature measured with optical pyrometer.	Measured in argon or vacuum. Data taken from curves.
23	De Vos	X	50-micron thick tungsten ribbon formed into triangular cross-section tube. Surface clean and annealed at 2400 K for 100 hours. Composition per cent: Fe 0.014 - 0.015 Si 0.004 - 0.008 Mn 0.001 - 0.003 Mg 0.0003 - 0.0006 Ni absent	Normal spectral emittance. Hole-in-tube method. Monochromator, photomultiplier, Cs phototube, and lead sulphide detectors. Temperatures measured with calibrated optical pyrometer.	Measured in vacuum and argon. Note: Tube heated in vacuum to 2400 K, then H₂, to reduce oxides. Measured at 2420, 2780, 3140, 3500, 3860, 4220, and 4580 F. Only the first and last are shown.
22	Larrabee	●	Purity - 99.9 per cent or better.	Normal spectral emittance. Monochromator and photomultiplier detector. Temperatures measured with calibrated optical pyrometer.	Measured in vacuum of 9.5 x 10^{-8} mm Hg or better. Measured at 2241, 2781, 3141, 3501, 3861 F. Only the first and last of these shown.
25	Forsythe and Adams	●	"A good surface" after aging at 3000 K in inert gas for 2 hours. Data given for 80 and 3140 F.	A compilation of earlier data obtained by Worthing, Weniger and Pfund, Hulbert, and Zwikker.	(Although not original test data, it was thought that some of the earlier work should be included for reference.)

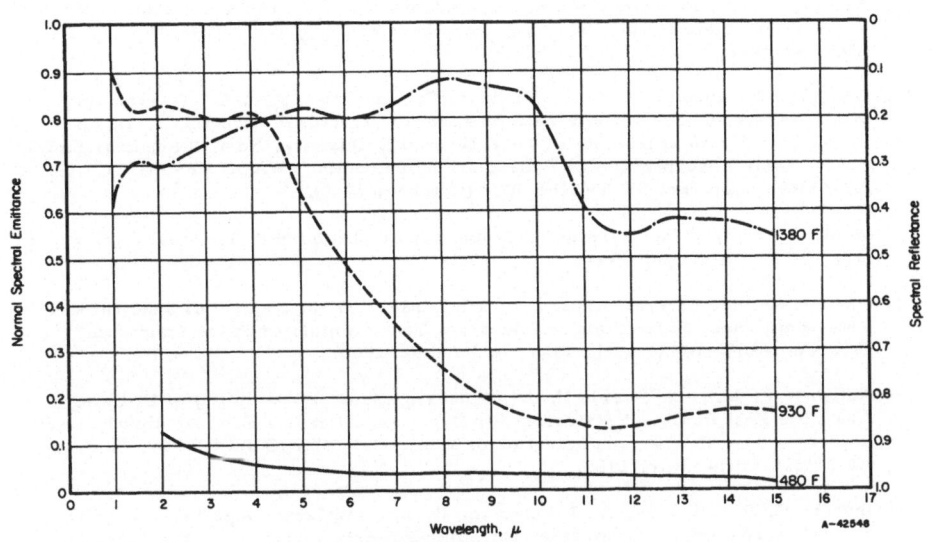

NORMAL SPECTRAL EMITTANCE OF TUNGSTEN

NORMAL SPECTRAL EMITTANCE OF TUNGSTEN--REFERENCE INFORMATION

Reference	Investigator	Symbol	Composition and Surface Condition	Test Method	Remarks
25	Adams, J. G.		As received Measured at 480 F Measured at 930 F Measured at 1380 F	Normal spectral emittance. Furnace-heated disk specimen. Comparison blackbody (Hohlraun). Spectrometer-monochromator with photomultiplier, lead sulphide, and thermocouple detectors. emperatures measured with thermocouples.'	Measured in air.

REFERENCES

(1) Barnes, E. T., Forsythe, W. E., and Adams, E. Q., "The Total Emissivity of Various Materials at 100-500 C", J. Opt. Soc. Am., 37 (10), 804-807 (October, 1947).

(2) Betz, H. T., Olson, O. H., Schurin, B. D., and Morris, J. C., "Determination of Emissivity and Reflectivity Data on Aircraft Structural Materials, Part II: Techniques for Measurement of Total Normal Emissivity, Normal Spectral Emissivity, Solar Absorptivity and Presentation of Results", WADC TR-56-222, ASTIA AD 202493, Contract AF 33(616)-3002 (October, 1958).

(3) Wahlin, H. B., "The Thermionic Properties of Chromium", Physical Review, 75 (12), 1458-1459 (June, 1948).

(4) Pratt & Whitney Aircraft Report No. PWA-1863, "Measurement of Spectral and Total Emittance of Materials and Surfaces Under Simulated Space Conditions", Contract NASW-104 (July 1, 1959, to June 30, 1960).

(5) Anthony, F. M., and Pearl, H. A., "Investigations of Feasibility of Utilizing Available Heat Resistant Materials for Hypersonic Leading Edge Applications, III, Screening Test Results and Selection of Materials", WADC TR 59-744, Contract AF 33(616)-6034 (July, 1960).

(6) Butler, C. P., Jenkins, R. J., Rudkin, R. L., and Laughridge, F. I., "High Temperature Surface Parameters for Solar Power", Coatings for the Aerospace Environment, WADD TR-60-773, 139-162 (November 9-10, 1960).

(7) Glasier, L. F., Jr., Allen, R. D., and Saldinger, I. L., "Mechanical and Physical Properties of the Refractory Metals, Tungsten, Tantalum, and Molybdenum Above 4000 F", GWR 356704-411 (April, 1959).

(8) Rudkin, R. L., "Thermal Properties of Molybdenum Between 2700°R and 4000°R, USNRDL-TR-433 (June 20, 1960).

(9) Rudkin, R. L., Parker, W. J., and Westover, R. W., "Measurements of the Thermal Properties of Metals at Elevated Temperatures", R&D Technical Report USNRDL-TR-419 (May 11, 1960).

(10) Coffman, J. A., Coulson, K. L., and Kibler, T. M., General Electric Company, Cincinnati, Ohio, preliminary information under an Air Force contract.

(11) Wade, William R., "Measurements of Total Hemispherical Emissivity of Several Stably Oxidized Metals and Some Refractory Oxide Coatings", NASA Memo 1-20-59L [AD No. 209192] (January, 1959).

(12) Pratt & Whitney Aircraft, preliminary information under an NASA contract.

(13) Taylor, J. E., "The Variation With Wavelength of the Spectral Emissivity of Iron and Molybdenum", J. Opt. Soc. Am., 42 (1), 33-36 (January, 1952).

(14) Price, D. J. , "The Emissivity of Hot Metals in the Infra-Red", Proc. Phys. Soc. (London), 59, 118-131 (1947).

(15) Gier, J. T. , Dunkle, R. V. , and Bevans, J. T. , "Measurement of Absolute Spectral Reflectivity From 1. 0 to 15 Microns", J. Opt. Soc. Am. , 44 (7), 558-562 (July, 1954).

(16) Betz, Howard T. , Olson, O. H. , Schurin, B. D. , and Morris, James C. , "Determination of Emissivity and Reflectivity Data on Aircraft Structural Materials", WADC TR-56-222, Part I (October, 1956), ASTIA Doc. No. AD 110458, pp 1-18, 40-43.

(17) Fieldhouse, I. B. , Lang, J. I. , and Blau, H. H. , Jr. , "Investigation of Feasibility of Utilizing Available Heat Resistant Materials for Hypersonic Leading Edge Applications, Vol IV — Thermal Properties of Molybdenum Alloy and Graphite", WADC Technical Report 59-744, Contract AF 33(616)-6034 (December, 1960).

(18) Fiske, Milan D. , "The Temperature Scale, Thermionics, and Thermatomics of Tantalum", Physical Review, 61 (April 1 and 15, 1942).

(19) Riethof, T. R. , "High Temperature Spectral Emissivity Studies", General Electric MSVD, Space Sciences Laboratory, R 61SD004 (January, 1961).

(20) Fabre, D. , and Romand, J. , "Reflection Measurements in the Far Ultraviolet With Evaporated Films of Tantalum, Tungsten, and Zirconium", Comp. rend. , 242, 893-896 (1960).

(21) Allen, R. D. , Glasier, L. F. , and Jordan, P. L. , "Spectral Emissivity, Total Emissivity, and Thermal Conductivity of Molybdenum, Tantalum, and Tungsten Above 2300 K", J. Appl. Phys. , 31 (8), 1382-1387 (August, 1960).

(22) Larrabee, Robert D. , "The Spectral Emissivity and Optical Properties of Tungsten", MIT TR 328 [ASTIA AD 156602] (May 21, 1957).

(23) DeVos, J. C. , "A New Determination of the Emissivity of Tungsten Ribbon", Physica, 20, 690-714 (1954).

(24) Wade, W. R. , "Measurements of Total Hemispherical Emissivity of Various Oxidized Metals at High Temperature", NACA TN-4206 (October, 1957).

(25) Adams, J. G. , "The Determination of Spectral Emissivities, Reflectivities, and Absorptivities of Materials and Coatings", Northrup Corporation Report No. NOR-61-189 (August 3, 1961).

RADIATIVE PROPERTY DATA

Coated Materials Suitable for Elevated-Temperature Use

TABLE OF CONTENTS

TABLE OF CONTENTS
(Continued)

TABLE OF CONTENTS
(Continued)

TABLE OF CONTENTS
(Continued)

TABLE OF CONTENTS
(Continued)

TABLE OF CONTENTS
(Continued)

TABLE OF CONTENTS
(Continued)

TABLE OF CONTENTS
(Continued)

TABLE OF CONTENTS
(Continued)

TABLE OF CONTENTS
(Continued)

TABLE OF CONTENTS
(Continued)

TABLE OF CONTENTS
(Continued)

TABLE OF CONTENTS
(Continued)

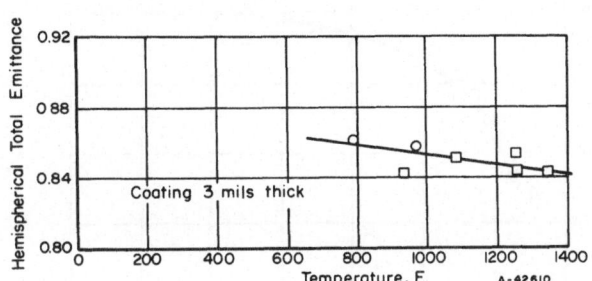

HEMISPHERICAL TOTAL EMITTANCE OF CRYSTALLINE BORON ON COLUMBIUM

HEMISPHERICAL TOTAL EMITTANCE OF CRYSTALLINE BORON ON COLUMBIUM--REFERENCE INFORMATION

Reference	Investigator	Symbol	Composition and Surface Condition	Test Method	Remarks
15	Pratt & Whitney Aircraft		3-mil-thick coating prepared by Linde-Plasmarc process on columbium tube.	Hemispherical total emittance. Resistance-heated tube specimen. Power dissipated in measured area. Temperatures measured with thermocouples.	Measured in vacuum. Data taken from curve.

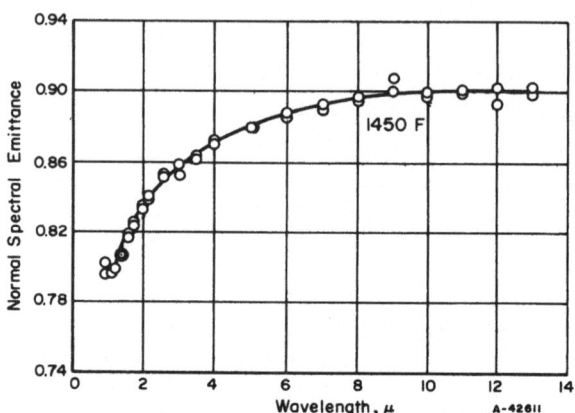

NORMAL SPECTRAL EMITTANCE OF CRYSTALLINE BORON ON COLUMBIUM

NORMAL SPECTRAL EMITTANCE OF CRYSTALLINE BORON ON COLUMBIUM--REFERENCE INFORMATION

Reference	Investigator	Symbol	Composition and Surface Condition	Test Method	Remarks
15	Pratt & Whitney Aircraft		3-mil-thick coating prepared by Linde Plasmarc process on columbium tube.	Normal spectral emittance. Electrically Heated tubular coated specimen. Integral blackbody slot in specimen tube. Temperatures measured with thermocouples and optical pyrometer.	Measured in vacuum. Data taken from curve.

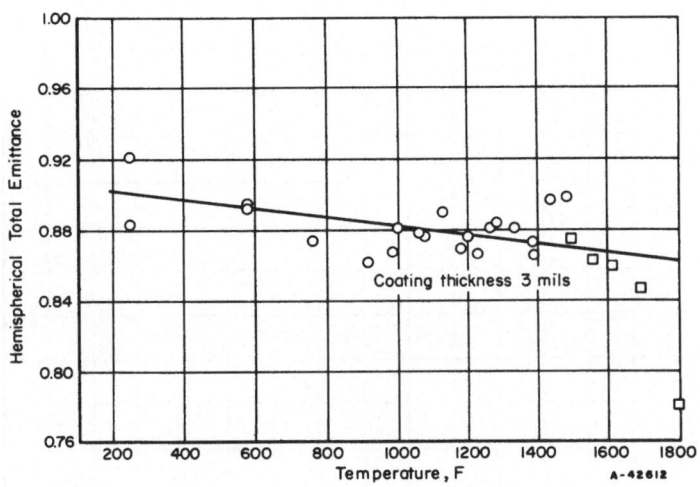

HEMISPHERICAL TOTAL EMITTANCE OF CRYSTALLINE BORON ON MOLYBDENUM

HEMISPHERICAL TOTAL EMITTANCE OF CRYSTALLINE BORON ON MOLYBDENUM--REFERENCE INFORMATION

Reference	Investigator	Symbol	Composition and Surface Condition	Test Method	Remarks
15	Pratt & Whitney Aircraft		Crystalline boron flame sprayed by Linde Plasmarc process on molybdenum strip. Coating 3 mils thick (coated both sides). Note: coating loosened from molybdenum.	Hemispherical total emittance. Resistance-heated strip specimen. Power dissipated in measured area. Temperatures measured with thermocouples.	Measured in vacuum. Data taken from curve.

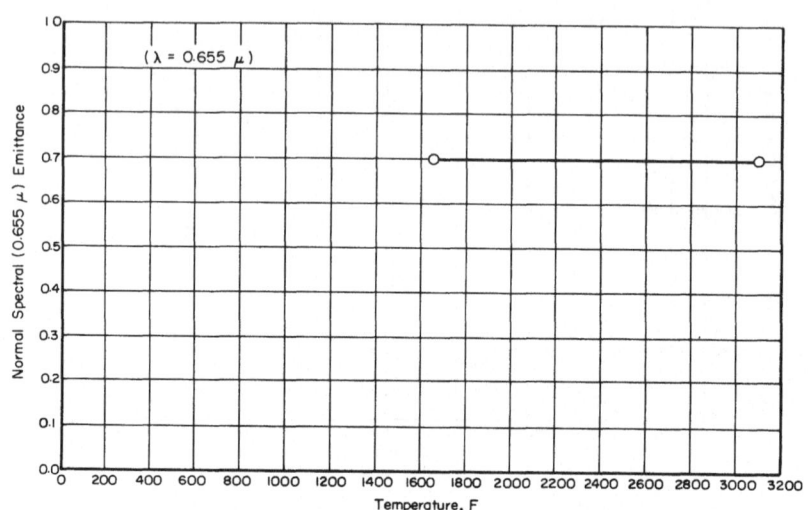

NORMAL SPECTRAL EMITTANCE OF TANTALUM BORIDE ON TUNGSTEN AND TANTALUM

NORMAL SPECTRAL EMITTANCE OF TANTALUM BORIDE ON TUNGSTEN AND TANTALUM—REFERENCE INFORMATION

Reference	Investigator	Symbol	Composition and Surface Condition	Test Method	Remarks
6	Morgan, F. H.	o	Purity or coating method not defined. Coating thickness not given.	Two methods used: (1) Coated-tungsten-strip heater. Temperatures measured with thermocouples. Brightness temperatures measured with optical pyrometer. (2) Hole-in-tube method. Tantalum tube coated with test material.	Measured in vacuum. Data taken from table and discussion. Data appear to be average of hole-in-tube and strip heater methods.

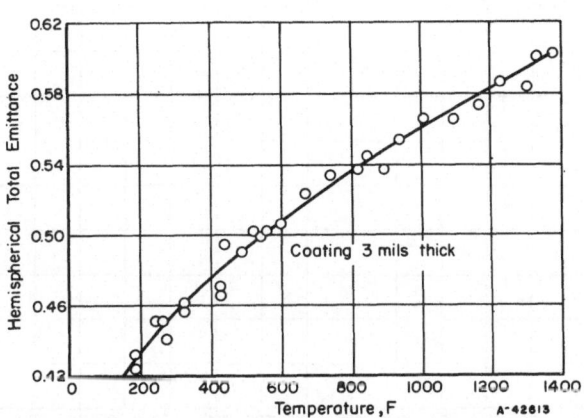

HEMISPHERICAL TOTAL EMITTANCE OF ZIRCONIUM BORIDE ON MOLYBDENUM

HEMISPHERICAL TOTAL EMITTANCE OF ZIRCONIUM BORIDE ON MOLYBDENUM--REFERENCE INFORMATION

Reference	Investigator	Symbol	Composition and Surface Condition	Test Method	Remarks
15	Pratt & Whitney Aircraft		3-mil-thick coating of ZrB$_2$ applied by the Linde Plasmarc process to a molybdenum strip.	Hemispherical total emittance. Resistance-heated strip specimen. Power dissipated in measured area. Temperatures measured with thermocouples.	Measured in vacuum. Data taken from curve.

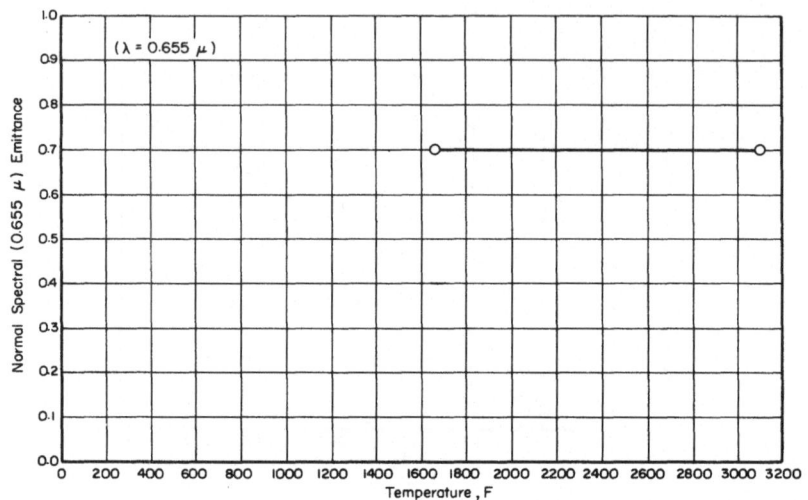

$(\lambda = 0.655\ \mu)$

NORMAL SPECTRAL EMITTANCE OF ZIRCONIUM BORIDE ON TANTALUM AND TUNGSTEN

NORMAL SPECTRAL EMITTANCE OF ZIRCONIUM BORIDE ON TANTALUM AND TUNGSTEN—REFERENCE INFORMATION

Reference	Investigator	Symbol	Composition and Surface Condition	Test Method	Remarks
6	Morgan, F. H.	○	Purity or coating method not defined. Coating thickness not given.	Two methods used: (1) Coated-tungsten-strip heater. Temperatures measured with thermocouples. Brightness temperatures measured with optical pyrometer. (2) Hole-in-tube method. Tantalum tube coated with test material.	Measured in vacuum. Data taken from table and discussion. Data appear to be average of hole-in-tube and strip heater methods.

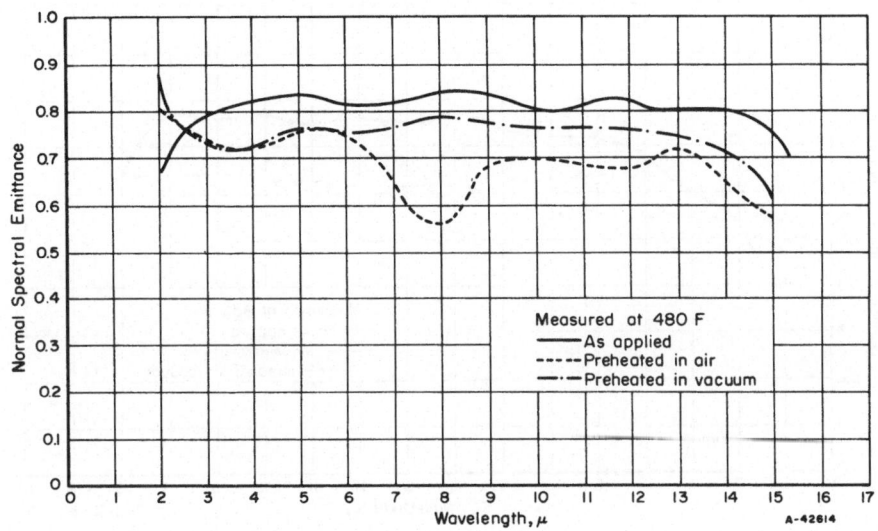

NORMAL SPECTRAL EMITTANCE OF BORON CARBIDE ON INCONEL X AT 480 F

NORMAL SPECTRAL EMITTANCE OF BORON CARBIDE ON INCONEL X AT 480 F--REFERENCE INFORMATION

Reference	Investigator	Symbol	Composition and Surface Condition	Test Method	Remarks
14	Adams, J. G.		Flame sprayed on Inconel X As applied - untreated Heated 30 minutes in air at 1500 F Heated 30 minutes in 6.8 x 10^{-5} mm Hg pressure at 1500 F	Normal spectral emittance. Furnace-heated disk specimen. Comparison blackbody (Hohlraun). Spectrometer-monochromator with photomultiplier, lead sulphide, and thermocouple detectors. Temperatures measured with thermocouples.	Measured in air.

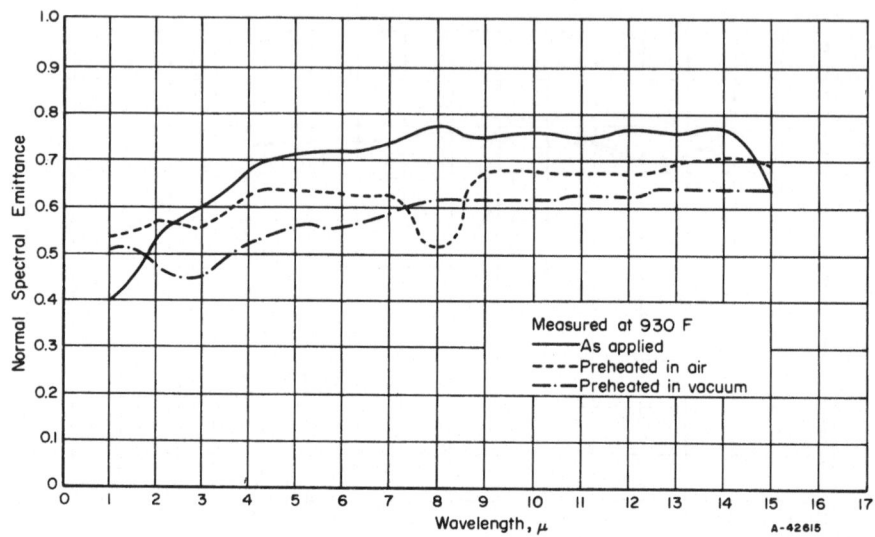

NORMAL SPECTRAL EMITTANCE OF BORON CARBIDE ON INCONEL X AT 930 F

NORMAL SPECTRAL EMITTANCE OF BORON CARBIDE ON INCONEL X AT 930 F--REFERENCE INFORMATION

Reference	Investigator	Symbol	Composition and Surface Condition	Test Method	Remarks
14	Adams, J. G.		Flame sprayed on Inconel X As applied - untreated Heated 30 minutes in air at 1500 F Heated 30 minutes in 6.8 x 10^{-5} mm Hg pressure at 1500 F	Normal spectral emittance. Furnace-heated disk specimen. Comparison blackbody (Hohlraun). Spectrometer-monochromator with photomultiplier, lead sulphide, and thermocouple detectors. Temperatures measured with thermocouples.	Measured in air.

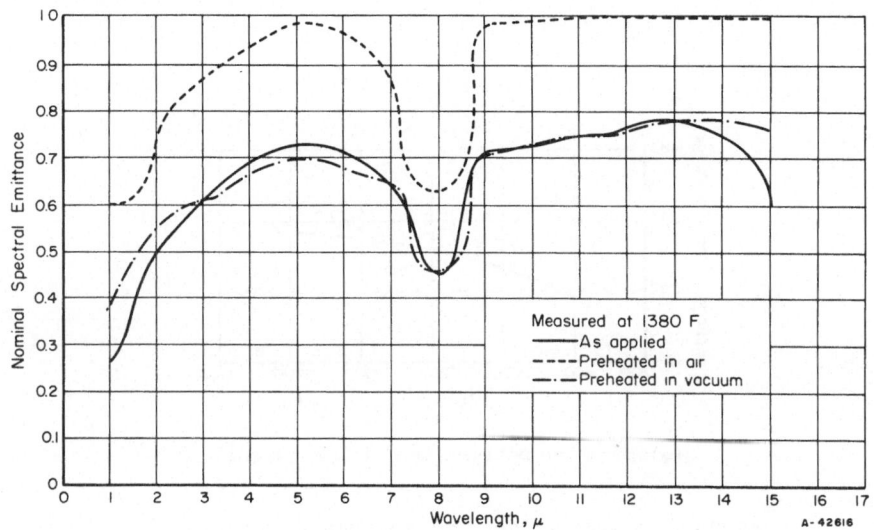

NORMAL SPECTRAL EMITTANCE OF BORON CARBIDE ON INCONEL X AT 1380 F

NORMAL SPECTRAL EMITTANCE OF BORON CARBIDE ON INCONEL X AT 1380 F--REFERENCE INFORMATION

Reference	Investigator	Symbol	Composition and Surface Condition	Test Method	Remarks
14	Adams, J. G.		Flame sprayed on Inconel X As applied - untreated Heated 30 minutes in air at 1500 F Heated 30 minutes in 6.8 x 10^{-5} mm Hg pressure at 1500 F	Normal spectral emittance. Furnace-heated disk specimen. Comparison blackbody (Hohlraun). Spectrometer-monochromator with photomultiplier, lead sulphide, and thermocouple detectors. Temperatures measured with thermocouples.	Measured in air.

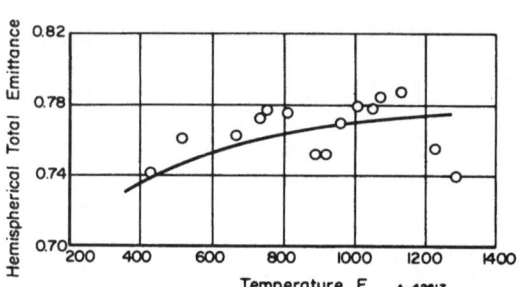

HEMISPHERICAL TOTAL EMITTANCE OF BORON CARBIDE ON MOLYBDENUM

HEMISPHERICAL TOTAL EMITTANCE OF BORON CARBIDE ON MOLYBDENUM--REFERENCE INFORMATION

Reference	Investigator	Symbol	Composition and Surface Condition	Test Method	Remarks
15	Pratt & Whitney Aircraft		2-mil-thick coating applied by the Linde Plasmarc process to both sides of a molybdenum strip.	Hemispherical total emittance. Resistance-heated strip specimen. Power dissipated in measured area. Temperatures measured with thermocouples.	Measured in vacuum. Data taken from curve.

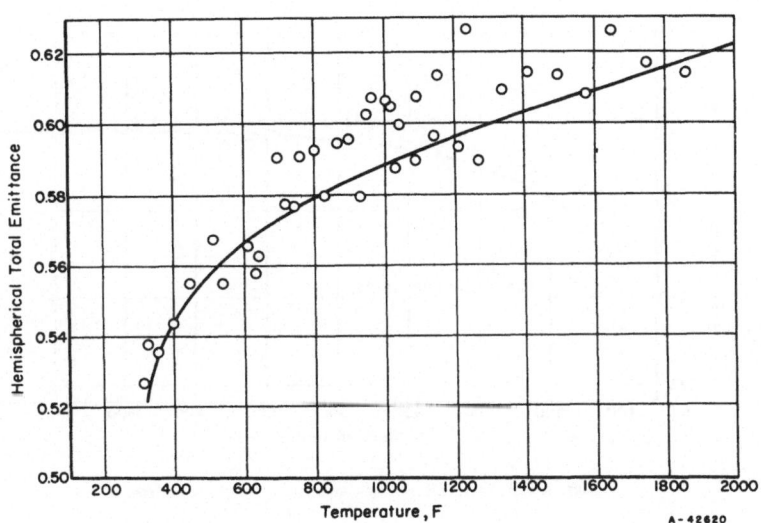

HEMISPHERICAL TOTAL EMITTANCE OF HAFNIUM CARBIDE ON MOLYBDENUM

HEMISPHERICAL TOTAL EMITTANCE OF HAFNIUM CARBIDE ON MOLYBDENUM—REFERENCE INFORMATION

Reference	Investigator	Symbol	Composition and Surface Condition	Test Method	Remarks
15	Pratt & Whitney Aircraft		3-mil-thick coating applied by the Linde Plasmarc process to both sides of a molybdenum strip.	Hemispherical total emittance. Resistance-heated strip specimen. Power dissipated in measured area. Temperatures measured with thermocouples.	Measured in vacuum. Data taken from curve.

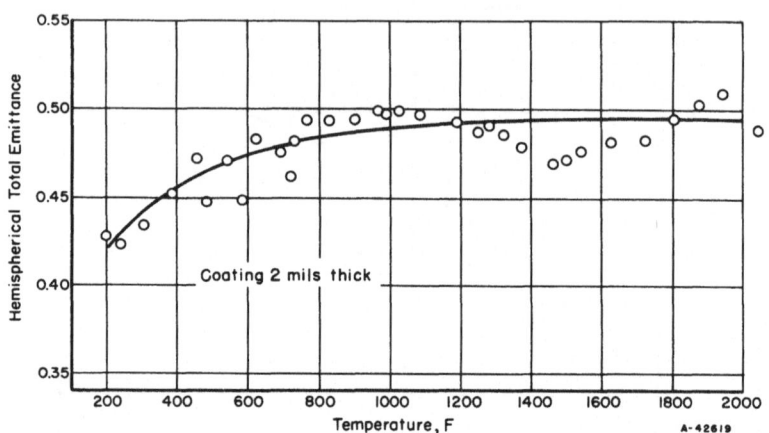

HEMISPHERICAL TOTAL EMITTANCE OF MOLYBDENUM CARBIDE ON MOLYBDENUM

HEMISPHERICAL TOTAL EMITTANCE OF MOLYBDENUM CARBIDE ON MOLYBDENUM--REFERENCE INFORMATION

Reference	Investigator	Symbol	Composition and Surface Condition	Test Method	Remarks
15	Pratt & Whitney Aircraft		2-mil-thick MoC coating applied by the Linde Plasmarc process to both sides of a molybdenum strip.	Hemispherical total emittance. Resistance-heated strip specimen. Power dissipated in measured area. Temperatures measured with thermocouples.	Measured in vacuum. Data taken from curve.

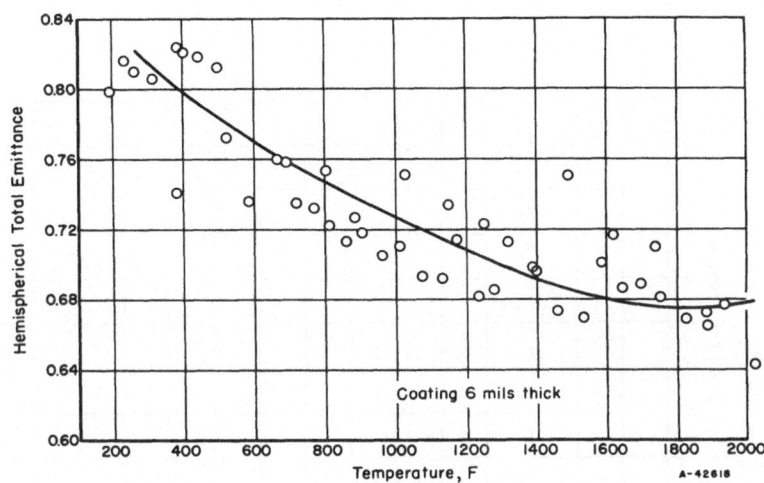

HEMISPHERICAL TOTAL EMITTANCE OF SILICON CARBIDE ON MOLYBDENUM

HEMISPHERICAL TOTAL EMITTANCE OF SILICON CARBIDE ON MOLYBDENUM--REFERENCE INFORMATION

Reference	Investigator	Symbol	Composition and Surface Condition	Test Method	Remarks
15	Pratt & Whitney Aircraft		6-mil-thick coating applied by an electrophoretic process and coated with an acrylic resin.	Hemispherical total emittance. Resistance-heated strip specimen. Power dissipated in measured area. Temperatures measured with thermocouples.	Measured in vacuum. Data taken from curve.

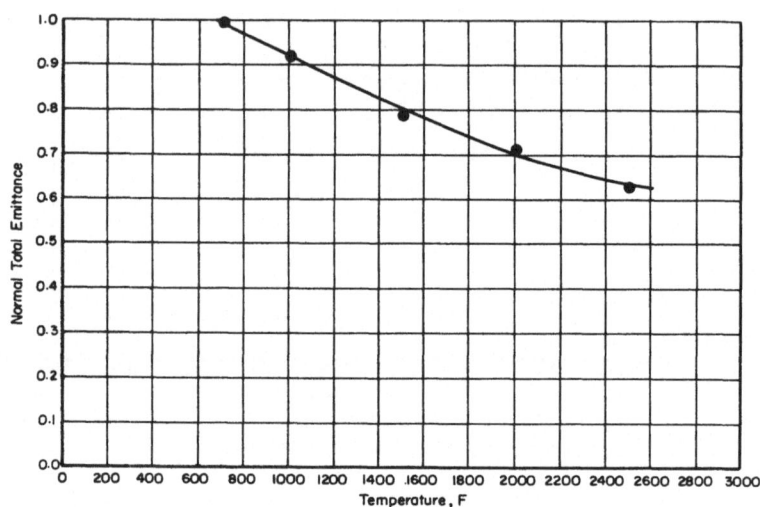

NORMAL TOTAL EMITTANCE OF SILICON CARBIDE ON GRAPHITE

NORMAL TOTAL EMITTANCE OF SILICON CARBIDE ON GRAPHITE--REFERENCE INFORMATION

Reference	Investigator	Symbol	Composition and Surface Condition	Test Method	Remarks
7	Anthony and Pearl	●	As received. Coating thickness not given.	Normal total emittance. Induction-heated specimen. Thermopile detector. Comparison blackbody. Temperatures measured with thermocouples and optical pyrometer.	Measured in continuous purge of helium gas.

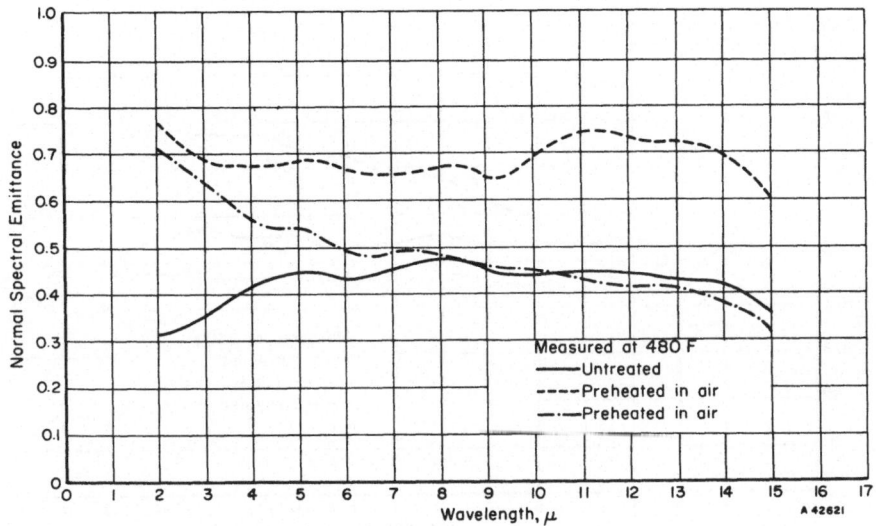

NORMAL SPECTRAL EMITTANCE OF TANTALUM CARBIDE ON INCONEL X AT 480 F

NORMAL SPECTRAL EMITTANCE OF TANTALUM CARBIDE ON INCONEL X AT 480 F--REFERENCE INFORMATION

Reference	Investigator	Symbol	Composition and Surface Condition	Test Method	Remarks
14	Adams, J. G.		Flame sprayed on Inconel X As applied - untreated Heated 30 minutes in air at 1500 F Heated 30 minutes in 6.9 x 10^{-5} mm Hg pressure at 1500 F	Normal spectral emittance. Furnace-heated disk specimen. Comparison blackbody (Hohlraun). Spectrometer-monochromator with photomultiplier, lead sulphide, and thermocouple detectors. Temperatures measured with thermocouples.	Measured in air.

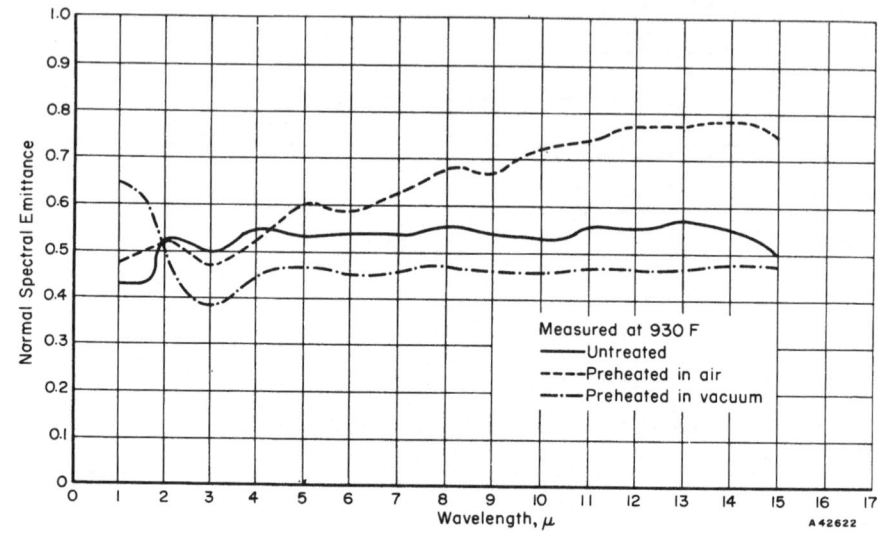

NORMAL SPECTRAL EMITTANCE OF TANTALUM CARBIDE ON INCONEL X AT 930 F

NORMAL SPECTRAL EMITTANCE OF TANTALUM CARBIDE ON INCONEL X AT 930 F—REFERENCE INFORMATION

Reference	Investigator	Symbol	Composition and Surface Condition	Test Method	Remarks
14	Adams, J. G.		Flame sprayed on Inconel X As applied – untreated Heated 30 minutes in air at 1500 F Heated 30 minutes in 6.9 x 10⁻5 mm Hg pressure at 1500 F	Normal spectral emittance. Furnace-heated disk specimen. Comparison blackbody (Hohlraun). Spectrometer-monochromator with photomultiplier, lead sulphide, and thermocouple detectors. Temperatures measured with thermocouples.	Measured in air.

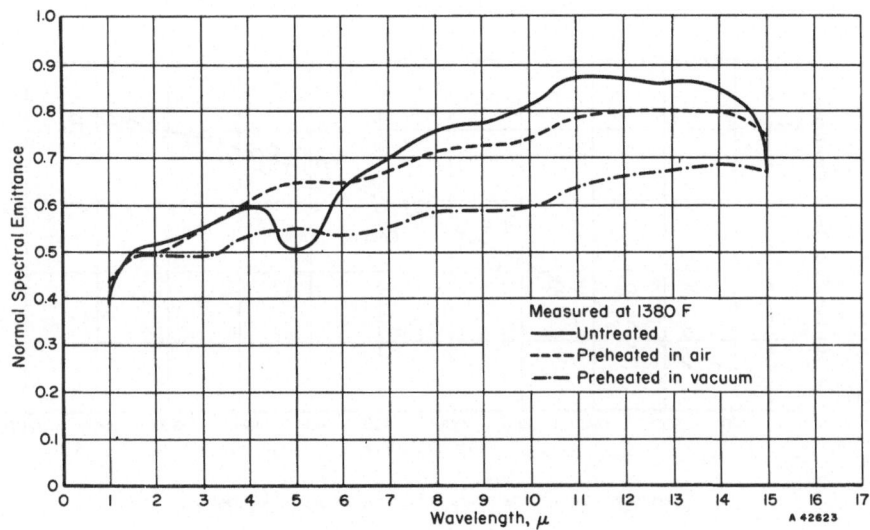

NORMAL SPECTRAL EMITTANCE OF TANTALUM CARBIDE ON INCONEL X AT 1380 F

NORMAL SPECTRAL EMITTANCE OF TANTALUM CARBIDE ON INCONEL X AT 1380 F--REFERENCE INFORMATION

Reference	Investigator	Symbol	Composition and Surface Condition	Test Method	Remarks
14	Adams, J. G.		Flame sprayed on Inconel X As applied - untreated Heated 30 minutes in air at 1500 F Heated 30 minutes in 6.9 x 10^{-5} mm Hg pressure at 1500 F	Normal spectral emittance. Furnace-heated disk specimen. Comparison blackbody (Hohlraun). Spectrometer-monochromator with photomultiplier, lead sulphide, and thermocouple detectors. Temperatures measured with thermocouples.	Measured in air.

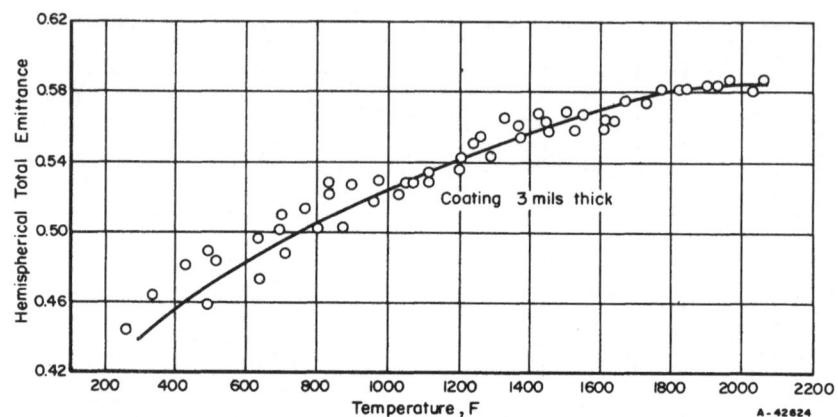

HEMISPHERICAL TOTAL EMITTANCE OF TANTALUM CARBIDE ON MOLYBDENUM

HEMISPHERICAL TOTAL EMITTANCE OF TANTALUM CARBIDE ON MOLYBDENUM—REFERENCE INFORMATION

Reference	Investigator	Symbol	Composition and Surface Condition	Test Method	Remarks
15	Pratt & Whitney Aircraft		3-mil-thick coating applied by the Linde Plasmarc process to a molybdenum strip.	Hemispherical total emittance. Resistance-heated strip specimen. Power dissipated in measured area. Temperatures measured with thermocouples.	Measured in vacuum.

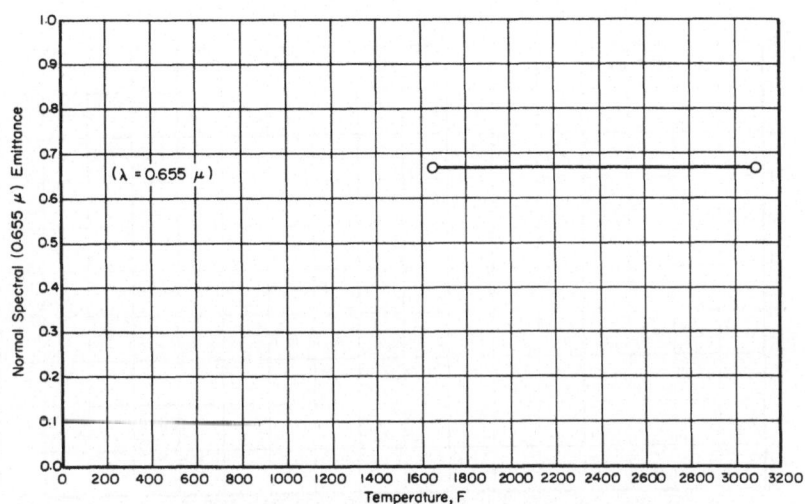

NORMAL SPECTRAL EMITTANCE OF TANTALUM CARBIDE ON TANTALUM AND TUNGSTEN

NORMAL SPECTRAL EMITTANCE OF TANTALUM CARBIDE ON TANTALUM AND TUNGSTEN--REFERENCE INFORMATION

Reference	Investigator	Symbol	Composition and Surface Condition	Test Method	Remarks
6	Morgan, F. H.	o	Purity or coating method not defined. Coating thickness not given.	Two methods used: (1) Coated-tungsten-strip heater. Temperatures measured with thermocouples. Brightness temperatures measured with optical pyrometer. (2) Hole-in-tube method. Tantalum tube coated with test material.	Measured in vacuum. Data taken from table and discussion. Data appear to be average of hole-in-tube and strip heater methods.

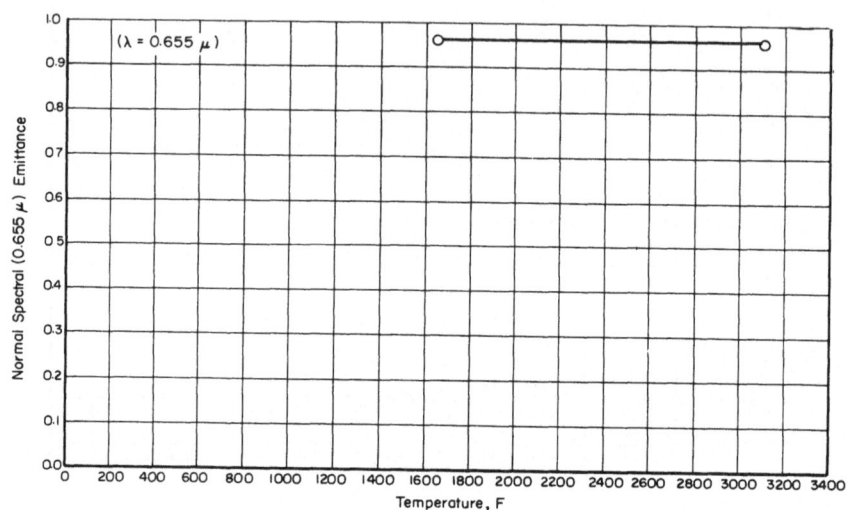

NORMAL SPECTRAL EMITTANCE OF TITANIUM CARBIDE ON TANTALUM AND TUNGSTEN

NORMAL SPECTRAL EMITTANCE OF TITANIUM CARBIDE ON TANTALUM AND TUNGSTEN--REFERENCE INFORMATION

Reference	Investigator	Symbol	Composition and Surface Condition	Test Method	Remarks
6	Morgan, F. H.	o	Purity or coating method not defined. Coating thickness not given.	Two methods used: (1) Coated-tungsten-strip heater. Temperatures measured with thermocouples. Brightness temperature measured with optical pyrometer, and (2) Hole-in-tube method. Tantalum tube coated with test material.	Measured in vacuum. Data taken from table and discussion. Data appear to be average of hole-in-tube and strip heater methods.

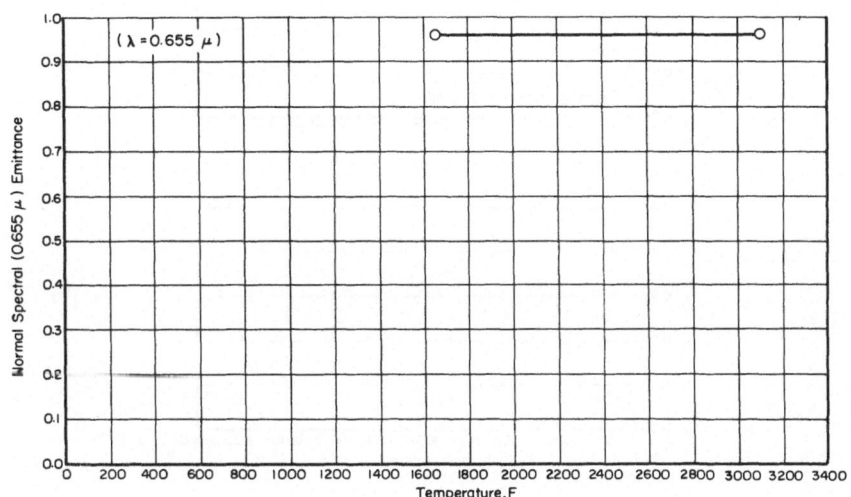

NORMAL SPECTRAL EMITTANCE OF ZIRCONIUM CARBIDE ON TANTALUM AND TUNGSTEN

NORMAL SPECTRAL EMITTANCE OF ZIRCONIUM CARBIDE ON TANTALUM AND TUNGSTEN--REFERENCE INFORMATION

Reference	Investigator	Symbol	Composition and Surface Condition	Test Method	Remarks
6	Morgan, F. H.	o	Purity or coating method not defined. Coating thickness not given.	Two methods used: (1) Coated-tungsten-strip heater with temperature measured with thermocouples, optical pyrometer for brightness temperatures and (2) Hole-in-tube. Tantalum tube coated with test material.	Measured in vacuum. Data taken from table and discussion. Data appear to be average of hole-in-tube and strip heater methods.

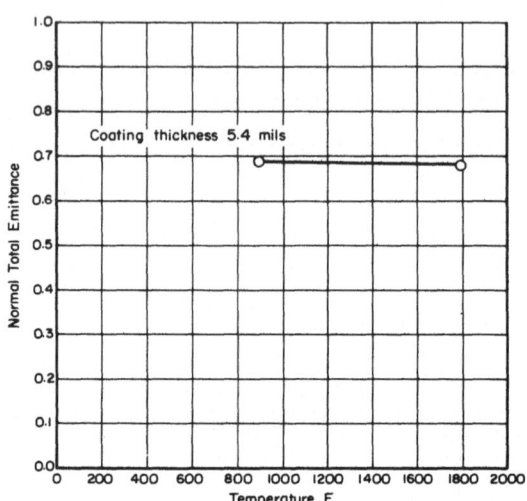

NORMAL TOTAL EMITTANCE OF A-418 ENAMEL ON INCONEL

NORMAL TOTAL EMITTANCE OF A-418 ENAMEL ON INCONEL--REFERENCE INFORMATION

Reference	Investigator	Symbol	Composition and Surface Condition	Test Method	Remarks
1	Burgess, Jasperse, Marcus, Martin, and Flint	o	A-418 Enamel on Inconel. Coating thickness 5.4 mils.	Normal total emittance. Rotating, hollow, cylindrical, Globar heating element. Blackbody hole. Specimen mounted in heating element flush with wall. Temperatures measured with thermocouples. Infrared spectrometer with prism replaced by plane mirror. Thermocouple detector.	Measured in air. Data taken from tables.

Coating Composition by Weight

NBS Frit No. 332 -- 70 per cent
Cr_2O_3 - 30 per cent

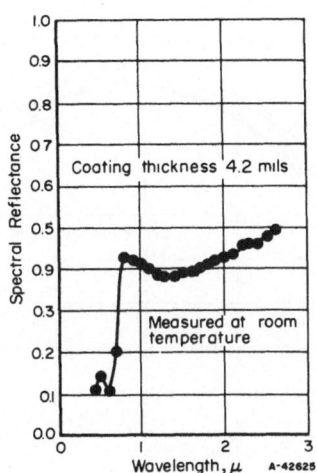

SPECTRAL REFLECTANCE OF A—418 ENAMEL ON INCONEL

SPECTRAL EMITTANCE OF A—418 ENAMEL ON INCONEL--REFERENCE INFORMATION

Reference	Investigator	Symbol	Composition and Surface Condition	Test Method	Remarks
1	Burgess, Jasperse, Marcus, Martin, and Flint	●	Enamel A-418 on Inconel. Coating thickness 4.2 mils.	Spectral reflectance. Commercial reflect-ometer and spectro-photometer. Quartz prism mono-chromator. MgO standard. (Normal viewing-diffuse reflection)	Measured in air at room temperature. Data taken from table.

Coating Composition by Weight

NBS Frit No. 332 - 70 per cent
Cr_2O_3 - 30 per cent

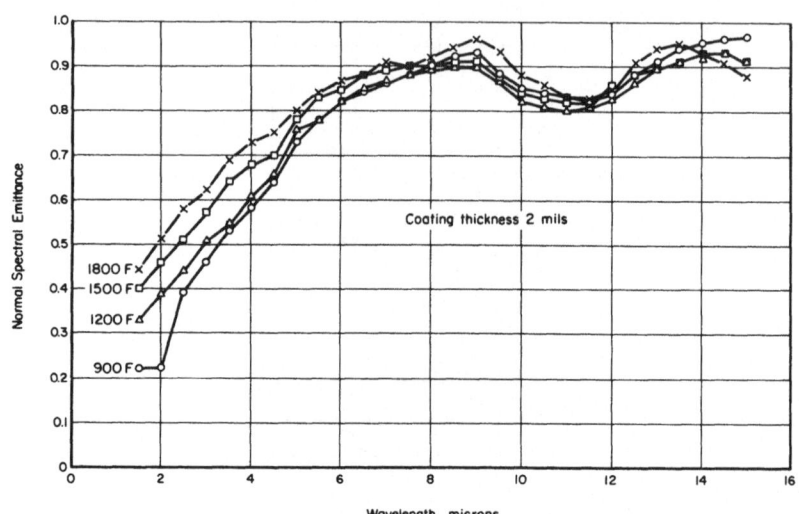

NORMAL SPECTRAL EMITTANCE OF A-418 ENAMEL ON INCONEL

NORMAL SPECTRAL EMITTANCE OF A-418 ENAMEL ON INCONEL--REFERENCE INFORMATION

Reference	Investigator	Symbol	Composition and Surface Condition	Test Method	Remarks
2	Richmond and Stewart		A-418 consists of alkali-free barium beryllium silicate frit with addition of chromic oxide. Coating thickness 2 mils. Coated on Inconel. Runs made at the following temperatures:	Normal spectral emittance. Double-beam infrared spectrometer with sodium chloride prism. Secondary standard [silicon carbide (Globar)] calibrated against laboratory blackbody. Temperatures measured with thermocouples.	Measured in air. Data taken from table.
		o	900 F		
		Δ	1200 F		
		□	1500 F		
		×	1800 F		

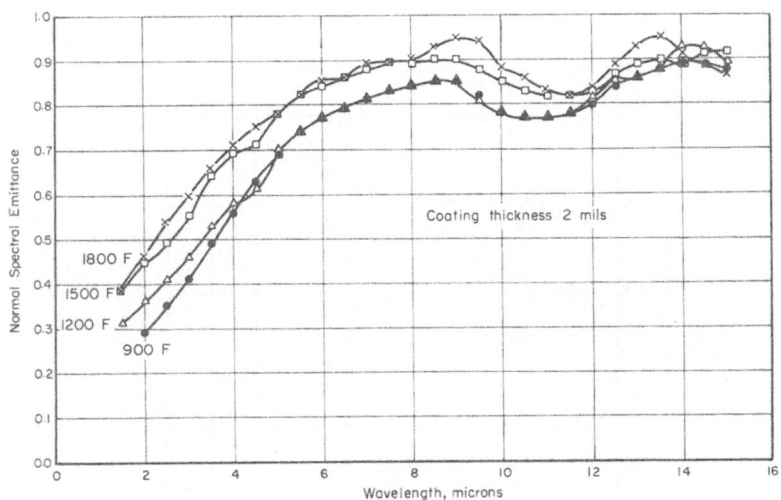

NORMAL SPECTRAL EMITTANCE OF A-418 ENAMEL ON TYPE 321 STAINLESS STEEL

NORMAL SPECTRAL EMITTANCE OF A-418 ENAMEL ON TYPE 321 STAINLESS STEEL--REFERENCE INFORMATION

Reference	Investigator	Symbol	Composition and Surface Condition	Test Method	Remarks
2	Richmond and Stewart		A-418 consists of alkali-free barium beryllium silicate frit with addition of chromic oxide. Coating thickness 2 mils. Coated on Inconel. Runs made at the following temperatures:	Normal spectral emittance. Double-beam infrared spectrometer with sodium chloride prism. Secondary standard [silicon carbide (Globar)] calibrated against laboratory blackbody. Temperatures measured with thermocouples.	Measured in air. Data taken from table.
		●	900 F		
		Δ	1200 F		
		□	1500 F		
		×	1800 F		

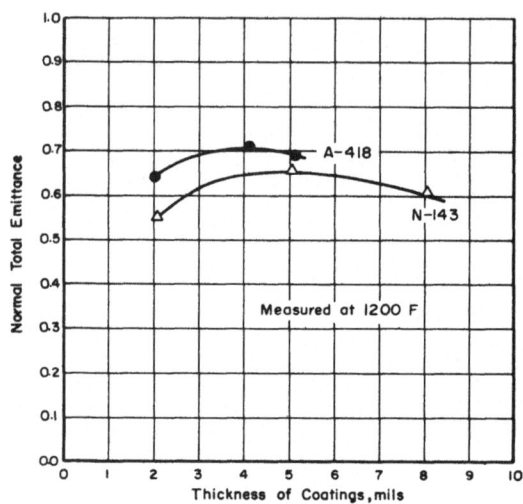

VARIATION OF NORMAL TOTAL EMITTANCE WITH THICKNESS OF A-418 AND N-143 ENAMELS ON INCONEL AT 1200 F

VARIATION OF NORMAL TOTAL EMITTANCE WITH THICKNESS OF A-418 AND N-143 ENAMELS
ON INCONEL AT 1200 F—REFERENCE INFORMATION

Reference	Investigator	Symbol	Composition and Surface Condition	Test Method	Remarks
2	Richmond and Stewart		Inconel coated with NBS coatings:	Normal total emittance. Thermopile detector. Comparison blackbody. Temperatures measured with thermocouples.	Measured in air. Data taken from curves.
		●	A-418		
		Δ	N-143		

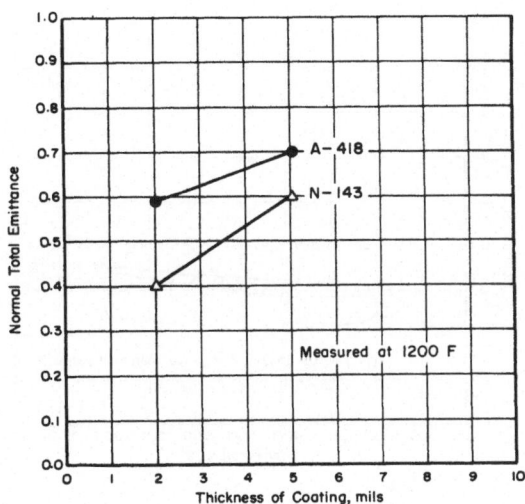

VARIATION OF NORMAL TOTAL EMITTANCE VERSUS COATING THICKNESS OF A-418 AND N-143 ENAMELS
ON TYPE 321 STAINLESS STEEL AT 1200 F

VARIATION OF NORMAL TOTAL EMITTANCE VERSUS COATING THICKNESS OF A-418 AND N-143 ENAMELS
ON TYPE 321 STAINLESS STEEL AT 1200 F--REFERENCE INFORMATION

Reference	Investigator	Symbol	Composition and Surface Condition	Test Method	Remarks
2	Richmond and Stewart		Type 321 stainless steel with NBS coatings:	Normal total emittance. Thermopile detector. Comparison blackbody.	Measured in air. Data taken from curves.
		●	A-418	Temperatures measured	
		Δ	N-143	with thermocouples.	

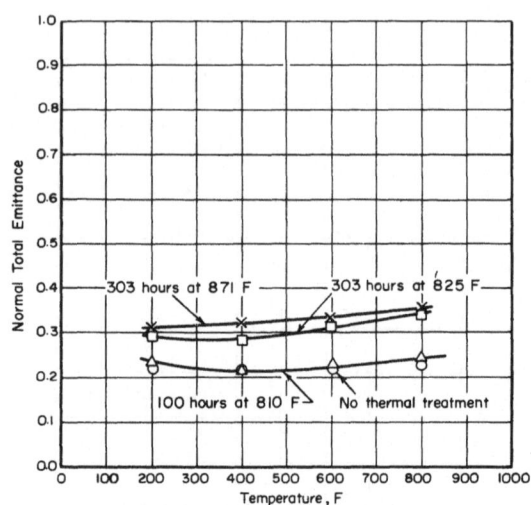

NORMAL TOTAL EMITTANCE OF ALUMINIZED SILICONE PAINT ON Ti-75A TITANIUM

NORMAL TOTAL EMITTANCE OF ALUMINIZED SILICONE PAINT ON Ti-75A TITANIUM--REFERENCE INFORMATION

Reference	Investigator	Symbol	Composition and Surface Condition	Test Method	Remarks
3	Bevans, Gier, and Dunkle		Dow-Corning XP-310 aluminized-silicone paint, on Ti-75A titanium (Mat'l. Spec. AMS 4901). No thickness given.	Normal total emittance. Calibrated thermopile detector. Temperatures measured with thermocouples.	Measured in air. Data taken from tables.
		O	No thermal treatment.		
		Δ	100 hours at 810 F.		
		□	303 hours at 825 F.		
		×	303 hours at 871 F.		

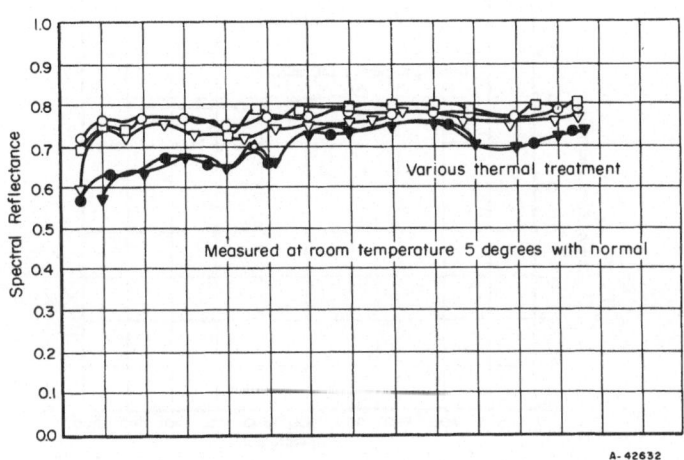

SPECTRAL REFLECTANCE OF ALUMINIZED SILICONE PAINT ON Ti-75A TITANIUM

SPECTRAL REFLECTANCE OF ALUMINIZED-SILICONE PAINT ON Ti-75A TITANIUM--REFERENCE INFORMATION

Reference	Investigator	Symbol	Composition and Surface Condition	Test Method	Remarks
3	Bevans, Gier, and Dunkle		Dow-Corning XP-310 aluminized-silicone paint on Ti-75A (Mat'l. Spec. AMS 4901). No thickness given.	Spectral reflectance at 5 degrees with normal. Gier-Dunkle reflectometer monochromator. Temperatures measured with thermocouples. (Diffuse illumination-normal viewing)	Measured in air at room temperature. Data taken from tables.
		△	No thermal treatment.		
		o	300 hours at 600 F.		
		□	100 hours at 810 F.		
		●	303 hours at 825 F.		
		▲	303 hours at 871 F.		

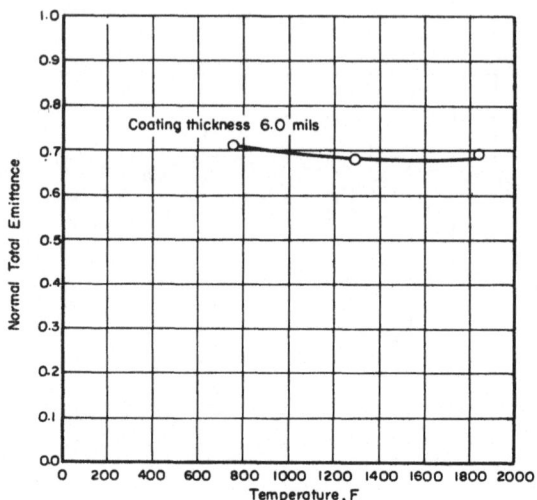

NORMAL TOTAL EMITTANCE OF B-1 ENAMEL ON INCONEL

NORMAL TOTAL EMITTANCE OF B-1 ENAMEL ON INCONEL--REFERENCE INFORMATION

Reference	Investigator	Symbol	Composition and Surface Condition	Test Method	Remarks
1	Burgess, Jasperse, Marcus, Martin, and Flint	o	B-1 Enamel on Inconel. Coating thickness 6.0 mils.	Normal total emittance. Rotating, hollow, cylindrical, Globar heating element. Blackbody hole. Specimen mounted in heating element flush with wall. Temperatures measured with thermocouples. Infrared spectrometer with prism replaced by plane mirror. Thermocouple detector.	Measured in air. Data taken from tables.

Coating Composition by Weight

NBS Frit No. 332 - 60 per cent
Black Stain* - 25 per cent
Cr_2O_3 - 15 per cent

*Co_2O_3, 28 per cent; Fe_2O_3, 37 per cent; Cr_2O_3, 10 per cent; MnO_2, 11 per cent; NiO, 14 per cent.

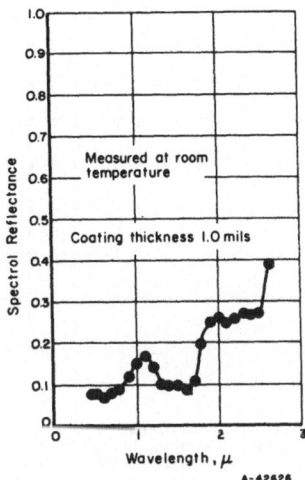

SPECTRAL REFLECTANCE OF INCONEL COATED WITH B-1 ENAMEL

SPECTRAL REFLECTANCE OF INCONEL COATED WITH B-1 ENAMEL--REFERENCE INFORMATION

Reference	Investigator	Symbol	Composition and Surface Condition	Test Method	Remarks
1	Burgess, Jasperse, Marcus, Martin, and Flint	●	Enamel B-1 on Inconel. Coating thickness 1.0 mil.	Spectral reflectance. Commercial reflect-ometer and spectro-photometer with quartz prism mono-chromator. MgO standard. (Normal viewing-diffuse reflection)	Measured in air at room temperature. Data taken from table.

Coating Composition by Weight

NBS Frit No. 332 - 60 per cent
Black Stain* - 25 per cent
Cr_2O_3 - 15 per cent

*Co_2O_3, 28 per cent; Fe_2O_3, 37 per cent; Cr_2O_3, 10 per cent; MnO_2, 11 per cent; NiO, 14 per cent.

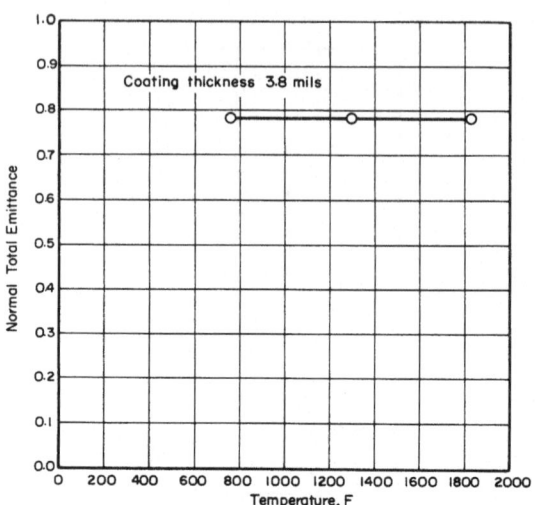

Coating thickness 3.8 mils

NORMAL TOTAL EMITTANCE OF B-4 ENAMEL ON INCONEL

NORMAL TOTAL EMITTANCE OF B-4 ENAMEL ON INCONEL--REFERENCE INFORMATION

Reference	Investigator	Symbol	Composition and Surface Condition	Test Method	Remarks
1	Burgess, Jasperse, Marcus, Martin, and Flint	o	B-4 Enamel on Inconel. Coating thickness 3.8 mils.	Normal total emittance. Rotating hollow cylindrical Globar heating element. Blackbody hole. Specimen mounted in heating element flush with wall. Temperatures measured with thermocouples. Infrared spectrometer with prism replaced by plane mirror. Thermocouple detector.	Measured in air. Data taken from tables.

Coating Composition by Weight

NBS Frit No. 332 - 60 per cent
Cr_2O_3 - 5 per cent
CoO - 15 per cent
Fe_2O_3 - 20 per cent

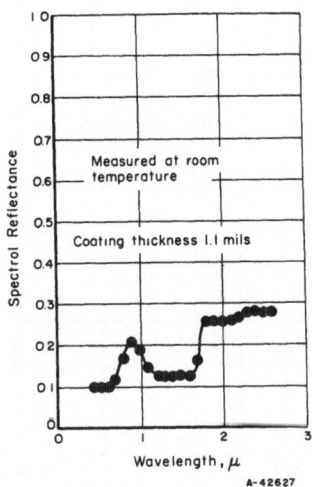

A-42627

SPECTRAL REFLECTANCE OF B-4 ENAMEL ON INCONEL

SPECTRAL REFLECTANCE OF B-4 ENAMEL ON INCONEL--REFERENCE INFORMATION

Reference	Investigator	Symbol	Composition and Surface Condition	Test Method	Remarks
1	Burgess, Jasperse, Marcus, Martin, and Flint	●	Enamel B-4 on Inconel. Coating thickness 1.1 mils.	Spectral reflectance. Commercial reflect-ometer with quartz prism monochromator. MgO standard. (Normal viewing-diffuse reflection)	Measured in air at room temperature.

Coating Composition by Weight

NBS Frit No. 332 – 60 per cent
  Cr_2O_3 – 5 per cent
  CoO – 15 per cent
  Fe_2O_3 – 20 per cent

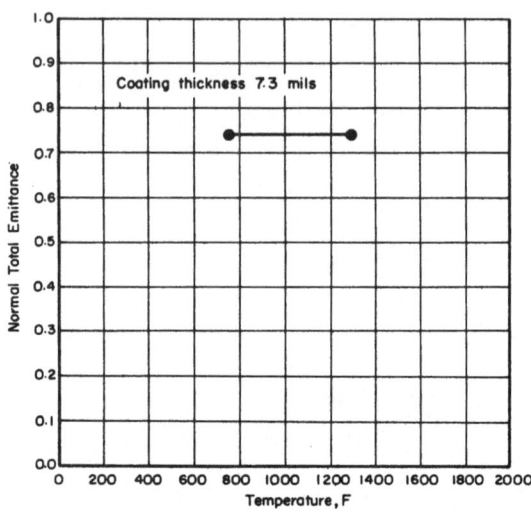

NORMAL TOTAL EMITTANCE OF B-7 ENAMEL ON INCONEL

NORMAL TOTAL EMITTANCE OF B-7 ENAMEL ON INCONEL--REFERENCE INFORMATION

Reference	Investigator	Symbol	Composition and Surface Condition	Test Method	Remarks
1	Burgess, Jasperse, Marcus, Martin, and Flint	●	B-7 Enamel on Inconel. Coating thickness 7.3 mils.	Normal total emittance. Rotating hollow cylindrical Globar heating element. Blackbody hole. Specimen mounted in heating element flush with wall. Temperatures measured with thermocouples. Infrared spectrometer with prism replaced by plane mirror. Thermocouple detector.	Measured in air. Data taken from tables.

Coating Composition by Weight

NBS Frit No. 332   - 60 per cent
  CoO·Cr$_2$O$_3$ spinel - 40 per cent

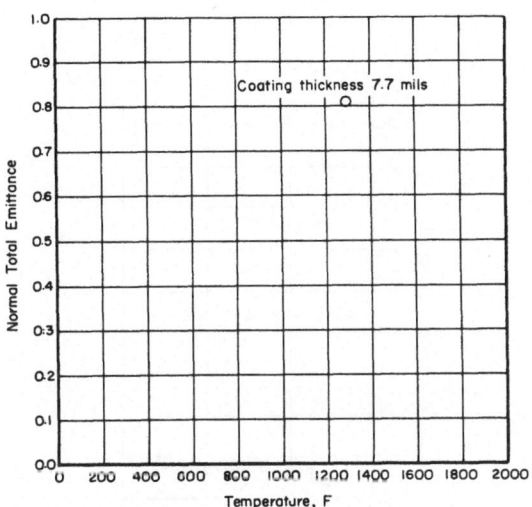

NORMAL TOTAL EMITTANCE OF B-8 ENAMEL ON INCONEL

NORMAL TOTAL EMITTANCE OF B-8 ENAMEL ON INCONEL--REFERENCE INFORMATION

Reference	Investigator	Symbol	Composition and Surface Condition	Test Method	Remarks
1	Burgess, Jasperse, Marcus, Martin, and Flint	O	B-8 Enamel on Inconel. Coating thickness 7.7 mils.	Normal total emittance. Rotating, hollow, cylindrical, Globar heating element. Blackbody hole. Specimen mounted in heating element flush with wall. Temperatures measured with thermocouples. Infrared spectrometer with prism replaced by plane mirror. Thermocouple detector.	Measured in air. Data taken from tables.

Coating Composition by Weight

NBS Frit No. 332   - 60 per cent
      NiO·Cr$_2$O$_3$ spinel - 40 per cent

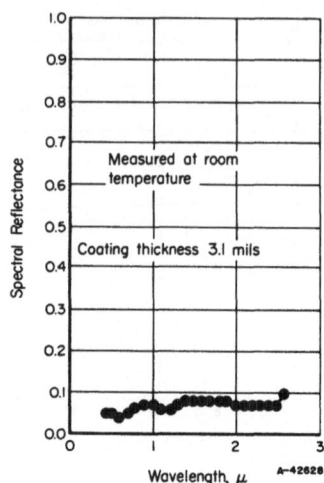

SPECTRAL REFLECTANCE OF B-8 ENAMEL ON INCONEL

SPECTRAL REFLECTANCE OF B-8 ENAMEL ON INCONEL—REFERENCE INFORMATION

Reference	Investigator	Symbol	Composition and Surface Condition	Test Method	Remarks
1	Burgess, Jasperse, Marcus, Martin, and Flint	●	Enamel B-8 on Inconel. Coating thickness 3.1 mils.	Spectral reflectance. Commercial reflectometer with quartz prism monochromator. MgO standard. (Normal viewing-diffuse reflection)	Measured in air at room temperature. Data taken from table.

Coating Composition by Weight

NBS Frit No. 332 - 60 per cent
$NiO \cdot Cr_2O_3$ - 40 per cent

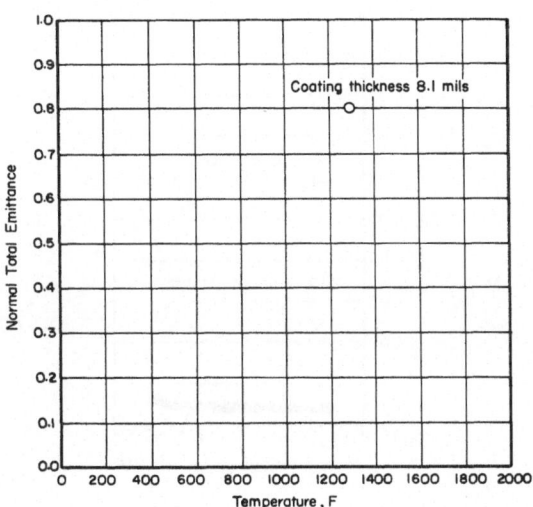

NORMAL TOTAL EMITTANCE OF B-9 ENAMEL ON INCONEL

NORMAL TOTAL EMITTANCE OF B-9 ENAMEL ON INCONEL--REFERENCE INFORMATION

Reference	Investigator	Symbol	Composition and Surface Condition	Test Method	Remarks
1	Burgess, Jasperse, Marcus, Martin, and Flint	O	B-9 Enamel on Inconel. Coating thickness 8.1 mils.	Normal total emittance. Rotating hollow cylindrical Globar heating element. Blackbody hole. Specimen mounted in heating element flush with wall. Temperatures measured with thermocouples.	Measured in air. Data taken from tables.

Coating Composition by Weight

NBS Frit No. 332   - 60 per cent
  NiO·Fe$_2$O$_3$ spinel - 40 per cent

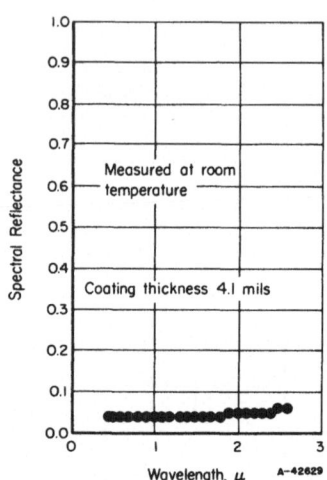

SPECTRAL REFLECTANCE OF B-9 ENAMEL ON INCONEL

SPECTRAL REFLECTANCE OF B-9 ENAMEL ON INCONEL—REFERENCE INFORMATION

Reference	Investigator	Symbol	Composition and Surface Condition	Test Method	Remarks
1	Burgess, Jasperse, Marcus, Martin, and Flint	●	Enamel B-9 on Inconel. Coating thickness 4.1 mils.	Spectral reflectance. Commercial reflect- ometer with quartz prism monochromator. MgO standard. (Normal viewing- diffuse reflection)	Measured in air at room temperature. Data taken from table.

Coating Composition by Weight

NBS Frit No. 332 - 60 per cent
NiO·Fe₂O₃ - 40 per cent

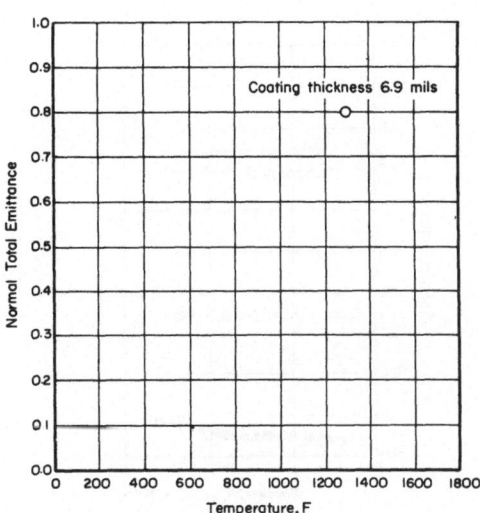

NORMAL TOTAL EMITTANCE OF B-11 ENAMEL ON INCONEL

NORMAL TOTAL EMITTANCE OF B-11 ENAMEL ON INCONEL--REFERENCE INFORMATION

Reference	Investigator	Symbol	Composition and Surface Condition	Test Method	Remarks
1	Burgess, Jasperse, Marcus, Martin, and Flint	o	B-11 Enamel on Inconel. Coating thickness 6.9 mils.	Normal total emittance. Rotating, hollow, cylindrical, Globar heating element. Blackbody hole. Specimen mounted in heating element flush with wall. Temperatures measured with thermocouples. Infrared spectrometer with prism replaced by plane mirror. Thermocouple detector.	Measured in air. Data taken from tables.

Coating Composition by Weight

NBS Frit No. 332   - 60 per cent
CoO·Fe$_2$O$_3$ spinel - 40 per cent

248

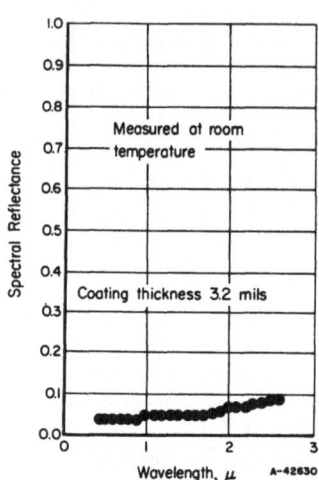

SPECTRAL REFLECTANCE OF B-11 ENAMEL ON INCONEL

SPECTRAL REFLECTANCE OF B-11 ENAMEL ON INCONEL--REFERENCE INFORMATION

Reference	Investigator	Symbol	Composition and Surface Condition	Test Method	Remarks
1	Burgess, Jasperse, Marcus, Martin, and Flint	●	Enamel B-11 on Inconel. Coating thickness 3.2 mils.	Spectral reflectance. Commercial reflect-ometer and spectro-photometer with quartz prism monochromator. MgO standard. (Normal viewing-diffuse reflection)	Measured in air at room temperature. Data taken from table.

Coating Composition by Weight

NBS Frit No. 332 - 60 per cent
CoO·Fe$_2$O$_3$ - 40 per cent

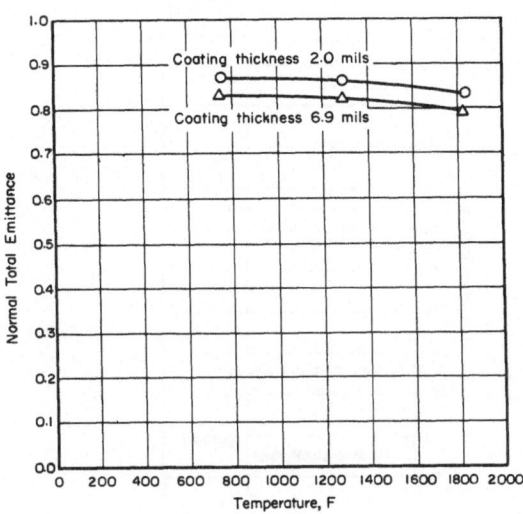

NORMAL TOTAL EMITTANCE OF B-12 ENAMEL ON INCONEL

NORMAL TOTAL EMITTANCE OF B-12 ENAMEL ON INCONEL--REFERENCE INFORMATION

Reference	Investigator	Symbol	Composition and Surface Condition	Test Method	Remarks
1	Burgess, Jasperse, Marcus, Martin, and Flint	O △	B-12 Enamel on Inconel. Coating thickness 2.0 mils. Coating thickness 6.9 mils.	Normal total emittance. Rotating, hollow, cylindrical, Globar heating element. Blackbody hole. Specimen mounted in heating element flush with wall. Temperatures measured with thermocouples. Infrared spectrometer with prism replaced by plane mirror. Thermocouple detector.	Measured in air. Data taken from tables.

Coating Composition by Weight

NBS Frit No. 332   - 60 per cent
CoO·Mn$_2$O$_3$ spinel - 40 per cent

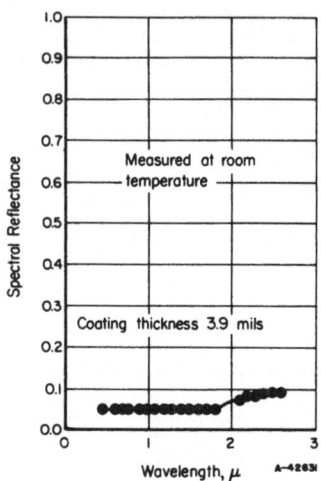

SPECTRAL REFLECTANCE OF B-12 ENAMEL ON INCONEL

SPECTRAL REFLECTANCE OF B-12 ENAMEL ON INCONEL--REFERENCE INFORMATION

Reference	Investigator	Symbol	Composition and Surface Condition	Test Method	Remarks
1	Burgess, Jasperse, Marcus, Martin, and Flint	●	Enamel B-12 on Inconel. Coating thickness 3.9 mils.	Spectral reflectance (normal viewing-diffuse reflection). Commercial reflect-ometer and spectro-photometer. Quartz prism mono-chromator. MgO standard.	Measured in air at room temperature. Data taken from table.

Coating Composition by Weight

NBS Frit No. 332 - 60 per cent
$CoO \cdot Mn_2O_3$ - 40 per cent

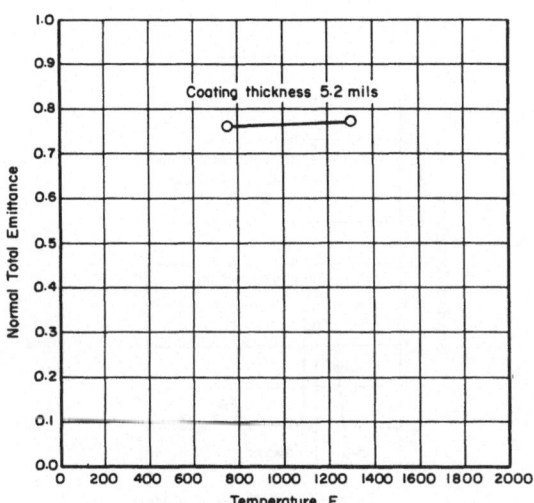

Coating thickness 5.2 mils

NORMAL TOTAL EMITTANCE OF B-13 ENAMEL ON INCONEL

NORMAL TOTAL EMITTANCE OF B-13 ENAMEL ON INCONEL--REFERENCE INFORMATION

Reference	Investigator	Symbol	Composition and Surface Condition	Test Method	Remarks
1	Burgess, Jasperse, Marcus, Martin, and Flint	○	B-13 Enamel on Inconel. Coating thickness 5.2 mils.	Normal total emittance. Rotating, hollow, cylindrical, Globar heating element. Blackbody hole. Specimen mounted in heating element flush with wall. Temperatures measured with thermocouples. Infrared spectrometer with prism replaced by plane mirror. Thermocouple detector.	Measured in air. Data taken from tables.

Coating Composition by Weight

NBS Frit No. 332   - 60 per cent
    $CoO \cdot Cr_2O_3$ spinel - 40 per cent

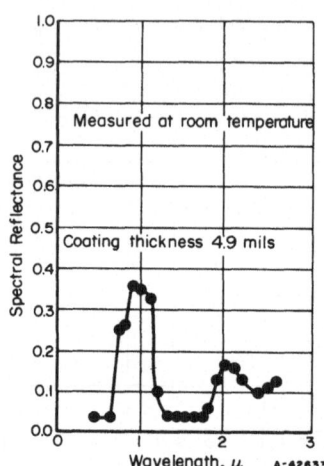

SPECTRAL REFLECTANCE OF B-13 ENAMEL ON INCONEL

SPECTRAL REFLECTANCE OF B-13 ENAMEL ON INCONEL—REFERENCE INFORMATION

Reference	Investigator	Symbol	Composition and Surface Condition	Test Method	Remarks
1	Burgess, Jasperse, Marcus, Martin, and Flint	●	Enamel B-13 on Inconel. Coating thickness 4.9 mils.	Spectral reflectance. (Normal viewing-diffuse reflection.) Commercial reflectometer and spectrophotometer. Quartz prism monochromator. MgO standard.	Measured in air at room temperature. Data taken from table.

Coating Composition by Weight

NBS Frit No. 332 – 50 per cent
$CoO \cdot Cr_2O_3$ – 50 per cent

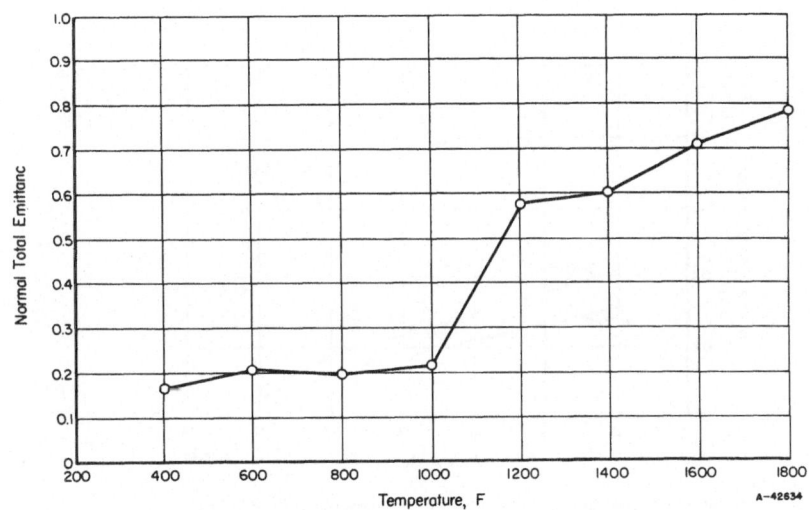

NORMAL TOTAL EMITTANCE OF ALUMINUM PAINT ON A-286 STEEL

NORMAL TOTAL EMITTANCE OF CHEM INDUSTRIES ALUMINUM PAINT ON A-286 STEEL--REFERENCE INFORMATION

Reference	Investigator	Symbol	Composition and Surface Condition	Test Method	Remarks
13	Gravina and Katz		Chem Industries high-temperature silicone-base aluminum paint. Coating thickness and surface condition not given.	Normal total emittance. Resistance-heated strip specimen. Thermistor-bolometer detector. Reference blackbody. Temperatures measured with thermocouples.	Measured in air. Data taken from curves.

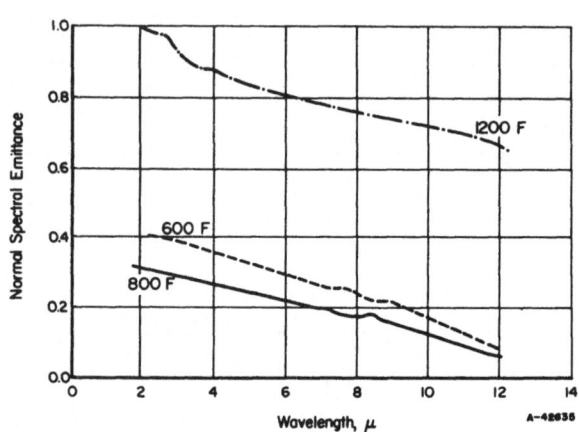

NORMAL SPECTRAL EMITTANCE OF CHEM INDUSTRIES ALUMINUM PAINT ON A-286 STEEL

NORMAL SPECTRAL EMITTANCE OF CHEM INDUSTRIES ALUMINUM PAINT ON A-286 STEEL--REFERENCE INFORMATION

Reference	Investigator	Symbol	Composition and Surface Condition	Test Method	Remarks
13	Gravina and Katz		Chem Industries high-temperature aluminum, silicone-base paint on A-286 steel. Thickness and surface condition not given. Measured at: 600 F 800 F 1200 F	Normal spectral emittance. Resistance-heated strip specimen. Thermistor-bolometer detector. Monochromator. Reference blackbody. Temperatures measured with thermocouples.	Measured in air. Data taken from curves.

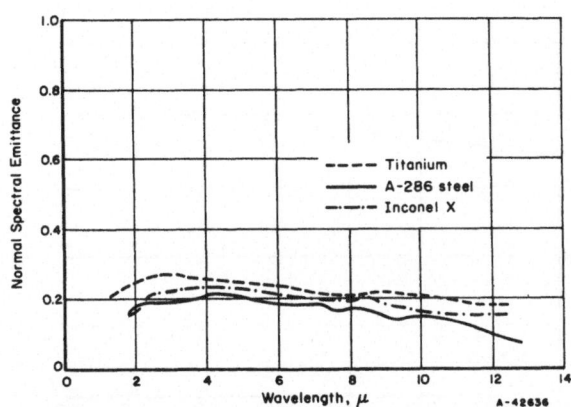

NORMAL SPECTRAL EMITTANCE OF CHEM INDUSTRIES ALUMINUM PAINT ON TITANIUM, STEEL, AND INCONEL X AT 800 F

NORMAL SPECTRAL EMITTANCE OF CHEM INDUSTRIES ALUMINUM PAINT ON TITANIUM, STEEL, AND INCONEL X—REFERENCE INFORMATION

Reference	Investigator	Symbol	Composition and Surface Condition	Test Method	Remarks
13	Gravina and Katz		Chem Industries high-temperature, silicone-base aluminum paint. Thickness and surface condition not given. Coated on: 6Al—4V Titanium Inconel X A-286 steel All measurements at 800 F.	Normal spectral emittance. Resistance-heated strip specimen. Thermistor-bolometer detector. Monochromator. Reference blackbody. Temperatures measured with thermocouples.	Measured in air. Data taken from curves.

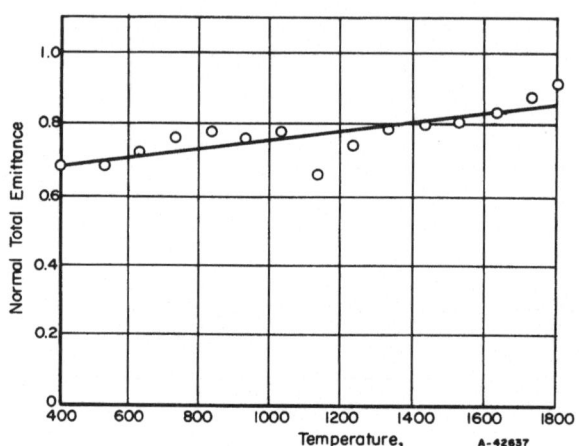

NORMAL TOTAL EMITTANCE OF DU LITE -0 ON A-286 STEEL

NORMAL TOTAL EMITTANCE OF DULITE 3-0 COATING ON A-286 STEEL--REFERENCE INFORMATION

Reference	Investigator	Symbol	Composition and Surface Condition	Test Method	Remarks
13	Gravina and Katz		DuLite 3-0, an oxide surface conversion coating. Composition or coating thickness not given.	Normal total emittance. Resistance-heated strip specimen. Thermistor-bolometer detector. Reference blackbody. Temperatures measured with thermocouples.	Measured in air. Data taken from curves.

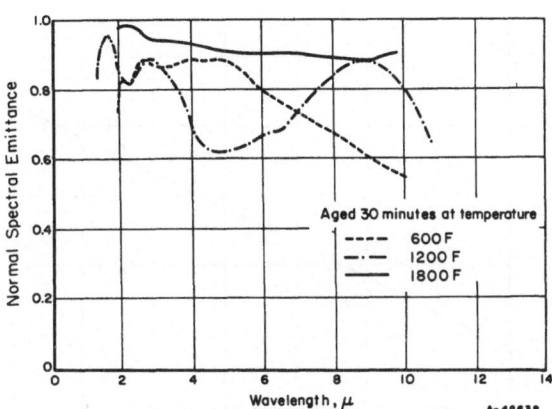

NORMAL SPECTRAL EMITTANCE OF DU LITE 3-0 ON A-286 STEEL

NORMAL SPECTRAL EMITTANCE OF DULITE 3-0 COATING ON A-286 STEEL--REFERENCE INFORMATION

Reference	Investigator	Symbol	Composition and Surface Condition	Test Method	Remarks
13	Gravina and Katz		DuLite 3-0, an oxide conversion coating. No thickness or composition given. Aged 30 minutes at temperature. Measured at: 600 F 1200 F 1800 F	Normal spectral emittance. Resistance-heated strip specimen. Thermistor-bolometer detector. Monochromator. Reference blackbody. Temperatures measured with thermocouples.	Measured in air. Data taken from curves.

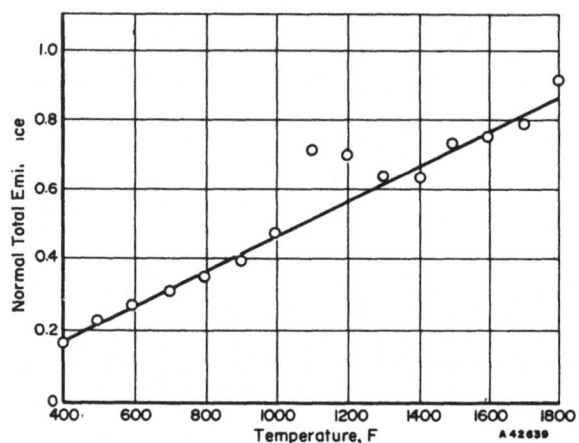

NORMAL TOTAL EMITTANCE OF DU LITE 3-0 ON INCONEL X

NORMAL TOTAL EMITTANCE OF DULITE 3-0 COATING ON INCONEL X—REFERENCE INFORMATION

eference	Investigator	Symbol	Composition and Surface Condition	Test Method	Remarks
13	Gravina and Katz		DuLite 3-0, an oxide surface conversion coating. Composition or coating thickness not given.	Normal total emittance. Resistance-heated strip specimen. Thermistor-bolometer detector. Reference blackbody. Temperatures measured with thermocouples.	Measured in air. Data taken from curves.

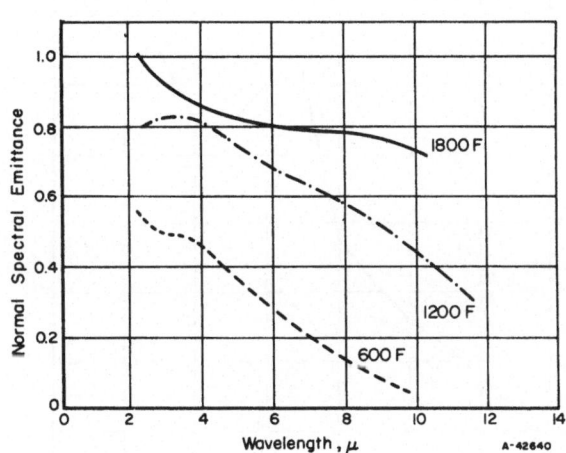

NORMAL SPECTRAL EMITTANCE OF DU LITE 3-0 ON INCONEL X

NORMAL SPECTRAL EMITTANCE OF DULITE 3-0 ON INCONEL X--REFERENCE INFORMATION

Reference	Investigator	Symbol	Composition and Surface Condition	Test Method	Remarks
13	Gravina and Katz		DuLite 3-0, an oxide conversion coating of the base metal. Thickness or surface condition not given. Measured at: 600 F 1200 F 1800 F	Normal spectral emittance. Resistance-heated strip specimen. Thermistor-bolometer detector. Monochromator. Reference blackbody. Temperatures measured with thermocouples.	Measured in air. Data taken from curves.

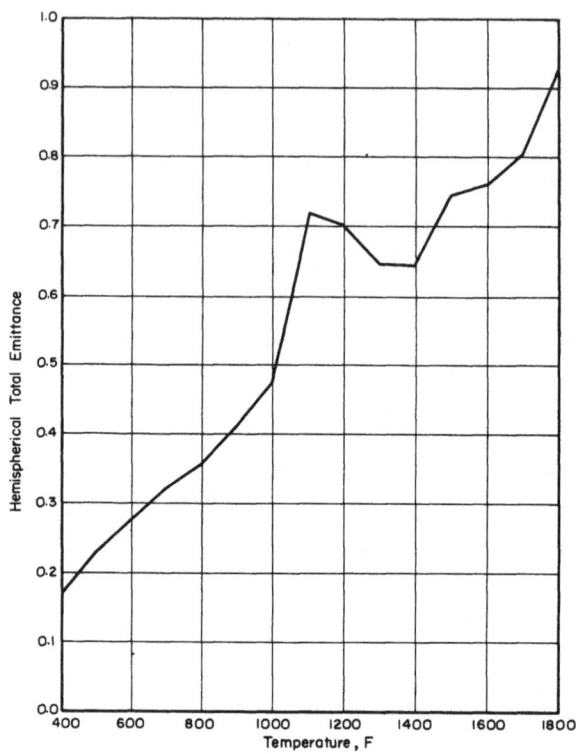

HEMISPHERICAL TOTAL EMITTANCE OF DuLITE 3-0 ON TITANIUM

HEMISPHERICAL TOTAL EMITTANCE OF DuLITE 3-0 ON TITANIUM--REFERENCE INFORMATION

Reference	Investigator	Symbol	Composition and Surface Condition	Test Method	Remarks
4	Dull, R. L.		DuLite 3-0 coating on titanium. No thickness given. (DuLite 3-0 is an oxide conversion coating of the base metal.) Note: Color of specimen surface changed considerably as the temperature increased. Original color - black.	Hemispherical total emittance. Resistance-heated strip. Specimens coated with test material. Measured power input to test section. Temperatures measured with thermocouples.	Measured in air. Data taken from curves.

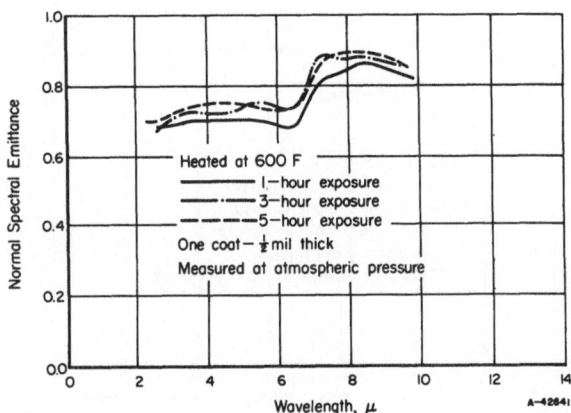

NORMAL SPECTRAL EMITTANCE OF GULTON CERAMIC COATING 6013 ON A-286 STEEL AT 600 F

NORMAL SPECTRAL EMITTANCE OF GULTON CERAMIC COATING 6013 ON A-286 STEEL AT 600 F—REFERENCE INFORMATION

Reference	Investigator	Symbol	Composition and Surface Condition	Test Method	Remarks
13	Gravina and Katz		Gulton ceramic coating 6013. A high chrome-bearing coating. Applied as a slip, dried, and fired. One coat, 1/2-mil thick, continuously heated at 600 F: 1 hour 3 hours 5 hours	Normal spectral emittance. Resistance-heated strip specimen. Thermistor-bolometer detector. Monochromator. Reference blackbody. Temperatures measured with thermocouples.	Measured in air. Data taken from curves.

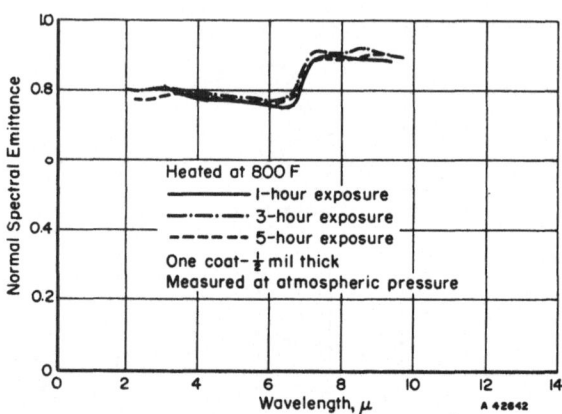

NORMAL SPECTRAL EMITTANCE OF GULTON CERAMIC COATING 6013 ON A-286 STEEL AT 800 F

NORMAL SPECTRAL EMITTANCE OF GULTON CERAMIC COATING 6013 ON A-286 STEEL AT 800 F--REFERENCE INFORMATION

Reference	Investigator	Symbol	Composition and Surface Condition	Test Method	Remarks
13	Gravina and Katz		Gulton ceramic coating 6013. A high chrome-bearing coating. Applied as a slip on sand-blasted A-286 steel, dried, and fired. One coat, 1/2 mil thick, continuously heated at 800 F: 1 hour 3 hours 5 hours	Normal spectral emittance. Resistance-heated strip specimen. Thermistor-bolometer detector. Monochromator. Reference blackbody. Temperatures measured with thermocouples.	Measured in air. Data taken from curves.

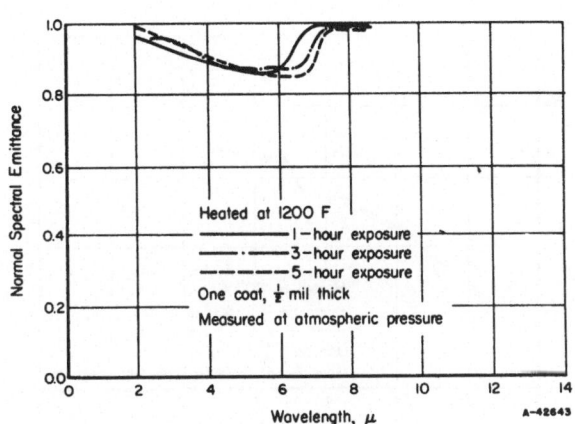

NORMAL SPECTRAL EMITTANCE OF GULTON CERAMIC COATING 6013 ON A-286 STEEL AT 1200 F

NORMAL SPECTRAL EMITTANCE OF GULTON CERAMIC COATING 6013 ON A-286 STEEL AT 1200 F--REFERENCE INFORMATION

Reference	Investigator	Symbol	Composition and Surface Condition	Test Method	Remarks
13	Gravina and Katz		Gulton ceramic coating 6013. A high-chrome-bearing coating. Applied as a slip on sand-blasted A-286 steel, dried, and fired. 1 coat, 1/2-mil thick, continuously heated at 1200 F: 1 hour 3 hours 5 hours	Normal spectral emittance. Resistance-heated strip specimen. Thermistor-bolometer detector. Monochromator. Reference blackbody. Temperatures measured with thermocouples.	Measured in air. Data taken from curves.

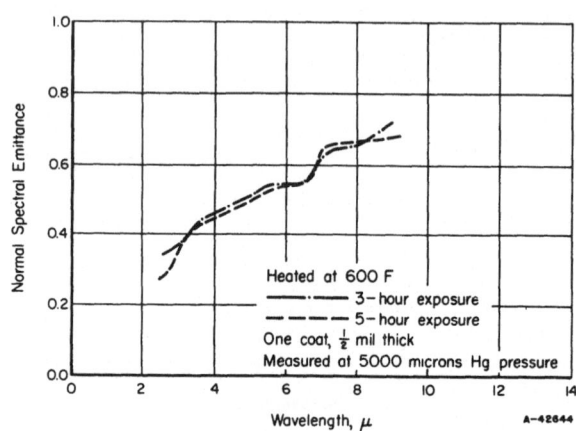

NORMAL SPECTRAL EMITTANCE OF GULTON CERAMIC COATING 6013 ON A-286 STEEL AT 600 F

NORMAL SPECTRAL EMITTANCE OF GULTON CERAMIC COATING 6013 ON A-286 STEEL AT 600 F--REFERENCE INFORMATION

Reference	Investigator	Symbol	Composition and Surface Condition	Test Method	Remarks
13	Gravina and Katz		Gulton ceramic coating 6013. A high-chrome-bearing coating. Applied as a slip on sand-blasted material, dried, and fired. 1 coat, 1/2 mil thick. Heated at 600 F: 3 hours 5 hours	Normal spectral emittance. Resistance-heated strip specimen. Thermistor-bolometer detector. Monochromator. Reference blackbody. Temperatures measured with thermocouples.	Measured in 5000 micron Hg pressure. Data taken from curves.

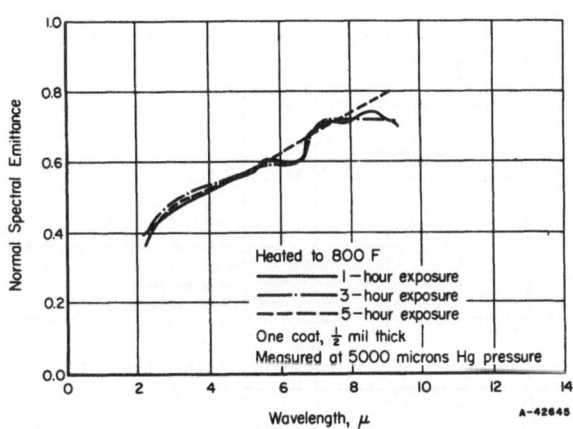

NORMAL SPECTRAL EMITTANCE OF GULTON CERAMIC COATING 6013 ON A-286 STEEL AT 800 F

NORMAL SPECTRAL EMITTANCE OF GULTON CERAMIC COATING 6013 ON A-286 STEEL AT 800 F--REFERENCE INFORMATION

Reference	Investigator	Symbol	Composition and Surface Condition	Test Method	Remarks
13	Gravina and Katz		Gulton ceramic coating 6013. A high-chrome-bearing coating. Applied as a slip on sand-blasted material, dried, and fired. 1 coat, 1/2 mil thick. Heated at 800 F: 1 hour 3 hours 5 hours	Normal spectral emittance. Resistance-heated strip specimen. Thermistor-bolometer detector. Monochromator. Reference blackbody. Temperatures measured with thermocouples.	Measured in 5000 micron Hg pressure. Data taken from curves.

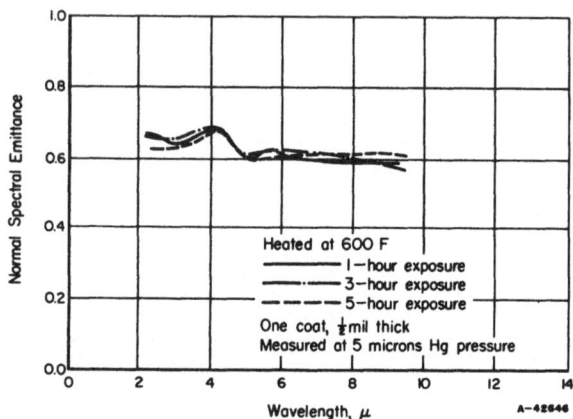

NORMAL SPECTRAL EMITTANCE OF GULTON CERAMIC COATING 6013 ON A-286 STEEL AT 600 F

NORMAL SPECTRAL EMITTANCE OF GULTON CERAMIC COATING 6013 ON A-286 STEEL AT 600 F—REFERENCE INFORMATION

Reference	Investigator	Symbol	Composition and Surface Condition	Test Method	Remarks
13	Gravina and Katz		Gulton ceramic coating 6013. A high-chrome-bearing coating. Applied as a slip on sand-blasted material, dried, and fired. 1 coat, 1/2-mil thick. Heated at 600 F: 1 hour 3 hours 5 hours	Normal spectral emittance. Resistance-heated strip specimen. Thermistor-bolometer detector. Monochromator. Reference blackbody. Temperatures measured with thermocouples.	Measured in 5 micron Hg pressure. Data taken from curves.

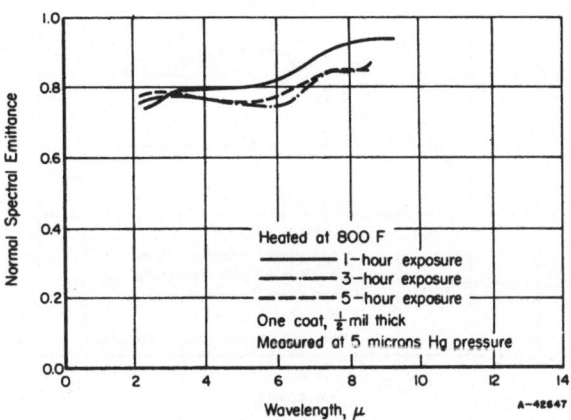

NORMAL SPECTRAL EMITTANCE OF GULTON CERAMIC COATING 6013 ON A-286 STEEL AT 800 F

NORMAL SPECTRAL EMITTANCE OF GULTON CERAMIC COATING 6013 ON A-286 STEEL AT 800 F--REFERENCE INFORMATION

Reference	Investigator	Symbol	Composition and Surface Condition	Test Method	Remarks
13	Gravina and Katz		Gulton ceramic coating 6013. A high-chrome-bearing coating. Applied as a slip on sand-blasted material, dried, and fired. 1 coat, 1/2-mil thick. Heated at 800 F: 1 hour 3 hours 5 hours	Normal spectral emittance. Resistance-heated strip specimen. Thermistor-bolometer detector. Monochromator. Reference blackbody. Temperatures measured with thermocouples.	Measured in 5 microns Hg pressure. Data taken from curves.

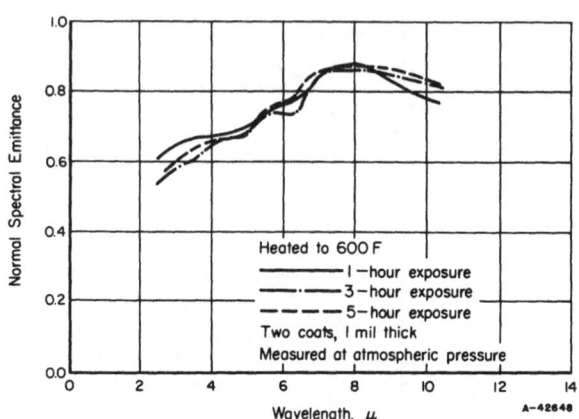

NORMAL SPECTRAL EMITTANCE OF GULTON CERAMIC COATING 6013 ON A-286 STEEL AT 600 F

NORMAL SPECTRAL EMITTANCE OF GULTON 6013 CERAMIC COATING ON A-286 STEEL AT 600 F--REFERENCE INFORMATION

Reference	Investigator	Symbol	Composition and Surface Condition	Test Method	Remarks
13	Gravina and Katz		Gulton ceramic coating 6013. A high-chrome-bearing coating. Applied as a slip to sand-blasted A-286 steel, dried, and fired. 2 coats, 1 mil thick. Continuously heated at 1600 F: 1 hour 3 hours 5 hours	Normal spectral emittance. Resistance-heated strip specimen. Thermistor-bolometer detector. Monochromator. Reference blackbody. Temperatures measured with thermocouples.	Measured in air. Data taken from curves.

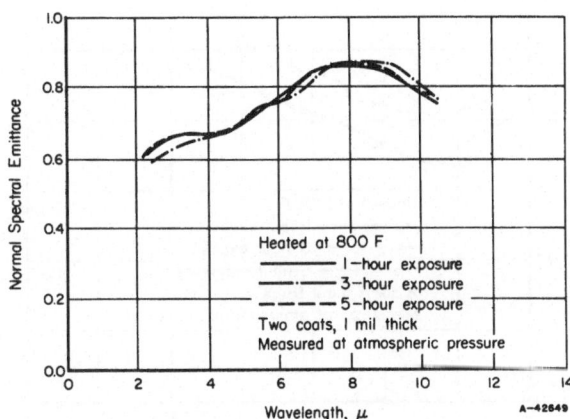

NORMAL SPECTRAL EMITTANCE OF GULTON CERAMIC COATING 6013 ON A-286 STEEL AT 800 F

NORMAL SPECTRAL EMITTANCE OF GULTON 6013 CERAMIC COATING ON A-286 STEEL AT 800 F---REFERENCE INFORMATION

Reference	Investigator	Symbol	Composition and Surface Condition	Test Method	Remarks
13	Gravina and Katz		Gulton ceramic coating 6013. A high-chrome-bearing coating. Applied as a slip on sand-blasted A-286 steel, dried, and fired. 2 coats, 1 mil thick. Continuously heated at 800 F: 1 hour 3 hours 5 hours	Normal spectral emittance. Resistance-heated strip specimen. Thermistor-bolometer detector. Monochromator. Reference blackbody. Temperatures measured with thermocouples.	Measured in air. Data taken from curves.

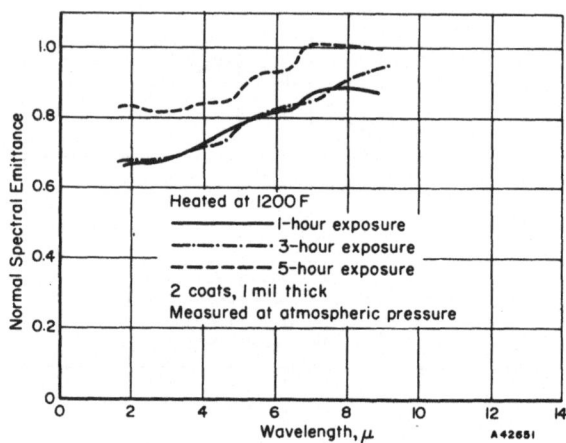

NORMAL SPECTRAL EMITTANCE OF GULTON CERAMIC COATING 6013 ON A-286 STEEL AT 1200 F

NORMAL SPECTRAL EMITTANCE OF GULTON 6013 CERAMIC COATING ON A-286 STEEL AT 1200 F--REFERENCE INFORMATION

Reference	Investigator	Symbol	Composition and Surface Condition	Test Method	Remarks
13	Gravina and Katz		Gulton ceramic coating 6013. A high-chrome-bearing coating. Applied as a slip on sand-blasted A-286 steel, dried, and fired. 2 coats, 1 mil thick. Heated at 1200 F: 1 hour 3 hours 5 hours	Normal spectral emittance. Resistance-heated strip specimen. Thermistor-bolometer detector. Monochromator. Reference blackbody. Temperatures measured with thermocouples.	Measured in air. Data taken from curves.

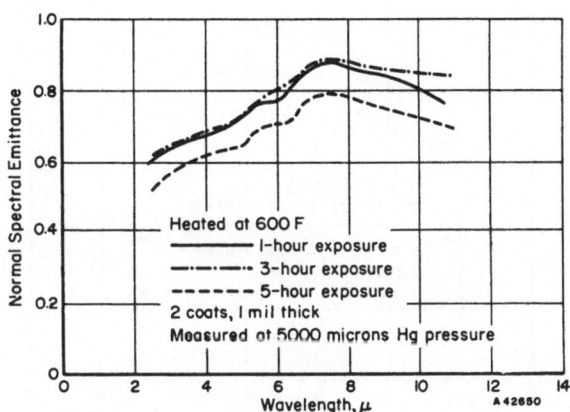

NORMAL SPECTRAL EMITTANCE OF GULTON CERAMIC COATING 6013 ON A-286 STEEL AT 600 F

NORMAL SPECTRAL EMITTANCE OF GULTON 6013 CERAMIC COATING ON A-286 STEEL AT 600 F--REFERENCE INFORMATION

Reference	Investigator	Symbol	Composition and Surface Condition	Test Method	Remarks
13	Gravina and Katz		Gulton ceramic coating 6013. A high-chrome-bearing coating. Applied as a slip on sand-blasted A-286 steel, dried, and fired. 2 coats, 1 mil thick. Heated at 600 F: 1 hour 3 hours 5 hours	Normal spectral emittance. Resistance-heated strip specimen. Thermistor-bolometer detector. Monochromator. Reference blackbody. Temperatures measured with thermocouples.	Measured in 5000 micron Hg pressure. Data taken from curves.

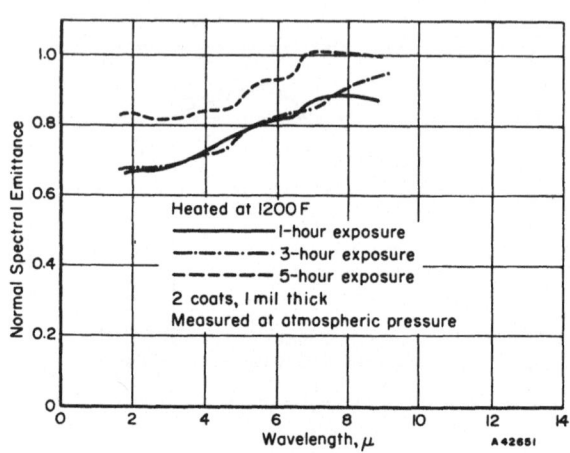

NORMAL SPECTRAL EMITTANCE OF GULTON CERAMIC COATING 6013 ON A-286 STEEL AT 1200 F

NORMAL SPECTRAL EMITTANCE OF GULTON 6013 CERAMIC COATING ON A-286 STEEL AT 1200 F--REFERENCE INFORMATION

Reference	Investigator	Symbol	Composition and Surface Condition	Test Method	Remarks
13	Gravina and Katz		Gulton ceramic coating 6013. A high-chrome-bearing coating. Applied as a slip on sand-blasted A-286 steel, dried, and fired. 2 coats, 1 mil thick. Heated at 1200 F: 1 hour 3 hours 5 hours	Normal spectral emittance. Resistance-heated strip specimen. Thermistor-bolometer detector. Monochromator. Reference blackbody. Temperatures measured with thermocouples.	Measured in air. Data taken from curves.

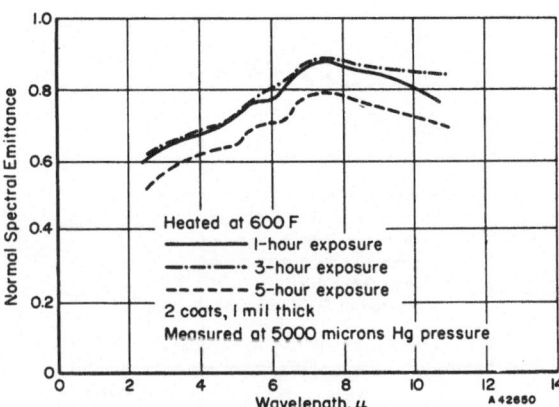

NORMAL SPECTRAL EMITTANCE OF GULTON CERAMIC COATING 6013 ON A-286 STEEL AT 600 F

NORMAL SPECTRAL EMITTANCE OF GULTON 6013 CERAMIC COATING ON A-286 STEEL AT 600 F--REFERENCE INFORMATION

Reference	Investigator	Symbol	Composition and Surface Condition	Test Method	Remarks
13	Gravina and Katz		Gulton ceramic coating 6013. A high-chrome-bearing coating. Applied as a slip on sand-blasted A-286 steel, dried, and fired. 2 coats, 1 mil thick. Heated at 600 F: 1 hour 3 hours 5 hours	Normal spectral emittance. Resistance-heated strip specimen. Thermistor-bolometer detector. Monochromator. Reference blackbody. Temperatures measured with thermocouples.	Measured in 5000 micron Hg pressure. Data taken from curves.

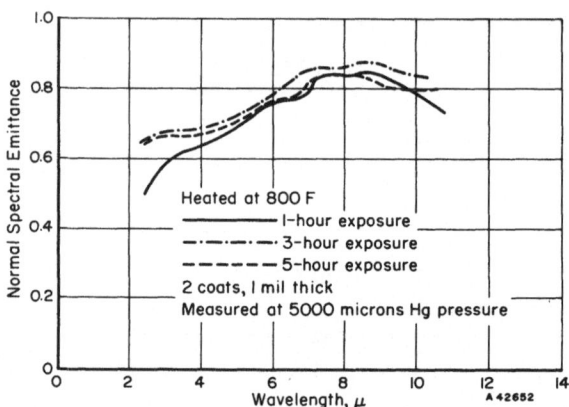

NORMAL SPECTRAL EMITTANCE OF GULTON CERAMIC COATING 6013 ON A-286 STEEL AT 800 F

NORMAL SPECTRAL EMITTANCE OF GULTON 6013 CERAMIC COATING ON A-286 STEEL AT 800 F--REFERENCE INFORMATION

Reference	Investigator	Symbol	Composition and Surface Condition	Test Method	Remarks
13	Gravina and Katz		Gulton ceramic coating 6013. A high-chrome-bearing coating. Applied as a slip on sand-blasted A-286 steel, dried, and fired. 2 coats, 1 mil thick. Heated at 800 F: 1 hour 3 hours 5 hours	Normal spectral emittance. Resistance-heated strip specimen. Thermistor-bolometer detector. Monochromator. Reference blackbody. Temperatures measured with thermocouples.	Measured in 5000 micron pressure. Data taken from curves.

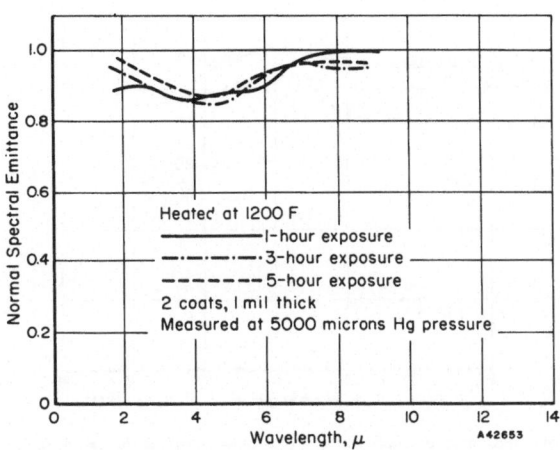

NORMAL SPECTRAL EMITTANCE OF GULTON CERAMIC COATING 6013 ON A-286 STEEL AT 1200 F

NORMAL SPECTRAL EMITTANCE OF GULTON 6013 CERAMIC COATING ON A-286 STEEL AT 1200 F--REFERENCE INFORMATION

Reference	Investigator	Symbol	Composition and Surface Condition	Test Method	Remarks
13	Gravina and Katz		Gulton ceramic coating 6013. A high-chrome-bearing coating. Applied as a slip on sand-blasted A-286 steel, dried, and fired. 2 coats, 1 mil thick. Heated at 1200 F: 1 hour 3 hours 5 hours	Normal spectral emittance. Resistance-heated strip specimen. Thermistor-bolometer detector. Monochromator. Reference blackbody. Temperatures measured with thermocouples.	Measured in 5000 micron pressure. Data taken from curves.

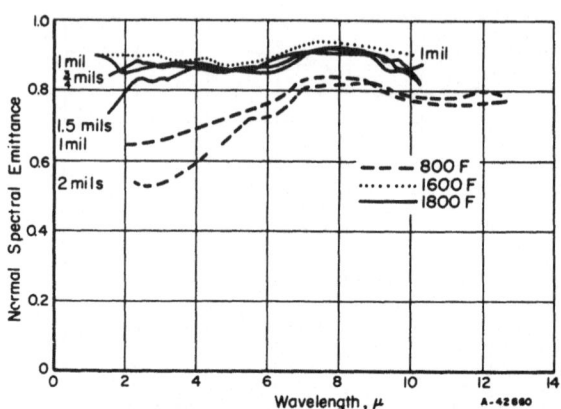

VARIATION OF THE NORMAL SPECTRAL EMITTANCE OF GULTON CERAMIC COATING 6013 ON INCONEL X
WITH COATING THICKNESS AND TEMPERATURE

NORMAL SPECTRAL EMITTANCE OF GULTON CERAMIC COATING 6013 ON INCONEL X--REFERENCE INFORMATION

Reference	Investigator	Symbol	Composition and Surface Condition	Test Method	Remarks
13	Gravina and Katz		Gulton ceramic coating 6013. A high-chrome-bearing coating. Applied as a slip on sand-blasted material, dried, and fired. Coating thicknesses, 3/4, 1, 1.5, and 2 mils Measured at: 800 F 1600 F 1800 F	Normal spectral emittance. Resistance-heated strip specimen. Thermistor-bolometer detector. Monochromator. Reference blackbody. Temperatures measured with thermocouples.	Measured in air. Data taken from curves.

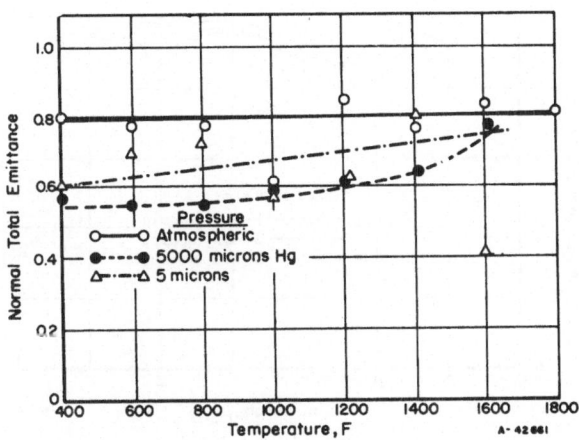

NORMAL TOTAL EMITTANCE OF NATIONAL LEAD BLACK PAINT 46H47 ON A-286 STEEL

NORMAL TOTAL EMITTANCE OF NATIONAL LEAD BLACK PAINT 46H47 ON A-286 STEEL—REFERENCE INFORMATION

Reference	Investigator	Symbol	Composition and Surface Condition	Test Method	Remarks
13	Gravina and Katz		National Lead 46H47 black paint. Composition or thickness not given. Measured at: Atmospheric pressure 5000 microns Hg 5 microns Hg.	Normal total emittance. Resistance-heated strip specimen. Thermistor-bolometer detector. Reference blackbody. Temperatures measured with thermocouples.	Measured in air and vacuum. Data taken from curves.

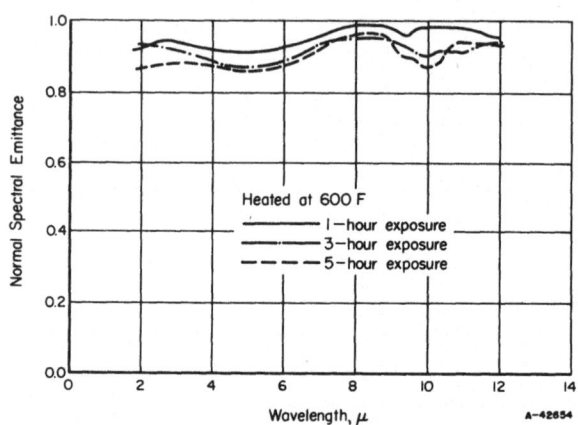

NORMAL SPECTRAL EMITTANCE OF NATIONAL LEAD BLACK PAINT 46H47 ON A-286 STEEL AT 600 F

NORMAL SPECTRAL EMITTANCE OF NATIONAL LEAD BLACK PAINT 46H47 ON A-286 STEEL AT 600 F--REFERENCE INFORMATION

Reference	Investigator	Symbol	Composition and Surface Condition	Test Method	Remarks
13	Gravina and Katz		National Lead 46H47 "high heat black" paint. Composition or thickness not given. Heated at 600 F: 1 hour 3 hours 5 hours	Normal spectral emittance. Resistance-heated strip specimen. Thermistor-bolometer detector. Monochromator. Reference blackbody. Temperatures measured with thermocouples.	Measured in air. Data taken from curves.

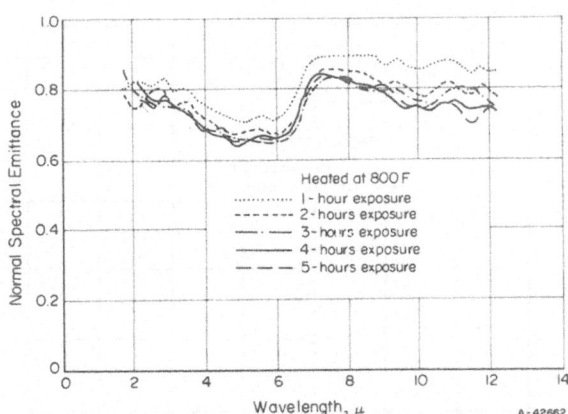

NORMAL SPECTRAL EMITTANCE OF NATIONAL LEAD BLACK PAINT 46H47 ON A-286 STEEL AT 800 F

NORMAL SPECTRAL EMITTANCE OF NATIONAL LEAD BLACK PAINT 46H47 ON A-286 STEELS AT 800 F—REFERENCE INFORMATION

Reference	Investigator	Symbol	Composition and Surface Condition	Test Method	Remarks
13	Gravina and Katz		National Lead 46H47 "high heat black" paint. Composition or thickness not given. Heated at 800 F: 1 hour 2 hours 3 hours 4 hours 5 hours	Normal spectral emittance. Resistance-heated strip specimen. Thermistor-bolometer detector. Monochromator. Reference blackbody. Temperatures measured with thermocouples.	Measured in air. Data taken from curves.

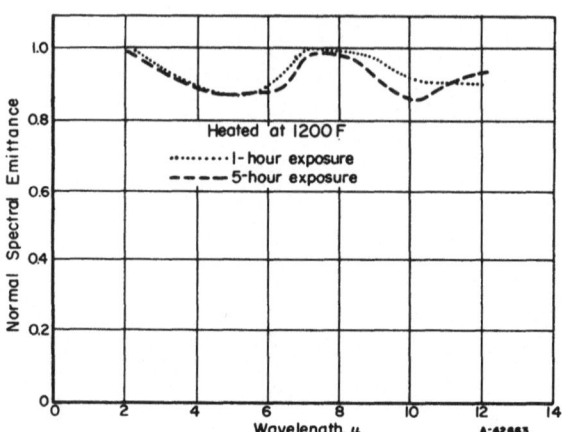

NORMAL SPECTRAL EMITTANCE OF NATIONAL LEAD BLACK PAINT 46H47 ON A-286 STEEL AT 1200 F

NORMAL SPECTRAL EMITTANCE OF NATIONAL LEAD BLACK PAINT 46H47 ON A-286 STEEL AT 1200 F--REFERENCE INFORMATION

Reference	Investigator	Symbol	Composition and Surface Condition	Test Method	Remarks
13	Gravina and Katz		National Lead 46H47 "high heat black" paint. Composition or thickness not given. Heated at 1200 F: 1 hour 5 hours	Normal spectral emittance. Resistance-heated strip specimen. Thermistor-bolometer detector. Monochromator. Reference blackbody. Temperatures measured with thermocouples.	Measured in air. Data taken from curves.

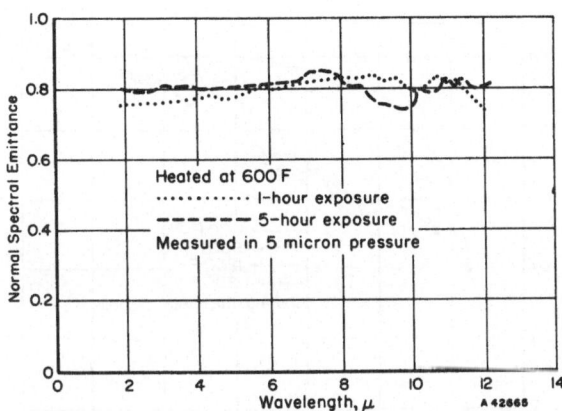

NORMAL SPECTRAL EMITTANCE OF NATIONAL LEAD BLACK PAINT 46H47 ON A-286 STEEL AT 600 F

NORMAL SPECTRAL EMITTANCE OF NATIONAL LEAD BLACK PAINT 46H47 ON A-286 STEEL AT 600 F AND
5 MICRONS PRESSURE--REFERENCE INFORMATION

Reference	Investigator	Symbol	Composition and Surface Condition	Test Method	Remarks
13	Gravina and Katz		National Lead 46H47 "high heat black" paint. Composition or thickness not given. Heated at 600 F: 1 hour 5 hours	Normal spectral emittance. Resistance-heated strip specimen. Thermistor-bolometer detector. Monochromator. Reference blackbody. Temperatures measured with thermocouples.	Measured in 5 microns Hg pressure. Data taken from curves.

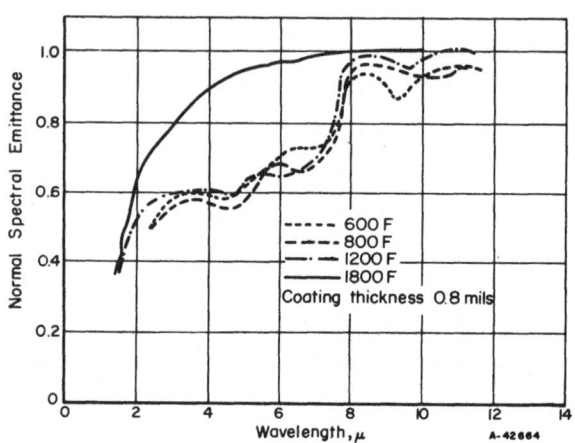

NORMAL SPECTRAL EMITTANCE OF VITA VAR PV100 ON A-286 STEEL

NORMAL SPECTRAL EMITTANCE OF VITA VAR PV 100 PAINT ON A-286 STEEL--REFERENCE INFORMATION

Reference	Investigator	Symbol	Composition and Surface Condition	Test Method	Remarks
13	Gravina and Katz		Vita Var PV 100 paint. A titanium dioxide pigment in silicone vehicle. Coating thickness 0.8 mil. Measured at: 600 F 800 F 1200 F 1800 F	Normal spectral emittance. Resistance-heated strip specimen. Thermistor-bolometer detector. Monochromator. Reference blackbody. Temperatures measured with thermocouples.	Measured in air. Data taken from curves.

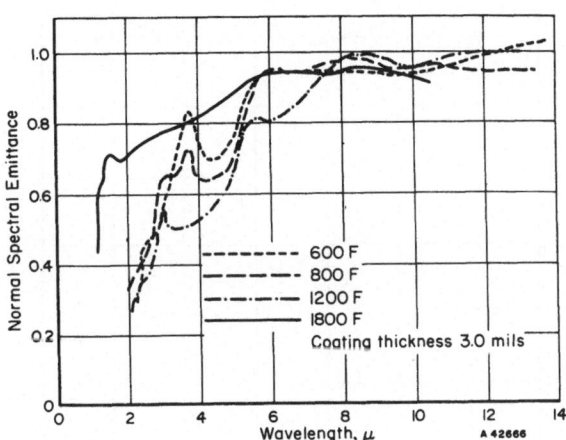

NORMAL SPECTRAL EMITTANCE OF VITA VAR PV 100 ON A-286 STEEL

NORMAL SPECTRAL EMITTANCE OF VITA VAR PV 100 PAINT ON A-286 STEEL—REFERENCE INFORMATION

Reference	Investigator	Symbol	Composition and Surface Condition	Test Method	Remarks
13	Gravina and Katz		Vita Var PV 100 paint. A titanium dioxide pigment in silicone vehicle. Coating thickness 3.0 mils. Measured at: 600 F 800 F 1200 F 1800 F	Normal spectral emittance. Resistance-heated strip specimen. Thermistor-bolometer detector. Monochromator. Reference blackbody. Temperatures measured with thermocouples.	Measured in air. Data taken from curves.

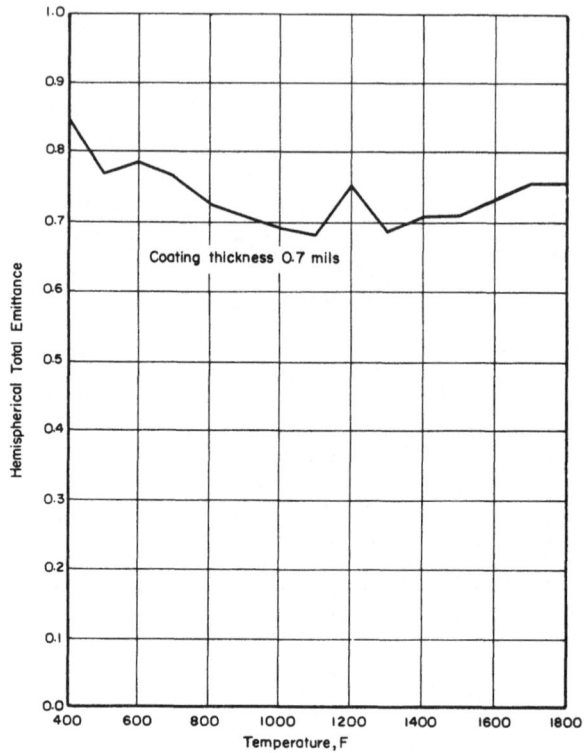

HEMISPHERICAL TOTAL EMITTANCE OF VITA VAR PV100 PAINT ON TITANIUM

HEMISPHERICAL TOTAL EMITTANCE OF VITA VAR PV100 PAINT ON TITANIUM--REFERENCE INFORMATION

Reference	Investigator	Symbol	Composition and Surface Condition	Test Method	Remarks
4	Dull, R. L.		Vita Var PV100 coating on titanium. Coating thickness 0.7 mil. (Vita Var PV100 is a white paint with a silicone vehicle and titanium dioxide pigment.) Note: Color began to change at 400 F and varied through yellow, tan, white, cream until brown and flaking at 1800 F.	Hemispherical total emittance. Resistance-heated strip specimen coated with test material. Measured power input to test section. Temperatures measured with thermocouples.	Measured in air. Data taken from curves.

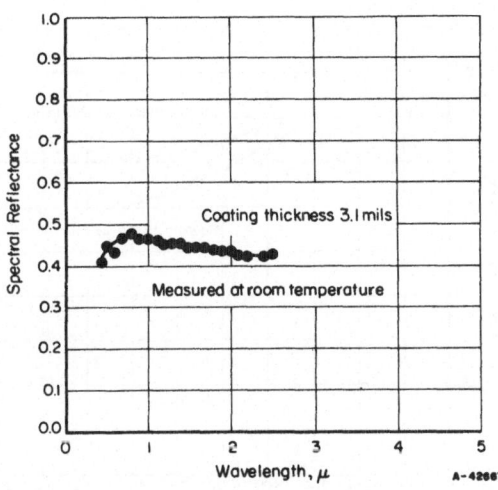

SPECTRAL REFLECTANCE OF W-1 WHITE ENAMEL ON INCONEL

SPECTRAL REFLECTANCE OF W-1 WHITE ENAMEL ON INCONEL--REFERENCE INFORMATION

Reference	Investigator	Symbol	Composition and Surface Condition	Test Method	Remarks
1	Burgess, Jasperse, Marcus, Martin, and Flint	●	W-1 white enamel on Inconel. Coating thickness 3.1 mils.	Spectral reflectance. Integrating sphere reflectometer. Commercial spectrophotometer, monochromator, lead sulphide detector. Hemispherical viewing. Illumination not clear from description-- whether diffuse or normal.	Measured in air at room temperature. Data taken from table.

Coating Composition by Weight

NBS Frit No. 332 - 60 per cent
CeO$_2$ - 30 per cent
MgO - 10 per cent

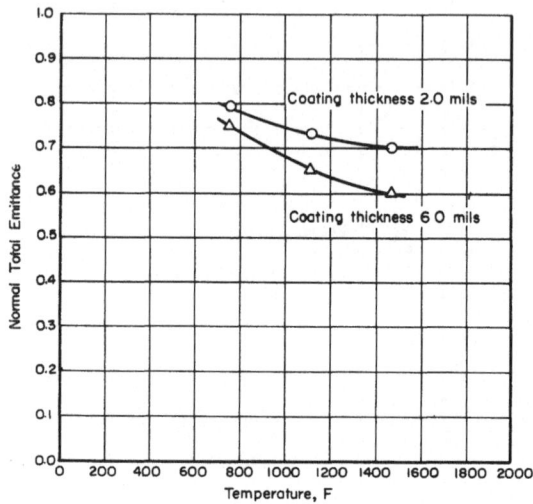

NORMAL TOTAL EMITTANCE OF W-3 WHITE ENAMEL ON INCONEL

NORMAL TOTAL EMITTANCE OF W-3 WHITE ENAMEL ON INCONEL--REFERENCE INFORMATION

Reference	Investigator	Symbol	Composition and Surface Condition	Test Method	Remarks
1	Burgess, Jasperse, Marcus, Martin, and Flint		W-3 white enamel on Inconel.	Normal total emittance. Sample recessed (flush) in wall of hollow, cylindrical, Globar heater. Comparison blackbody, hole. Infrared spectrometer with prism replaced by plane mirror. Thermocouple detector. Temperatures measured with thermocouples	Measured in air. Data taken from table.
		O	Coating thickness 2.0 mils.		
		Δ	Coating thickness 6.0 mils.		

Coating Composition by Weight

NBS Frit No. 332 - 60 per cent
CeO_2 - 20 per cent
SnO_2 - 20 per cent

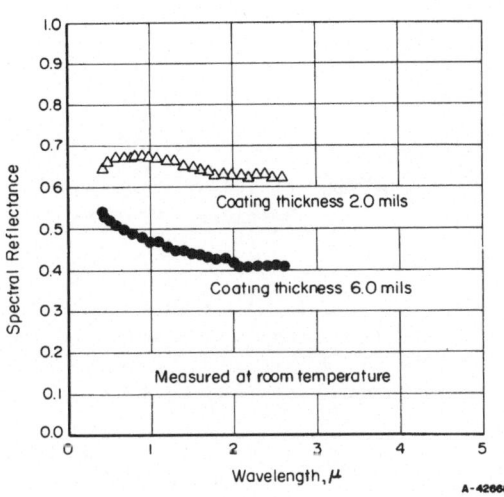

SPECTRAL REFLECTANCE OF W-3 WHITE ENAMEL ON INCONEL

SPECTRAL REFLECTANCE OF W-3 WHITE ENAMEL ON INCONEL--REFERENCE INFORMATION

Reference	Investigator	Symbol	Composition and Surface Condition	Test Method	Remarks
1	Burgess, Jasperse, Marcus, Martin, and Flint	● Δ	W-3 white enamel on Inconel. Coating thickness 2.0 mils. Coating thickness 6.0 mils.	Spectral reflectance. Integrating sphere reflectometer. Commercial spectrophotometer, monochromator, and lead sulphide detector. Hemispherical viewing. Illumination not clear from description--whether diffuse or normal.	Measured in air at room temperature. Data taken from table.

Coating Composition by Weight

NBS Frit No. 332 - 60 per cent
CeO_2 - 20 per cent
SnO_2 - 20 per cent

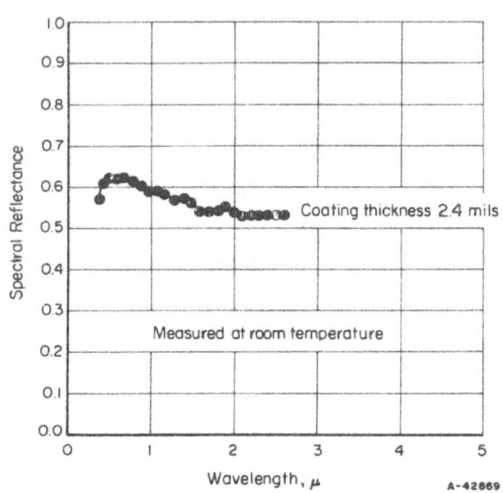

SPECTRAL REFLECTANCE OF W-4 WHITE ENAMEL ON INCONEL

SPECTRAL REFLECTANCE OF W-4 WHITE ENAMEL ON INCONEL--REFERENCE INFORMATION

Reference	Investigator	Symbol	Composition and Surface Condition	Test Method	Remarks
1	Burgess, Jasperse, Marcus, Martin, and Flint		W-4 white enamel on Inconel. Coating thickness 2.4 mils.	Spectral reflectance. Integrating sphere reflectometer. Commercial spectrophotometer, monochromator, lead sulphide detector. Hemispherical viewing. Illumination not clear from description--whether diffuse or normal.	Measured in air at room temperature. Data taken from table.

Coating Composition by Weight

NBS Frit No. 332 - 60 per cent
CeO$_2$ - 20 per cent
ZrO$_2$ - 20 per cent

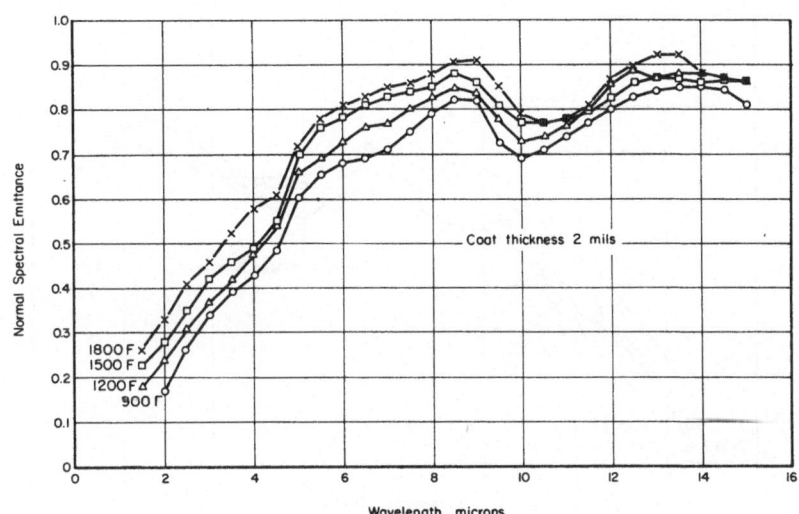

NORMAL SPECTRAL EMITTANCE OF N-143 ENAMEL ON INCONEL

NORMAL SPECTRAL EMITTANCE OF N-143 ENAMEL ON INCONEL--REFERENCE INFORMATION

Reference	Investigator	Symbol	Composition and Surface Condition	Test Method	Remarks
2	Richmond and Stewart		N-143 consists of boron-free barium beryllium silicate frit with addition of cerium oxide. Coating thickness 2 mils. Coated on Inconel. Runs made at the following temperatures:	Normal spectral emittance. Double-beam infrared spectrometer with sodium chloride prism. Secondary standard [silicon carbide (Globar)] cali-brated against laboratory black-body. Temperatures meas-ured with thermo-couples.	Measured in air. Data taken from tables.
		o	900 F		
		Δ	1200 F		
		□	1500 F		
		x	1800 F		

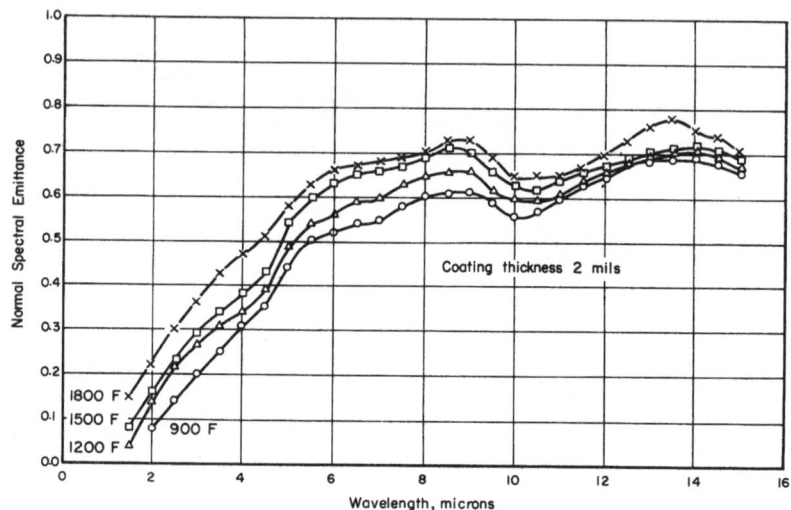

NORMAL SPECTRAL EMITTANCE OF N-143 ENAMEL ON TYPE 321 STAINLESS STEEL

NORMAL SPECTRAL EMITTANCE OF N-143 ENAMEL ON TYPE 321 STAINLESS STEEL-REFERENCE INFORMATION

Reference	Investigator	Symbol	Composition and Surface Condition	Test Method	Remarks
2	Richmond and Stewart		N-143 consists of boron-free barium beryllium silicate frit with addition of cerium oxide. Coating thickness 2 mils. Coated on Inconel. Runs made at the following temperatures:	Normal spectral emittance. Double-beam infrared spectrometer with sodium chloride prism. Secondary standard [silicon carbide (Globar)] calibrated against laboratory black-body. Temperatures measured with thermocouples.	Measured in air. Data taken from tables.
		O	900 F		
		△	1200 F		
		□	1500 F		
		×	1800 F		

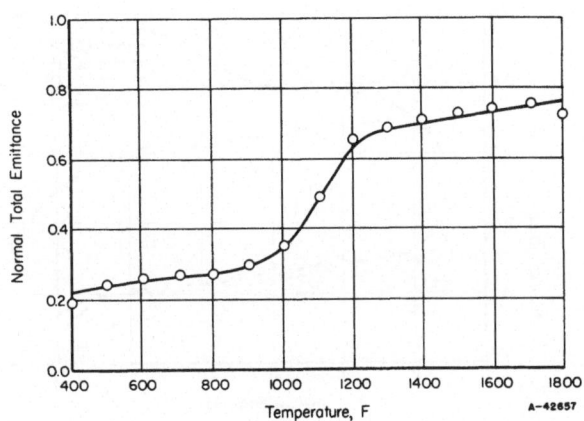

NORMAL TOTAL EMITTANCE FOR PRATT AND LAMBERT 91-1524 PAINT ON INCONEL X

NORMAL TOTAL EMITTANCE OF PRATT AND LAMBERT 91-1524 PAINT ON INCONEL X—REFERENCE INFORMATION

Reference	Investigator	Symbol	Composition and Surface Condition	Test Method	Remarks
13	Gravina and Katz		Pratt and Lambert 91-1524 coating. Butyl titanate paint with aluminum pigment.	Normal total emittance. Resistance-heated strip specimen. Thermistor-bolometer detector. Reference blackbody. Temperatures measured with thermocouples.	Measured in air. Data taken from curves.

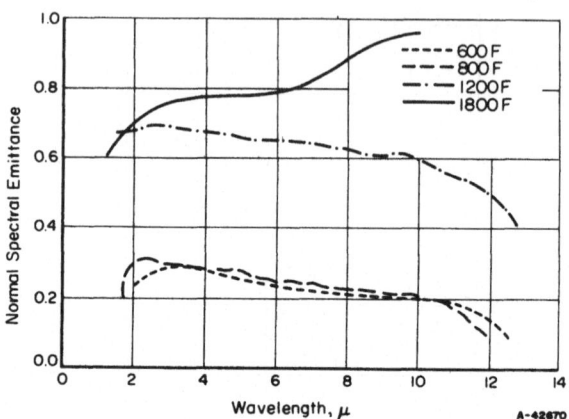

NORMAL SPECTRAL EMITTANCE OF PRATT AND LAMBERT 91-1524 PAINT ON INCONEL X

NORMAL SPECTRAL EMITTANCE OF PRATT AND LAMBERT 91-1524 PAINT ON INCONEL X—REFERENCE INFORMATION

Reference	Investigator	Symbol	Composition and Surface Condition	Test Method	Remarks
13	Gravina and Katz		Pratt and Lambert 91-1524, a butyl titanate paint with aluminum pigment. Measured at: 600 F 800 F 1200 F 1800 F	Normal spectral emittance. Resistance-heated strip specimen. Thermistor-bolometer detector. Monochromator. Reference blackbody. Temperatures measured with thermocouples.	Measured in air. Data taken from curves.

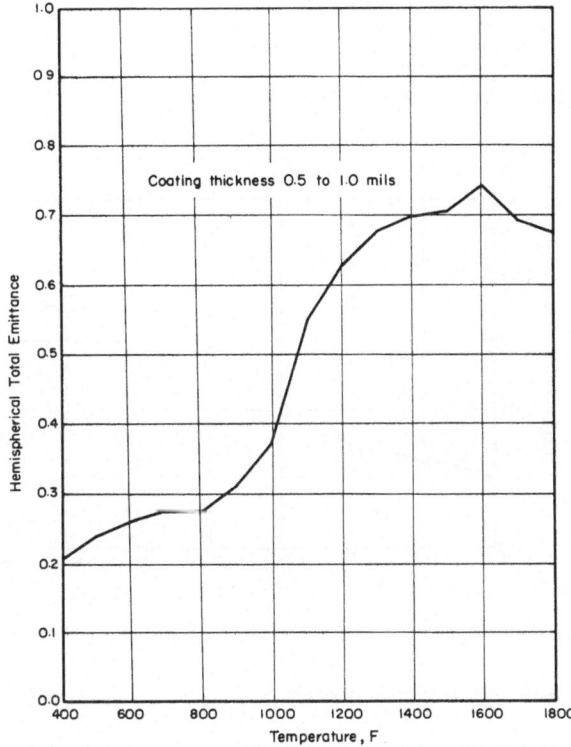

Coating thickness 0.5 to 1.0 mils

HEMISPHERICAL TOTAL EMITTANCE OF PRATT AND LAMBERT 91-1524 PAINT ON TITANIUM

HEMISPHERICAL TOTAL EMITTANCE OF PRATT AND LAMBERT 91-1524 PAINT ON TITANIUM--REFERENCE INFORMATION

Reference	Investigator	Symbol	Composition and Surface Condition	Test Method	Remarks
4	Dull, R. L.		Pratt and Lambert coating No. 91-1524 on titanium. Coating thickness 0.5 to 1.0 mil. (Coating is a butyl titanate paint with aluminum pigment.) Note: Surface began minute blistering at 1500 F and turned to dark brown, peeling flakes at 1800 F.	Hemispherical total emittance. Resistance-heated strip specimens coated with test material. Measured power input to test section. Temperatures measured with thermocouples.	Measured in air. Data taken from curves.

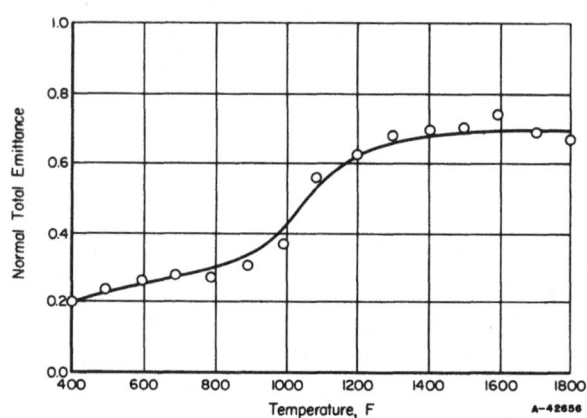

NORMAL TOTAL EMITTANCE FOR PRATT AND LAMBERT 91-1524 PAINT ON TITANIUM

NORMAL TOTAL EMITTANCE OF PRATT AND LAMBERT 91-1524 PAINT ON TITANIUM—REFERENCE INFORMATION

Reference	Investigator	Symbol	Composition and Surface Condition	Test Method	Remarks
13	Gravina and Katz		Pratt and Lambert 91-1524 coating. Butyl titanate paint with aluminum pigment.	Normal total emittance. Resistance-heated strip specimen. Thermistor-bolometer detector. Reference blackbody. Temperatures measured with thermocouples.	Measured in air. Data taken from curves.

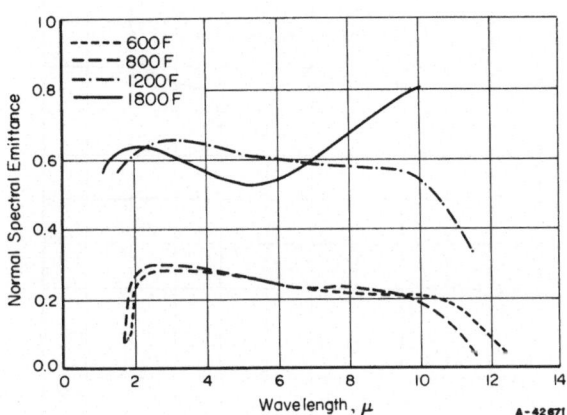

NORMAL SPLCTRAL EMITTANCE OF PRATT AND LAMBERT 91-1524 PAINT ON TITANIUM

NORMAL SPECTRAL EMITTANCE OF PRATT AND LAMBERT 91-1524 PAINT ON TITANIUM--REFERENCE INFORMATION

Reference	Investigator	Symbol	Composition and Surface Condition	Test Method	Remarks
13	Gravina and Katz		Pratt and Lambert 91-1524, a butyl titanate paint with aluminum pigment. Measured at: 600 F 800 F 1200 F 1800 F	Normal spectral emittance. Resistance-heated strip specimen. Thermistor-bolometer detector. Monochromator. Reference blackbody. Temperatures measured with thermocouples.	Measured in air. Data taken from curves.

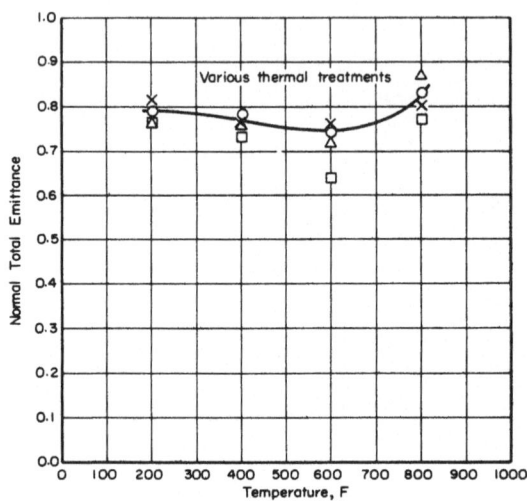

NORMAL TOTAL EMITTANCE OF RINSHED-MASON H12144 ENAMEL ON TYPE 321 STAINLESS STEEL

NORMAL TOTAL EMITTANCE OF RINSHED-MASON H12144 ENAMEL ON TYPE 321 STAINLESS STEEL--REFERENCE INFORMATION

Reference	Investigator	Symbol	Composition and Surface Condition	Test Method	Remarks
3	Bevans, Gier, and Dunkle		Rinshed-Mason black heat-resistant, air-dry enamel H12144, painted on Type 321 stainless steel (Mat'l. Spec. MIS-S-6721.) No thickness given.	Normal total emittance. Calibrated thermopile detector. Temperatures measured with thermocouples.	Measured in air. Data taken from tables.
		o	No thermal treatment.		
		△	300 hours at 497 F.		
		□	307 hours at 690 F.		
		✕	1000 hours at 705 F.		

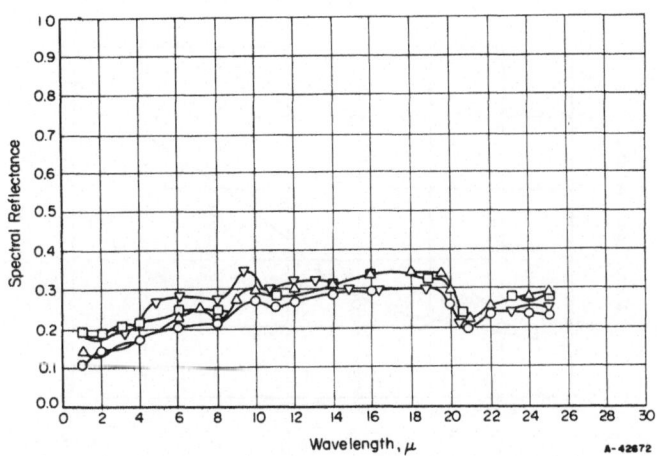

SPECTRAL REFLECTANCE OF RINSHED-MASON H12144 ENAMEL ON TYPE 321 STAINLESS STEEL

SPECTRAL REFLECTANCE OF RINSHED-MASON H12144 ENAMEL ON TYPE 321 STAINLESS STEEL--REFERENCE INFORMATION

Reference	Investigator	Symbol	Composition and Surface Condition	Test Method	Remarks
3	Bevans, Gier, and Dunkle		Rinshed-Mason black heat-resistant, air-dry Enamel H12144, painted on Type 321 stainless steel (Mat'l. Spec. MIS-S-6721.) No thickness given.	Spectral reflectance at 5 degrees with normal. Gier-Dunkle reflect-ometer-monochromator. Temperatures measured with thermocouples. (Diffuse illumination-normal viewing.)	Measured in air at room temperature. Data taken from tables.
		O	No thermal treatment.		
		△	300 hours at 497 F.		
		□	307 hours at 690 F.		
		x	1000 hours at 705 F.		

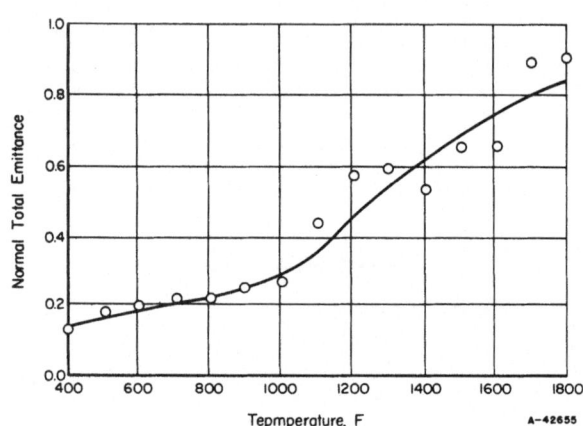

NORMAL SPECTRAL EMITTANCE OF RINSHED—MASON J—15934 PAINT ON A—286 STEEL

NORMAL TOTAL EMITTANCE OF RINSHED—MASON J—15934 PAINT ON A—286 STEEL—REFERENCE INFORMATION

Reference	Investigator	Symbol	Composition and Surface Condition	Test Method	Remarks
13	Gravina and Katz		Rinshed—Mason J—15934, silicone paint with aluminum pigment. Coating thickness not given.	Normal total emittance. Resistance-heated strip specimen. Thermistor-bolometer detector. Reference blackbody. Temperatures measured with thermocouples.	Measured in air. Data taken from curves.

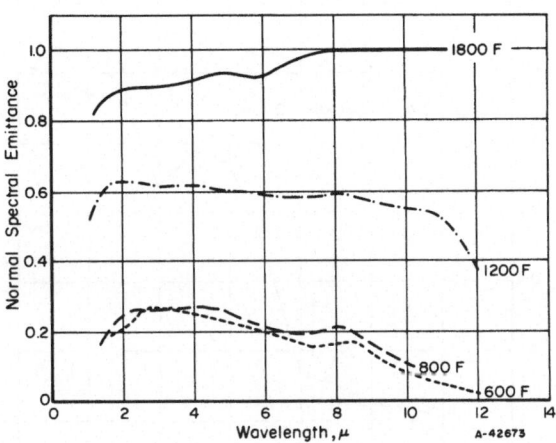

NORMAL TOTAL EMITTANCE OF RINSHED—MASON J-15934 PAINT ON A-286 STEEL

NORMAL SPECTRAL EMITTANCE OF RINSHED—MASON J-15934 PAINT ON A-286 STEEL--REFERENCE INFORMATION

Reference	Investigator	Symbol	Composition and Surface Condition	Test Method	Remarks
13	Gravina and Katz		Rinshed—Mason Paint J-15934, silicone vehicle, aluminum pigment. Coating thickness not given. Measured at: 600 F 800 F 1200 F 1800 F	Normal spectral emittance. Resistance-heated strip specimen. Thermistor-bolometer detector. Monochromator. Reference blackbody. Temperatures measured with thermocouples.	Measured in air. Data taken from curves.

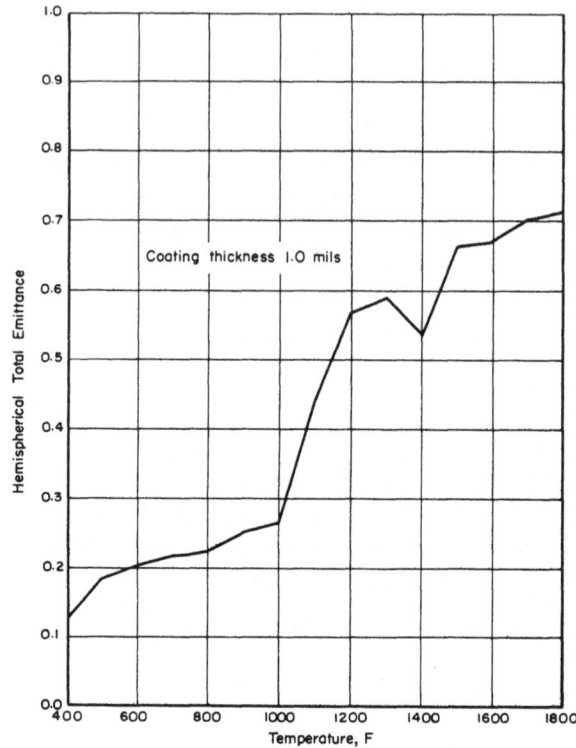

HEMISPHERICAL TOTAL EMITTANCE OF RINSHED-MASON J-15934 PAINT ON TITANIUM

HEMISPHERICAL TOTAL EMITTANCE OF RINSHED-MASON J-15934 PAINT ON TITANIUM--REFERENCE INFORMATION

Reference	Investigator	Symbol	Composition and Surface Condition	Test Method	Remarks
4	Dull, R. L.		Rinshed-Mason Coating J-15934 on titanium. Coating thickness 1.0 mil. (A silicone paint with aluminum pigment.) Note: Extensive peeling began at about 1300 F. Nearly all peeled off at 1800 F.	Hemispherical total emittance. Resistance-heated strip specimens coated with test material. Measured power input to test section. Temperatures measured with thermocouples.	Measured in air. Data taken from curves.

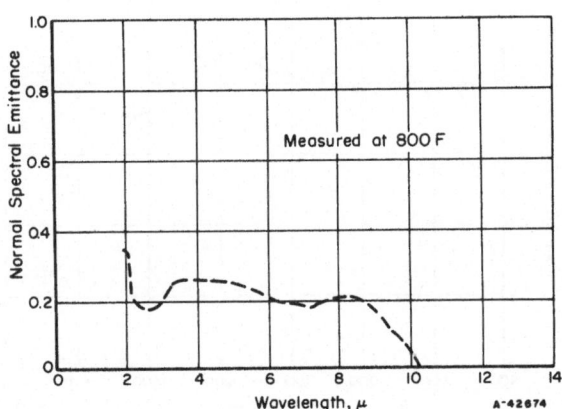

NORMAL SPECTRAL EMITTANCE OF RINSHED-MASON J-15934 PAINT ON TITANIUM

NORMAL SPECTRAL EMITTANCE OF RINSHED-MASON J-15934 PAINT ON TITANIUM—REFERENCE INFORMATION

Reference	Investigator	Symbol	Composition and Surface Condition	Test Method	Remarks
13	Gravina and Katz		Rinshed-Mason Paint J-15934, a silicone paint with aluminum pigment. Coating thickness not given. Measured at 800 F	Normal spectral emittance. Resistance-heated strip specimen. Thermistor-bolometer detector. Monochromator. Reference blackbody. Temperatures measured with thermocouples.	Measured in air. Data taken from curves.

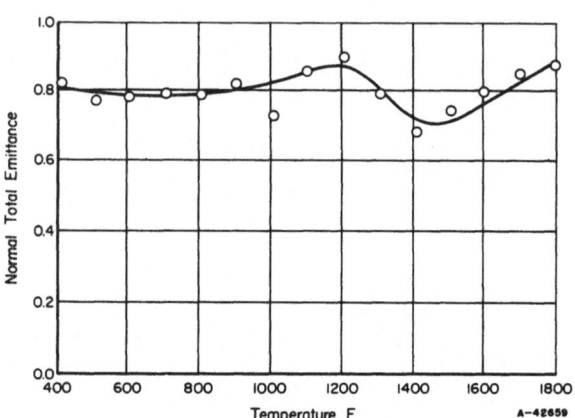

NORMAL TOTAL EMITTANCE OF RINSHED-MASON Q36K802 PAINT ON A-286 STEEL

NORMAL TOTAL EMITTANCE OF RINSHED-MASON Q36K802 PAINT ON A-286 STEEL--REFERENCE INFORMATION

Reference	Investigator	Symbol	Composition and Surface Condition	Test Method	Remarks
13	Gravina and Katz		Rinshed-Mason Q36K802, silicone paint with carbon black pigment. Coating thickness not given.	Normal total emittance. Resistance-heated strip specimen. Thermistor-bolometer detector. Reference blackbody. Temperatures measured with thermocouples.	Measured in air. Data taken from curves.

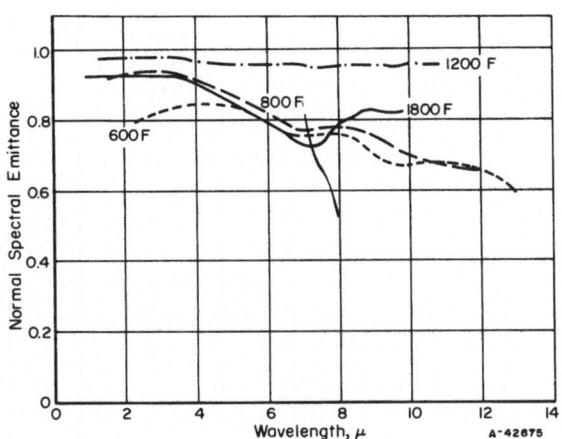

NORMAL SPECTRAL EMITTANCE OF RINSHED—MASON Q36K802 PAINT ON A-286 STEEL

NORMAL SPECTRAL EMITTANCE OF RINSHED—MASON Q36K802 PAINT ON A-286 STEEL—REFERENCE INFORMATION

Reference	Investigator	Symbol	Composition and Surface Condition	Test Method	Remarks
13	Gravina and Katz		Rinshed—Mason Q36K802 black paint is a silicone paint with carbon black pigment. Coating thickness not given. Measured at: 600 F 800 F 1200 F 1800 F	Normal spectral emittance. Resistance-heated strip specimen. Thermistor—bolometer detector. Monochromator. Reference blackbody. Temperatures measured with thermocouples.	Measured in air. Data taken from curves.

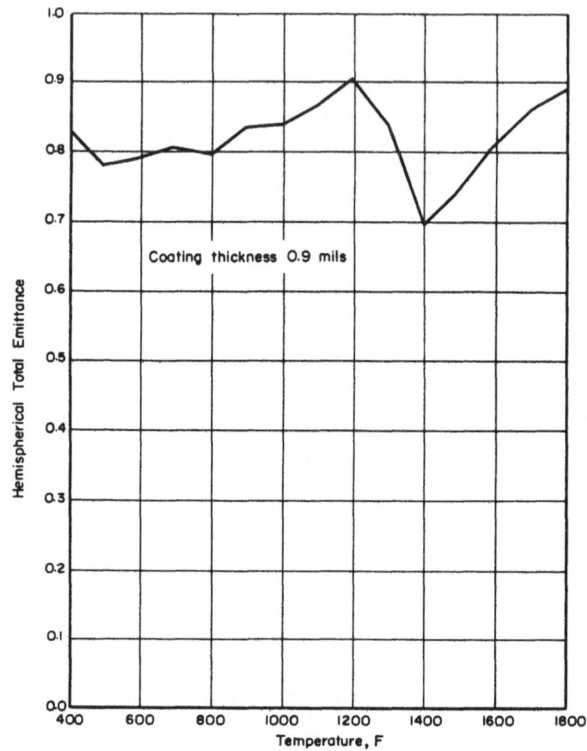

HEMISPHERICAL TOTAL EMITTANCE OF RINSHED—MASON Q-36K802 PAINT ON TITANIUM

HEMISPHERICAL TOTAL EMITTANCE OF RINSHED—MASON Q-36K802 PAINT ON TITANIUM--REFERENCE INFORMATION

Reference	Investigator	Symbol	Composition and Surface Condition	Test Method	Remarks
4	Dull, R. L.		Rinshed—Mason Q-36K802 coating on titanium. Coating thickness 0.9 mil. (A silicone paint with carbon black pigment.) Note: Discoloration began at 400 F. Blistering began at about 1600 F.	Hemispherical total emittance. Resistance-heated strip. Specimen coated with test material. Measured power input to test section. Temperatures measured with thermocouples.	Measured in air. Data taken from curves.

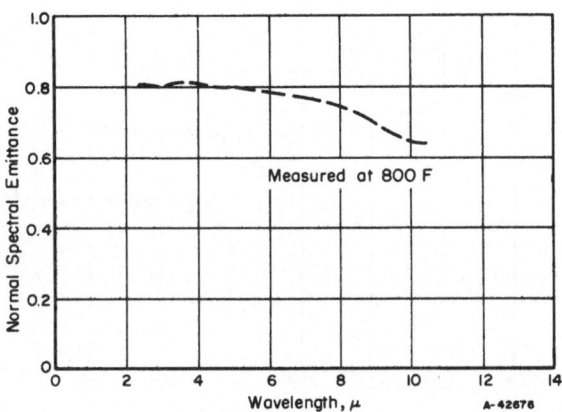

NORMAL SPECTRAL EMITTANCE OF RINSHED–MASON Q36K802 PAINT ON TITANIUM

NORMAL SPECTRAL EMITTANCE OF RINSHED–MASON Q36K802 PAINT ON TITANIUM—REFERENCE INFORMATION

Reference	Investigator	Symbol	Composition and Surface Condition	Test Method	Remarks
13	Gravina and Katz		Rinshed–Mason Q36K802 black paint. A sili-cone paint with carbon black pigment. Coating thickness not given. Measured at 800 F	Normal spectral emittance. Resistance–heated strip specimen. Thermistor–bolometer detector. Monochromator. Reference blackbody. Temperatures measured with thermocouples.	Measured in air. Data taken from curves.

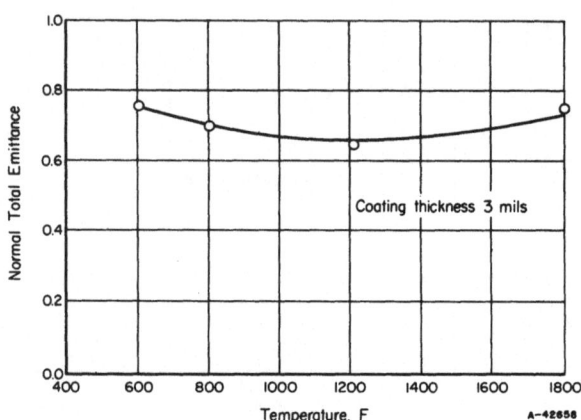

NORMAL TOTAL EMITTANCE OF VITA VAR PV100 ON A-286 STEEL

NORMAL TOTAL EMITTANCE OF VITA VAR PV 100 PAINT ON A-286 STEEL--REFERENCE INFORMATION

Reference	Investigator	Symbol	Composition and Surface Condition	Test Method	Remarks
13	Gravina and Katz		Vita Var PV 100, silicone paint with titanium dioxide pigment. Coating thickness 3 mils	Normal total emittance. Resistance-heated strip specimen. Thermistor-bolometer detector. Reference blackbody. Temperatures measured with thermocouples.	Measured in air. Data taken from curves.

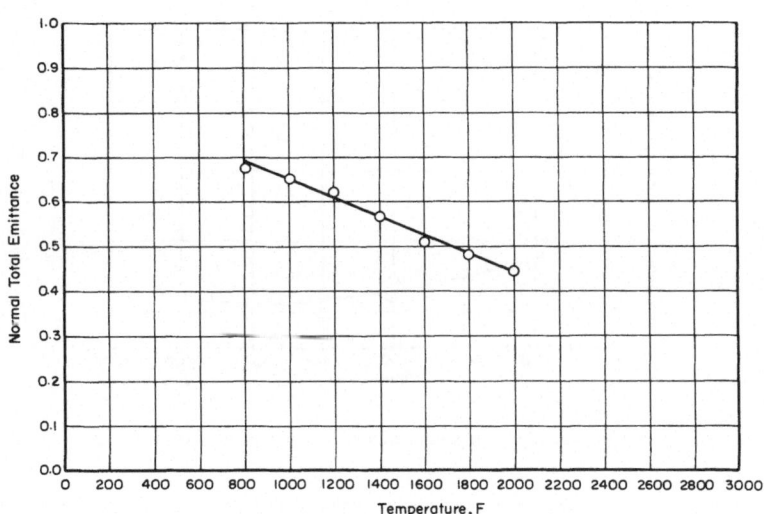

NORMAL TOTAL EMITTANCE OF ALUMINUM OXIDE ON INCONEL

NORMAL TOTAL EMITTANCE OF ALUMINUM OXIDE ON INCONEL—REFERENCE INFORMATION

Reference	Investigator	Symbol	Composition and Surface Condition	Test Method	Remarks
10	Wade, W. R.	o	Flame-sprayed alumina on Inconel heater strip. Thickness not given.	Normal total emittance. Thermopile detector. Resistance-heated Inconel strip with test material flame sprayed to "opaque" thickness.	Measured in air. Temperatures given are those of Inconel heater strip. Data taken from curve.

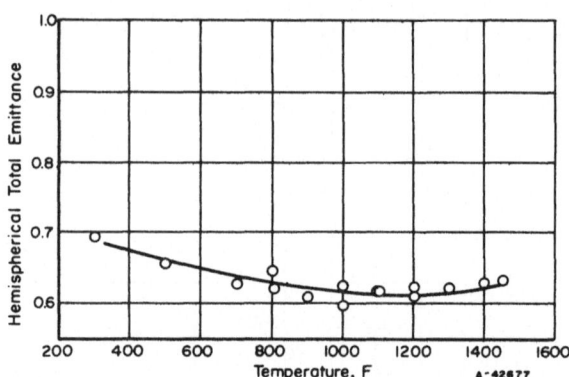

HEMISPHERICAL TOTAL EMITTANCE OF ALUMINUM OXIDE ON TYPE 310 STAINLESS STEEL

HEMISPHERICAL TOTAL EMITTANCE OF ALUMINUM OXIDE ON TYPE 310 STAINLESS STEEL--REFERENCE INFORMATION

Reference	Investigator	Symbol	Composition and Surface Condition	Test Method	Remarks
13	Pratt & Whitney Aircraft		Plasmadyne powder. Coated on both sides. Flame sprayed on Type 310 stainless strip.	Hemispherical total emittance. Resistance-heated strip specimen. Power dissipated in measured area. Temperatures measured with thermocouples.	Measured in vacuum. Data taken from curves.

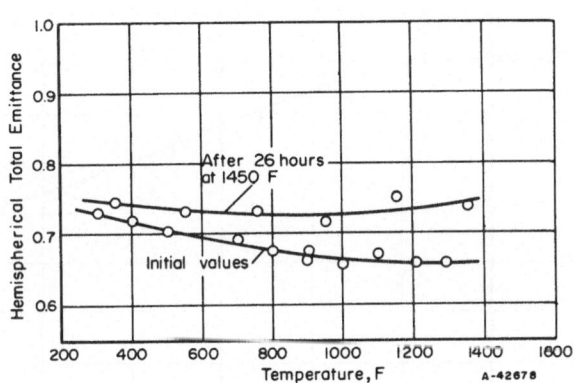

HEMISPHERICAL TOTAL EMITTANCE OF ALUMINUM OXIDE ON TYPE 310 STAINLESS STEEL

HEMISPHERICAL TOTAL EMITTANCE OF ALUMINUM OXIDE ON TYPE 310 STAINLESS STEEL--REFERENCE INFORMATION

Reference	Investigator	Symbol	Composition and Surface Condition	Test Method	Remarks
17	Pratt & Whitney Aircraft		Metco 101 powder. Coated on both sides. Flame sprayed. Initial runs After 26 hours at 1450 F	Hemispherical total emittance. Resistance-heated strip specimen. Power dissipated in measured area. Temperatures measured with thermocouples.	Measured in vacuum. Data taken from curves.

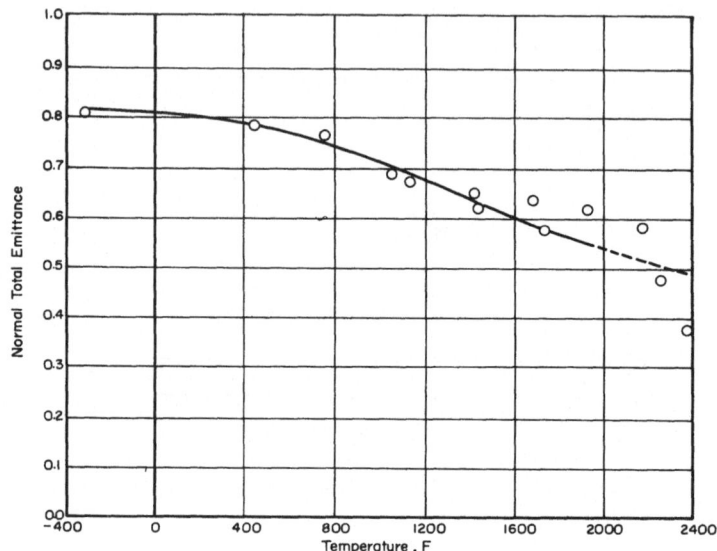

NORMAL TOTAL EMITTANCE OF ALUMINUM OXIDE (ROKIDE) ON TYPE 446 STAINLESS STEEL

NORMAL TOTAL EMITTANCE OF ALUMINUM OXIDE (ROKIDE) ON TYPE 446 STAINLESS STEEL--REFERENCE INFORMATION

Reference	Investigator	Symbol	Composition and Surface Condition	Test Method	Remarks
11	Olson and Morris	o	As received. Showed purple discoloration after test. Thickness or surface condition not given.	Normal total emittance. Comparison blackbody. Furnace heated specimens. Temperatures measured with thermocouples. Thermistor-bolometer detector.	Measured in air. Data taken from curves.

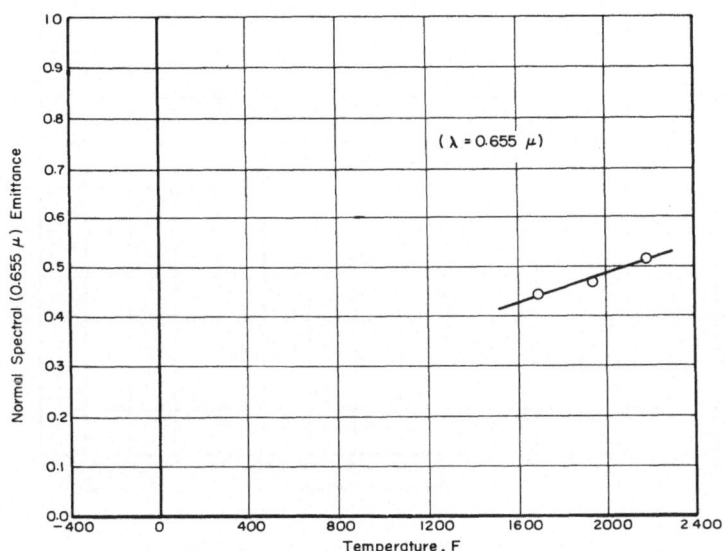

NORMAL SPECTRAL EMITTANCE OF ALUMINUM OXIDE (ROKIDE) ON TYPE 446 STAINLESS STEEL

NORMAL SPECTRAL EMITTANCE OF ALUMINUM OXIDE (ROKIDE) ON TYPE 446 STAINLESS STEEL--REFERENCE INFORMATION

Reference	Investigator	Symbol	Composition and Surface Condition	Test Method	Remarks
11	Olson and Morris	O	Aluminum Oxide (Rokide) on Type 446 stainless steel. Thickness or surface condition not given.	Normal spectral emittance. Furnace-heated specimen. Commercial sensing unit. Appropriate lenses and filters. Temperatures measured with thermocouples.	Measured in air. Data taken from curves.

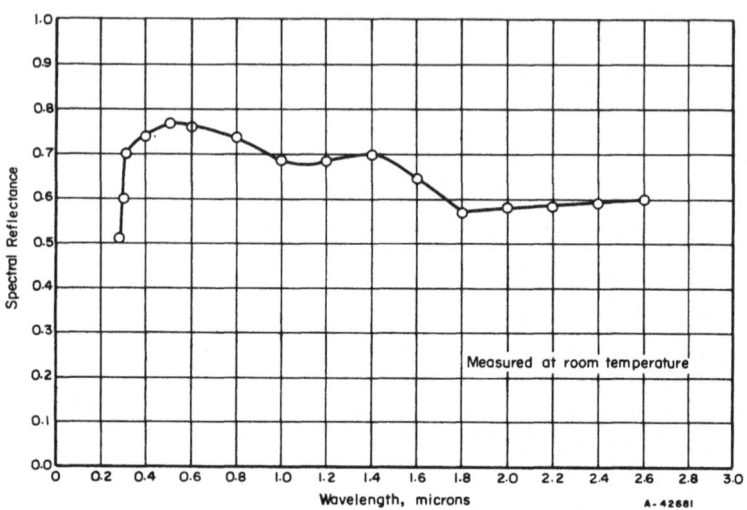

SPECTRAL REFLECTANCE OF ALUMINUM OXIDE (ROKIDE) ON TYPE 446 STAINLESS STEEL

SPECTRAL REFLECTANCE OF ALUMINUM OXIDE (ROKIDE) ON TYPE 446 STAINLESS STEEL—REFERENCE INFORMATION

Reference	Investigator	Symbol	Composition and Surface Condition	Test Method	Remarks
11	Olson and Morris	O	Aluminum oxide (Norton Co., Rokide A) on Type · 16 stainless steel. Thickness of surface condition not given.	Spectral reflectance at 9 degrees from normal (incident radiation). Recording spectrophotometer, integrating sphere reflectometer, and lead sulphide detector. (Normal illumination—hemispherical viewing)	Measured in air at room temperature. Data taken from curves.

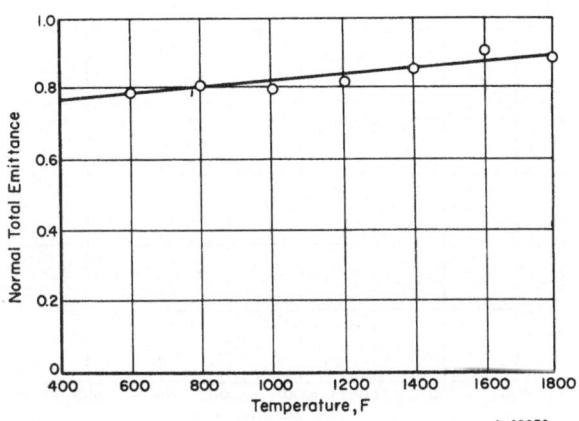

A-42679

NORMAL TOTAL EMITTANCE OF CERAMCO B-682P ON A-286 STEEL

NORMAL TOTAL EMITTANCE OF CERAMCO B-682P ON A-286 STEEL--REFERENCE INFORMATION

Reference	Investigator	Symbol	Composition and Surface Condition	Test Method	Remarks
13	Gravina and Katz		Ceramco B-682P, a proprietary black oxide.	Normal total emittance. Resistance-heated strip specimen. Thermistor-bolometer detector. Reference blackbody. Temperatures measured with thermocouples.	Measured at atmospheric pressure. Data taken from curves.

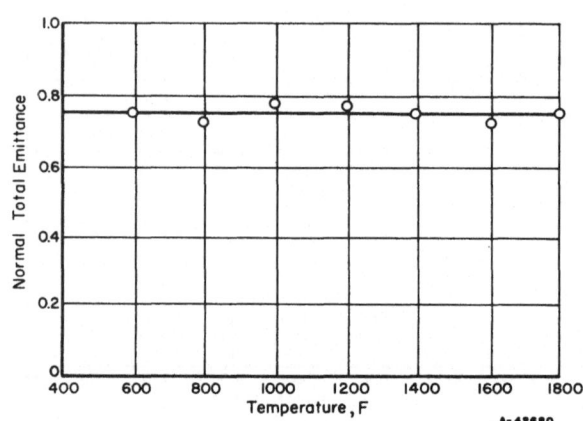

NORMAL TOTAL EMITTANCE OF CERAMCO G-683P ON A-286 STEEL

NORMAL TOTAL EMITTANCE OF CERAMCO G-683P ON A-286 STEEL--REFERENCE INFORMATION

Reference	Investigator	Symbol	Composition and Surface Condition	Test Method	Remarks
13	Gravina and Katz		Ceramco G-683P, a proprietary green oxide. Composition or thickness not given.	Normal total emittance. Resistance-heated strip specimen. Thermistor-bolometer detector. Reference blackbody. Temperatures measured with thermocouples.	Measured at atmospheric pressure. Data taken from curves.

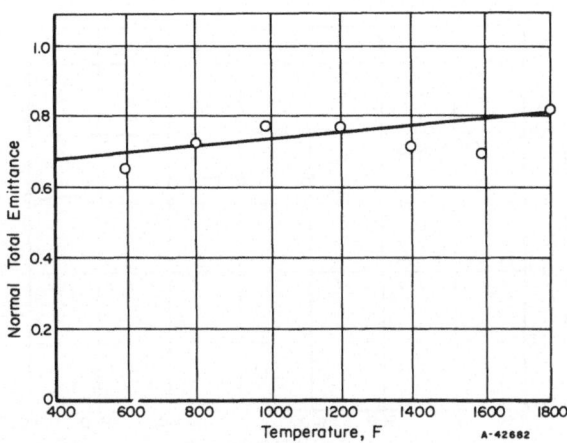

NORMAL TOTAL EMITTANCE OF CERAMCO G-684tc ON INCONEL X

NORMAL TOTAL EMITTANCE OF CERAMCO G-684tc ON INCONEL X--REFERENCE INFORMATION

Reference	Investigator	Symbol	Composition and Surface Condition	Test Method	Remarks
13	Gravina and Katz		Ceramco G-684tc, a proprietary green oxide. Composition on thickness not given.	Normal total emittance. Resistance-heated strip specimen. Thermistor-bolometer detector. Reference blackbody. Temperatures measured with thermocouples.	Measured at atmospheric pressure. Data taken from curves.

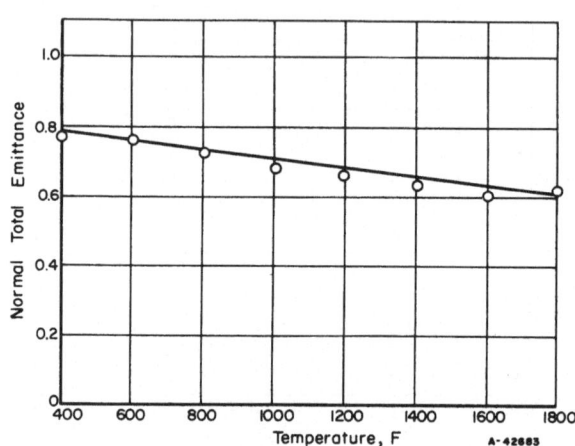

NORMAL TOTAL EMITTANCE OF CERAMCO W–683P ON A–286 STEEL

NORMAL TOTAL EMITTANCE OF CERAMCO W–683P ON A–286 STEEL––REFERENCE INFORMATION

Reference	Investigator	Symbol	Composition and Surface Condition	Test Method	Remarks
13	Gravina and Katz		Ceramco W–683P, a proprietary white oxide.	Normal total emittance. Resistance–heated strip specimen. Thermistor–bolometer detector. Reference blackbody. Temperatures measured with thermocouples.	Measured at atmospheric pressure. Data taken from curves.

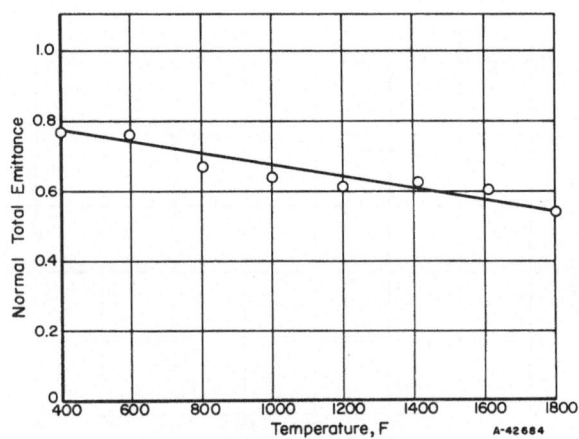

NORMAL TOTAL EMITTANCE OF CERAMCO W-683tc ON INCONEL X

NORMAL TOTAL EMITTANCE OF CERAMCO W-683tc ON INCONEL X--REFERENCE INFORMATION

Reference	Investigator	Symbol	Composition and Surface Condition	Test Method	Remarks
13	Gravina and Katz		Ceramco W-683tc, a proprietary white oxide. Composition or thickness not given.	Normal total emittance. Resistance-heated strip specimen. Thermistor-bolometer detector. Reference blackbody. Temperatures measured with thermocouples.	Measured at atmospheric pressure. Data taken from curves.

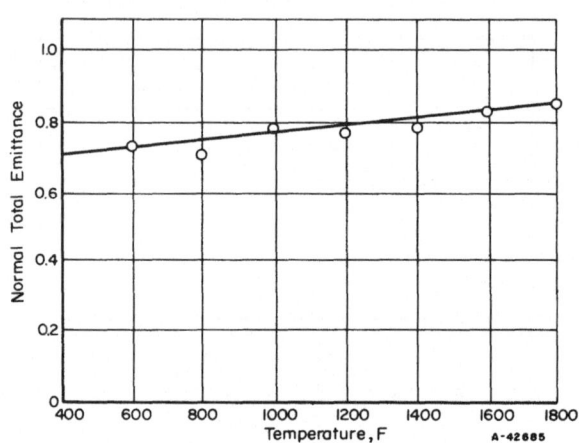

NORMAL TOTAL EMITTANCE OF CERAMCO WB-6832 ON A-286 STEEL

NORMAL TOTAL EMITTANCE OF CERAMCO WB-6832 ON A-286 STEEL--REFERENCE INFORMATION

Reference	Investigator	Symbol	Composition and Surface Condition	Test Method	Remarks
13	Gravina and Katz		Ceramco WB-6832, a proprietary brown oxide.	Normal total emittance. Resistance-heated strip specimen. Thermistor-bolometer detector. Reference blackbody. Temperatures measured with thermocouples.	Measured at atmospheric pressure. Data taken from curves.

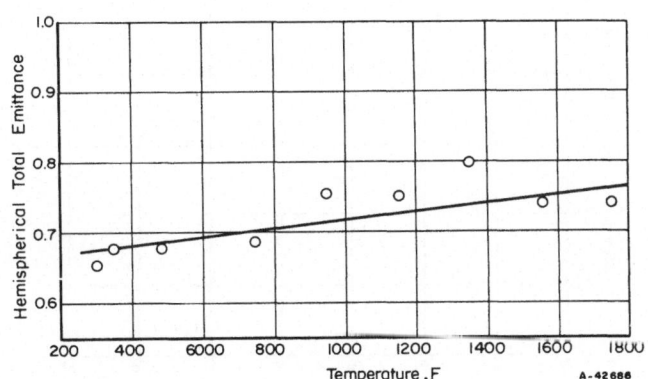

HEMISPHERICAL TOTAL EMITTANCE OF CERIC OXIDE ON TYPE 310 STAINLESS STEEL

HEMISPHERICAL TOTAL EMITTANCE OF CERIC OXIDE ON TYPE 310 STAINLESS STEEL--REFERENCE INFORMATION

Reference	Investigator	Symbol	Composition and Surface Condition	Test Method	Remarks
16	Pratt & Whitney Aircraft		Metco plasma flame spray powder XP-111.	Hemispherical total emittance. Resistance-heated strip specimen. Power dissipated in measured area. Temperatures measured with thermocouples.	Measured in vacuum. Data taken from curves.

318

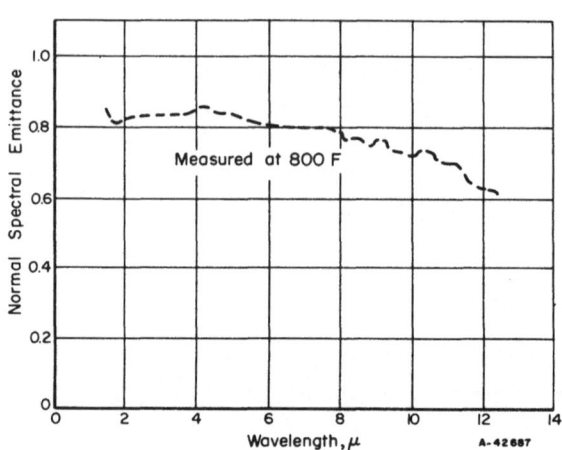

NORMAL SPECTRAL EMITTANCE OF NORTON LA 9696 ON 6A1-4V TITANIUM

NORMAL SPECTRAL EMITTANCE OF NORTON LA9683 ON 6A1-4V TITANIUM--REFERENCE INFORMATION

Reference	Investigator	Symbol	Composition and Surface Condition	Test Method	Remarks
13	Gravina and Katz		A refractory oxide. Composition not given. Flame sprayed on cleaned, grit-blasted surface. Coated with a 12-mil-thick Nichrome V under-coat. Coating thickness not given. Measured at 800 F.	Normal spectral emittance. Resistance-heated strip specimen. Thermistor-bolometer detector. Monochromator. Reference blackbody. Temperatures measured with thermocouples.	Measured in air. Data taken from curves.

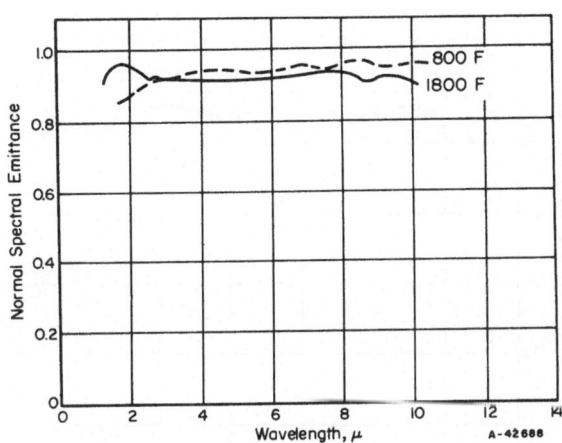

NORMAL SPECTRAL EMITTANCE OF NORTON LN 9684 ON INCONEL X

NORMAL SPECTRAL EMITTANCE OF NORTON LN9684 ON INCONEL X--REFERENCE INFORMATION

Reference	Investigator	Symbol	Composition and Surface Condition	Test Method	Remarks
13	Gravina and Katz		Norton LN9684, a very dark nickel oxide. Melting point about 3500 F. Porosity about 2 per cent. Flame sprayed on cleaned, grit-blasted surface coated with 12-mil-thick Nichrome V undercoat. Coating thickness not given. Measured at: 800 F 1800 F	Normal spectral emittance. Resistance-heated strip specimen. Thermistor-bolometer detector. Monochromator. Reference blackbody. Temperatures measured with thermocouples.	Measured in air. Data taken from curves.

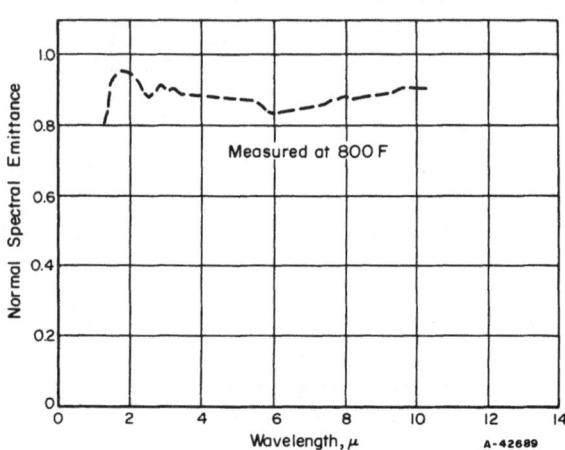

NORMAL SPECTRAL EMITTANCE OF NORTON LN 9684 ON 6A1-4V TITANIUM

NORMAL SPECTRAL EMITTANCE OF NORTON LN9684 CERAMIC COATING ON 6A1-4V TITANIUM--REFERENCE INFORMATION

Reference	Investigator	Symbol	Composition and Surface Condition	Test Method	Remarks
13	Gravina and Katz		Norton LN9684, a very dark nickel oxide. Melting point about 3500 F. Porosity about 2 per cent. Flame sprayed on cleaned, grit-blasted surface coated with 12-mil-thick Nichrome V undercoat. Coating thickness not given. Measured at 800 F	Normal spectral emittance. Resistance-heated strip specimen. Thermistor-bolometer detector. Monochromator. Reference blackbody. Temperatures measured with thermocouples.	Measured in air. Data taken from curves.

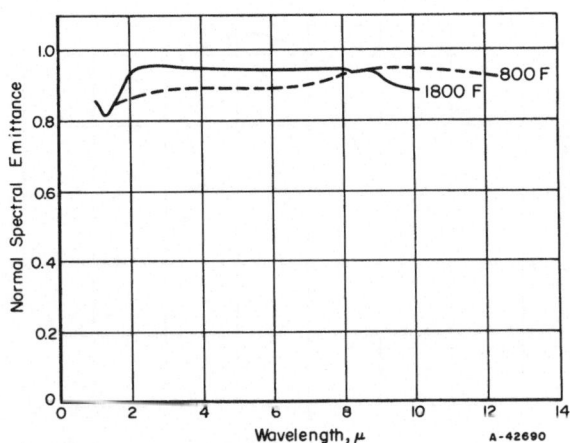

NORMAL SPECTRAL EMITTANCE OF NORTON LN 9684 ON A-286 STEEL

NORMAL SPECTRAL EMITTANCE OF NORTON LN9684 CERAMIC COATING ON A-286 STEEL--REFERENCE INFORMATION

Reference	Investigator	Symbol	Composition and Surface Condition	Test Method	Remarks
13	Gravina and Katz		Norton LN9684, a very dark nickel oxide. Melting point about 3500 F. Porosity about 2 per cent. Flame sprayed on cleaned, grit-blasted surface coated with 12-mil-thick Nichrome V undercoat. Coating thickness not given. Measured at: 800 F 1800 F	Normal spectral emittance. Resistance-heated strip specimen. Thermistor-bolometer detector. Monochromator. Reference blackbody. Temperatures measured with thermocouples.	Measured in air. Data taken from curves.

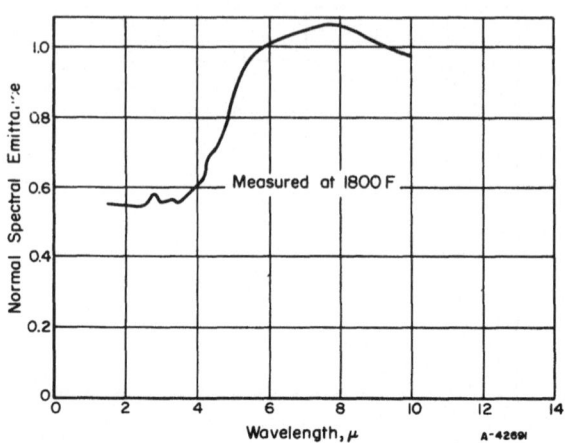

NORMAL SPECTRAL EMITTANCE OF NORTON LA 9696 ON INCONEL X

NORMAL SPECTRAL EMITTANCE OF NORTON LA-9696 ON INCONEL X--REFERENCE INFORMATION

Reference	Investigator	Symbol	Composition and Surface Condition	Test Method	Remarks
13	Gravina and Katz		Norton ceramic coating LA-9696, a tan alundum, 92 per cent Al_2O_3. Melting point about 3500 F. Porosity about 5 per cent. Flame sprayed on cleaned, grit-blasted surface coated with 12-mil-thick Nichrome V undercoat. Coating thickness not given. Measured at 1800 F	Normal spectral emittance. Resistance-heated strip specimen. Thermistor-bolometer detector. Monochromator. Reference blackbody. Temperatures measured with thermocouples.	Measured in air. Data taken from curves.

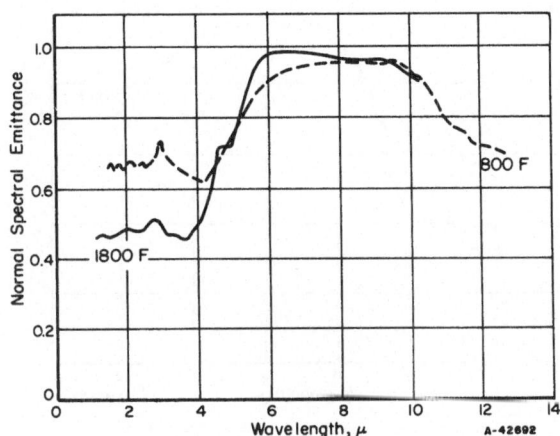

NORMAL SPECTRAL EMITTANCE OF NORTON LA 9683 ON TITANIUM

NORMAL SPECTRAL EMITTANCE OF NORTON LA-9696 ON 6A1-4V TITANIUM--REFERENCE INFORMATION

Reference	Investigator	Symbol	Composition and Surface Condition	Test Method	Remarks
13	Gravina and Katz		Norton ceramic coating LA-9696, a tan alundum, 92 per cent Al_2O_3. Melting point about 3500 F. Porosity about 5 per cent. Flame sprayed on cleaned, grit-blasted surface coated with 12-mil-thick Nichrome V undercoat. Coating thickness not given. Measured at: 800 F 1800 F	Normal spectral emittance. Resistance-heated strip specimen. Thermistor-bolometer detector. Monochromator. Reference blackbody. Temperatures measured with thermocouples.	Measured in air. Data taken from curves.

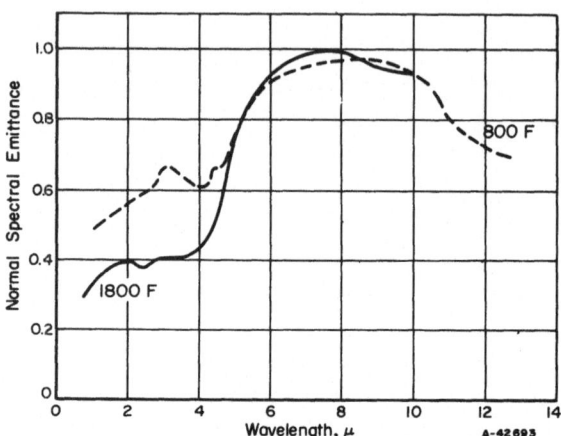

NORMAL SPECTRAL EMITTANCE OF NORTON LA 9696 ON A-286 STEEL

NORMAL SPECTRAL EMITTANCE OF NORTON LA-9696 ON A-286 STEEL—REFERENCE INFORMATION

Reference	Investigator	Symbol	Composition and Surface Condition	Test Method	Remarks
13	Gravina and Katz		Norton ceramic coating LA-9696, a tan alundum, 92 per cent Al_2O_3. Melting point about 3500 F. Porosity about 5 per cent. Flame sprayed on cleaned, grit-blasted surface coated with 12-mil-thick Nichrome V undercoat. Coating thickness not given. Measured at: 800 F 1800 F	Normal spectral emittance. Resistance-heated strip specimen. Thermistor-bolometer detector. Monochromator. Reference blackbody. Temperatures measured with thermocouples.	Measured in air. Data taken from curves.

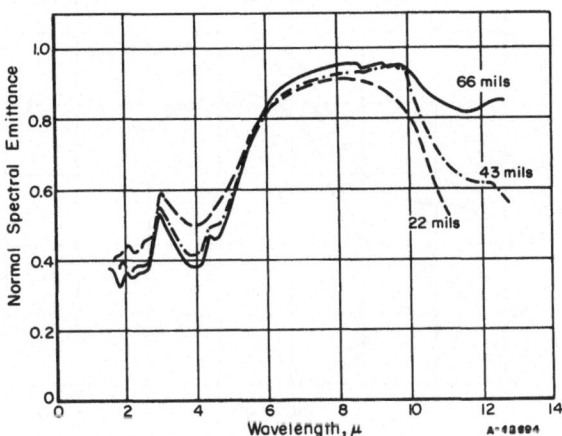

NORMAL SPECTRAL EMITTANCE OF ROKIDE A ON INCONEL X AT 800 F

NORMAL SPECTRAL EMITTANCE OF ROKIDE A ON INCONEL X AT 800 F--REFERENCE INFORMATION

Reference	Investigator	Symbol	Composition and Surface Condition	Test Method	Remarks
13	Gravina and Katz		Norton Rokide A, white 98.5 per cent alumina. Melting point about 3600 F. Porosity about 4 to 8 per cent. Flame sprayed on degreased, grit-blasted surface coated with 12-mil-thick Nichrome V undercoat. Coating thickness: 66 mils 43 mils 22 mils	Normal total emittance. Resistance-heated strip specimen. Thermistor-bolometer detector. Reference blackbody. Temperatures measured with thermocouples.	Measured in air. Data taken from curves.

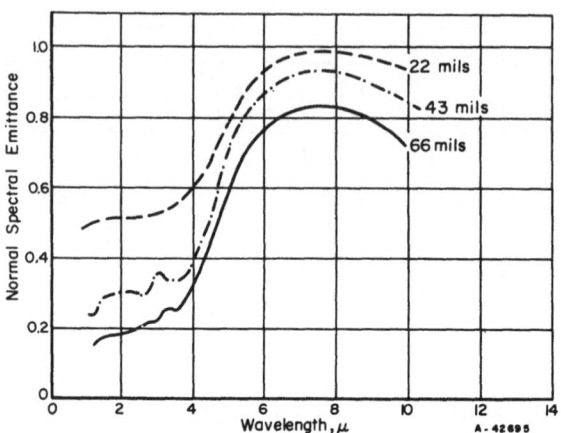

NORMAL SPECTRAL EMITTANCE OF ROKIDE A ON INCONEL X AT 1800 F

NORMAL SPECTRAL EMITTANCE OF ROKIDE A ON INCONEL X AT 1800 F--REFERENCE INFORMATION

Reference	Investigator	Symbol	Composition and Surface Condition	Test Method	Remarks
13	Gravina and Katz		Norton Rokide A, white 93.5 per cent alumina. Melting point about 3600 F. Porosity about 4 to 8 per cent. Flame sprayed on degreased, grit-blasted surface coated with 12-mil-thick Nichrome V undercoat. Coating thickness: 66 mils 43 mils 22 mils	Normal total emittance. Resistance-heated strip specimen. Thermistor-bolometer detector. Reference blackbody. Temperatures measured with thermocouples.	Measured in air. Data taken from curves.

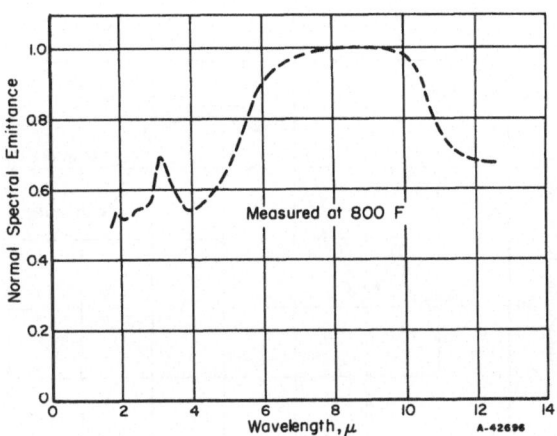

NORMAL SPECTRAL EMITTANCE OF ROKIDE A ON 6A1-4V TITANIUM

NORMAL SPECTRAL EMITTANCE OF ROKIDE A ON 6A1-4V TITANIUM—REFERENCE INFORMATION

Reference	Investigator	Symbol	Composition and Surface Condition	Test Method	Remarks
13	Gravina and Katz		Norton Rokide A, white 98.5 per cent alumina. Melting point about 3600 F. Porosity about 4 to 8 per cent. Flame sprayed on degreased, grit-blasted surface coated with 12-mil-thick Nichrome V undercoat. Coating thickness not given. Measured at 800 F	Normal spectral emittance. Resistance-heated strip specimen. Thermistor-bolometer detector. Monochromator. Reference blackbody. Temperatures measured with thermocouples.	Measured in air. Data taken from curves.

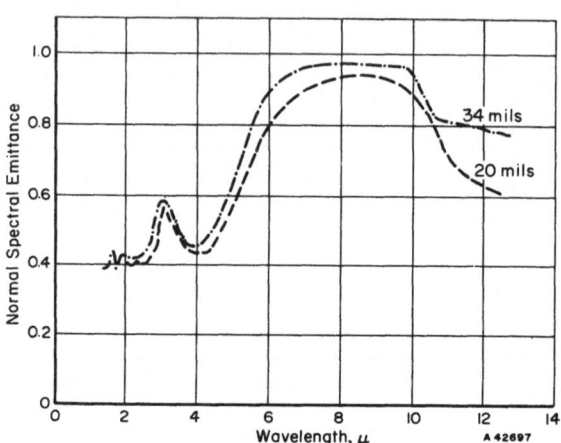

NORMAL SPECTRAL EMITTANCE OF ROKIDE A ON A-286 STEEL AT 800 F

NORMAL SPECTRAL EMITTANCE OF ROKIDE A ON A-286 STEEL AT 800 F--REFERENCE INFORMATION

Reference	Investigator	Symbol	Composition and Surface Condition	Test Method	Remarks
13	Gravina and Katz		Norton Rokide A, white 98.5 per cent alumina. Melting point about 3600 F. Porosity about 4 to 8 per cent. Flame sprayed on degreased, grit-blasted surface coated with 12-mil-thick Nichrome V undercoat. Coating thickness: 34 mils 20 mils	Normal spectral emittance. Resistance-heated strip specimen. Thermistor-bolometer detector. Monochromator. Reference blackbody. Temperatures measured with thermocouples.	Measured in air. Data taken from curves.

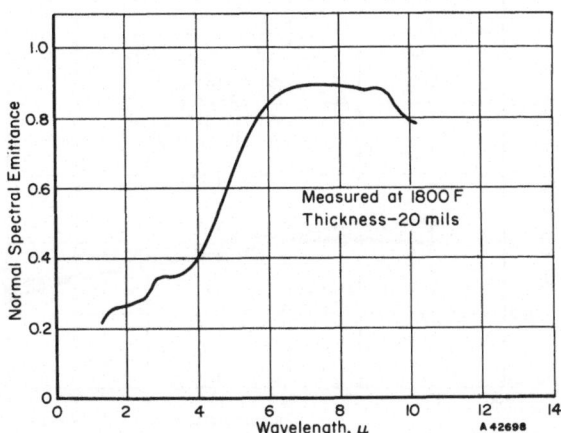

NORMAL SPECTRAL EMITTANCE OF ROKIDE A ON A-286 STEEL AT 1800 F

NORMAL SPECTRAL EMITTANCE OF ROKIDE A ON A-286 STEEL AT 1800 F--REFERENCE INFORMATION

Reference	Investigator	Symbol	Composition and Surface Condition	Test Method	Remarks
13	Gravina and Katz		Norton Rokide A, white 98.5 per cent alumina. Melting point about 3600 F. Porosity about 4 to 8 per cent. Flame sprayed on degreased, grit-blasted surface coated with 12-mil-thick Nichrome V undercoat. Coating thickness 20 mils.	Normal total emittance. Resistance-heated strip specimen. Thermistor-bolometer detector. Reference blackbody. Temperatures measured with thermocouples.	Measured in air. Data taken from curves.

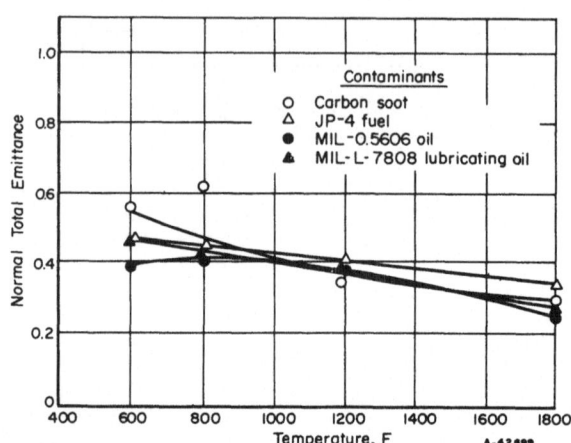

NORMAL TOTAL EMITTANCE OF ROKIDE A (CONTAMINATED) ON A-286 STEEL

NORMAL TOTAL EMITTANCE OF ROKIDE A (CONTAMINATED) ON A-286 STEEL--REFERENCE INFORMATION

Reference	Investigator	Symbol	Composition and Surface Condition	Test Method	Remarks
13	Gravina and Katz		Rokide A (Norton, 98.5 per cent alumina). Contaminated with carbon soot Contaminated with JP-4 fuel Contaminated with MIL-O-5606 oil Contaminated with MIL-L-7808 lubricating oil.	Normal total emittance. Resistance-heated strip specimen. Thermistor-bolometer detector. Reference blackbody. Temperatures measured with thermocouples.	Measured at atmospheric pressure. Data taken from curves.

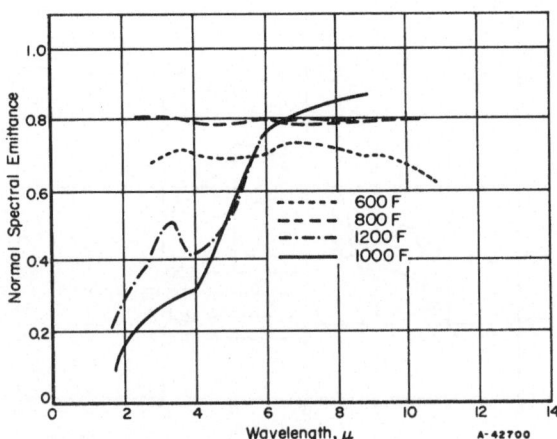

NORMAL SPECTRAL EMITTANCE OF ROKIDE A (CONTAMINATED) ON A-286 STEEL

NORMAL SPECTRAL EMITTANCE OF ROKIDE A (CONTAMINATED) ON A-286 STEEL—REFERENCE INFORMATION

Reference	Investigator	Symbol	Composition and Surface Condition	Test Method	Remarks
13	Gravina and Katz		Rokide A, Norton 98.5 per cent alumina. Flame sprayed onto 12-mil-thick Nichrome V undercoat. Contaminated with carbon deposits. Measured at: 600 F 800 F 1200 F 1800 F	Normal spectral emittance. Resistance-heated strip specimen. Thermistor-bolometer detector. Monochromator. Reference blackbody. Temperatures measured with thermocouples.	Measured in air. Data taken from curves.

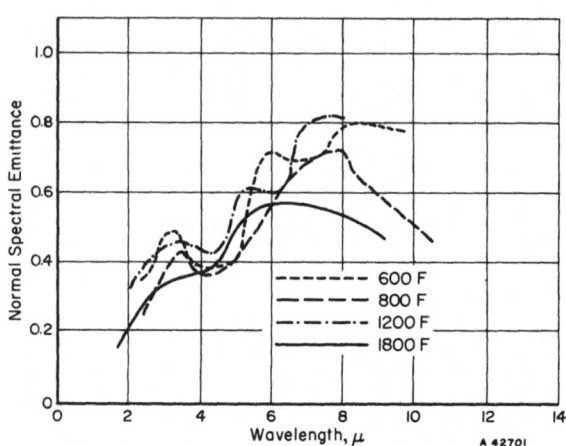

NORMAL SPECTRAL EMITTANCE OF ROKIDE A (CONTAMINATED) ON A-286 STEEL

NORMAL SPECTRAL EMITTANCE OF ROKIDE A (CONTAMINATED) ON A-286 STEEL--REFERENCE INFORMATION

Reference	Investigator	Symbol	Composition and Surface Condition	Test Method	Remarks
13	Gravina and Katz		Rokide A, Norton 98.5 per cent alumina. Flame sprayed onto 12-mil-thick Nichrome V under-coat. Contaminated with JP-4 fuel. Measured at: 600 F 800 F 1200 F 1800 F	Normal spectral emittance. Resistance-heated strip specimen. Thermistor-bolometer detector. Monochromator. Reference blackbody. Temperatures measured with thermocouples.	Measured in air. Data taken from curves.

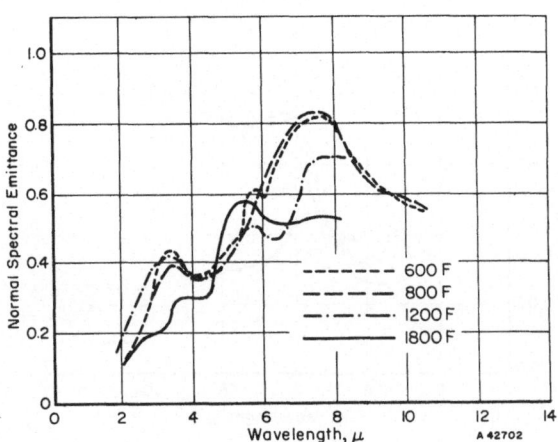

NORMAL SPECTRAL EMITTANCE OF ROKIDE A (CONTAMINATED) ON A-286 STEEL

NORMAL SPECTRAL EMITTANCE OF ROKIDE A (CONTAMINATED) ON TYPE A-286 STEEL--REFERENCE INFORMATION

Reference	Investigator	Symbol	Composition and Surface Condition	Test Method	Remarks
13	Gravina and Katz		Rokide A, Norton 98.5 per cent alumina. Flame sprayed onto 12-mil-thick Nichrome V undercoat. Contaminated with MIL-L-7808. Measured at: 600 F 800 F 1200 F 1800 F	Normal spectral emittance. Resistance-heated strip specimen. Thermistor-bolometer detector. Monochromator. Reference blackbody. Temperatures measured with thermocouples.	Measured in air. Data taken from curves.

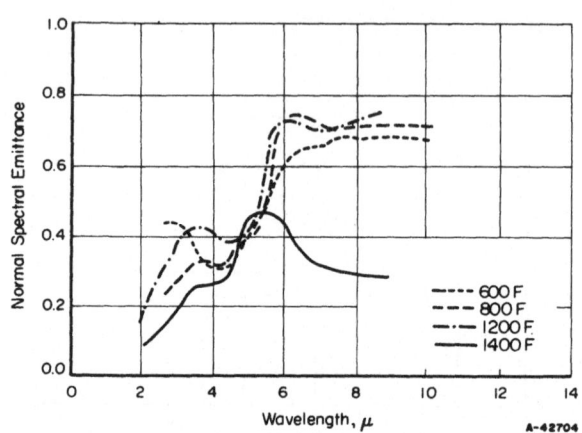

NORMAL SPECTRAL EMITTANCE OF ROKIDE A (CONTAMINATED) ON A-286 STEEL

NORMAL SPECTRAL EMITTANCE OF ROKIDE A (CONTAMINATED) ON TYPE A-286 STEEL—REFERENCE INFORMATION

Reference	Investigator	Symbol	Composition and Surface Condition	Test Method	Remarks
13	Gravina and Katz		Rokide A, Norton 98.5 per cent alumina. Flame sprayed onto 12-mil-thick Nichrome V under-coat. Contaminated with MIL-O-5606. Measured at: 600 F 800 F 1200 F 1800 F	Normal spectral emittance. Resistance-heated strip specimen. Thermistor-bolometer detector. Monochromator. Reference blackbody. Temperatures measured with thermocouples.	Measured in air. Data taken from curves.

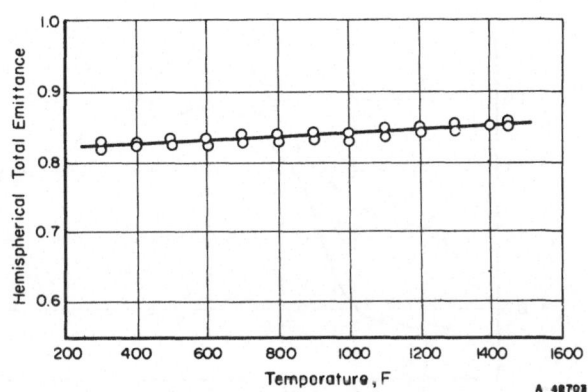

A 48703

HEMISPHERICAL TOTAL EMITTANCE OF ROKIDE C ON TYPE 310 STAINLESS STEEL

HEMISPHERICAL TOTAL EMITTANCE OF ROKIDE C ON TYPE 310 STAINLESS STEEL—REFERENCE INFORMATION

Reference	Investigator	Symbol	Composition and Surface Condition	Test Method	Remarks
16	Pratt & Whitney Aircraft		Rokide C applied to Type 310 stainless steel.	Hemispherical total emittance. Resistance-heated strip specimen. Power dissipated in measured area. Temperatures measured with thermocouples.	Measured in vacuum. Data taken from curves.

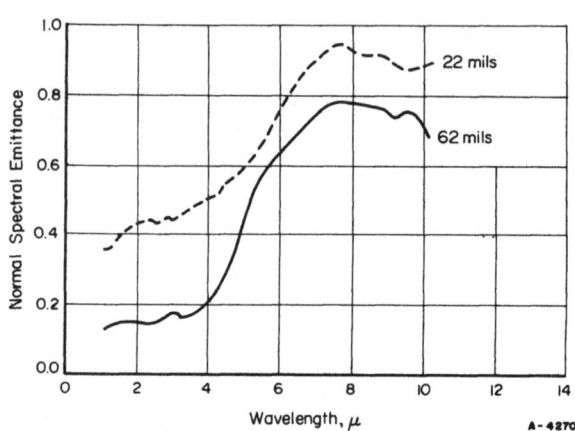

NORMAL SPECTRAL EMITTANCE OF ROKIDE Z ON INCONEL X

NORMAL SPECTRAL EMITTANCE OF ROKIDE Z ON INCONEL X--REFERENCE INFORMATION

Reference	Investigator	Symbol	Composition and Surface Condition	Test Method	Remarks
13	Gravina and Katz		Norton Rokide Z, stabilized ZrO_2. Melting point about 4500 F. Porosity—about 8 per cent total pores. Coating thickness: 62 mils 22 mils	Normal spectral emittance. Resistance-heated strip specimen. Thermistor-bolometer detector. Monochromator. Reference blackbody. Temperatures measured with thermocouples.	Measured in air. Data taken from curves.

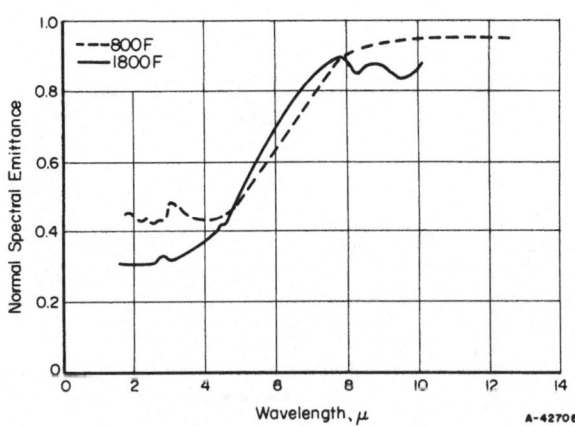

NORMAL SPECTRAL EMITTANCE OF ROKIDE Z ON 6A1-4V TITANIUM

NORMAL SPECTRAL EMITTANCE OF ROKIDE Z ON 6A1-4V TITANIUM--REFERENCE INFORMATION

Reference	Investigator	Symbol	Composition and Surface Condition	Test Method	Remarks
13	Gravina and Katz		Norton Rokide Z, stabilized ZrO_2. Melting point about 4500 F. Porosity about 8 per cent total pores. Measured at: 800 F 1800 F	Normal spectral emittance. Resistance-heated strip specimen. Thermistor-bolometer detector. Monochromator. Reference blackbody. Temperatures measured with thermocouples.	Measured in air. Data taken from curves.

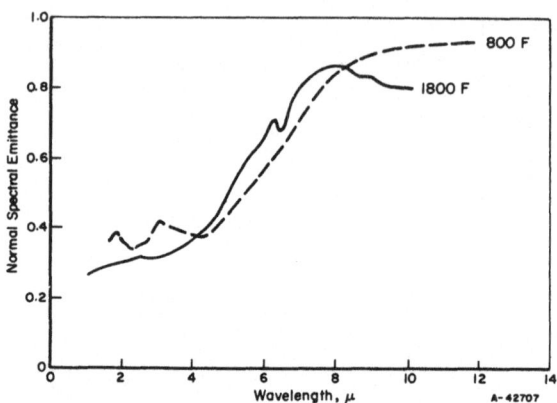

NORMAL SPECTRAL EMITTANCE OF ROKIDE Z ON A-286 STEEL

NORMAL SPECTRAL EMITTANCE OF ROKIDE Z ON A-286 STEEL—REFERENCE INFORMATION

Reference	Investigator	Symbol	Composition and Surface Condition	Test Method	Remarks
13	Gravina and Katz		Norton Rokide Z, stabilized ZrO_2. Melting point about 4500 F. Porosity about 8 per cent total pores. Measured at: 800 F 1800 F	Normal spectral emittance. Resistance-heated strip specimen. Thermistor-bolometer detector. Monochromator. Reference blackbody. Temperatures measured with thermocouples.	Measured in air. Data taken from curves.

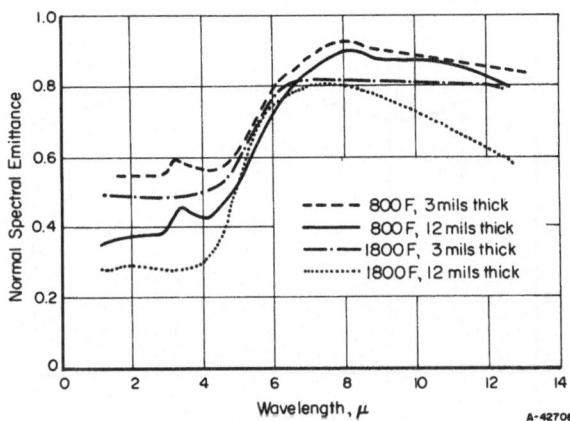

NORMAL SPECTRAL EMITTANCE OF SYLVESTER FCM-10 ON INCONEL X

NORMAL SPECTRAL EMITTANCE OF SYLVESTER FCM-10 ON INCONEL X—REFERENCE INFORMATION

Reference	Investigator	Symbol	Composition and Surface Condition	Test Method	Remarks
13	Gravina and Katz		Sylvester ceramic coating FCM-10, a dark gray mullite. Flame sprayed on degreased, sand blasted, preheated Inconel X. Surface roughness approximately 180 to 200 microinches. Measured at: 800 F, 3 mils thick 800 F, 12 mils thick 1800 F, 3 mils thick 1800 F, 12 mils thick	Normal spectral emittance. Resistance-heated strip specimen. Thermistor-bolometer detector. Monochromator. Reference blackbody. Temperatures measured with thermocouples.	Measured in air. Data taken from curves.

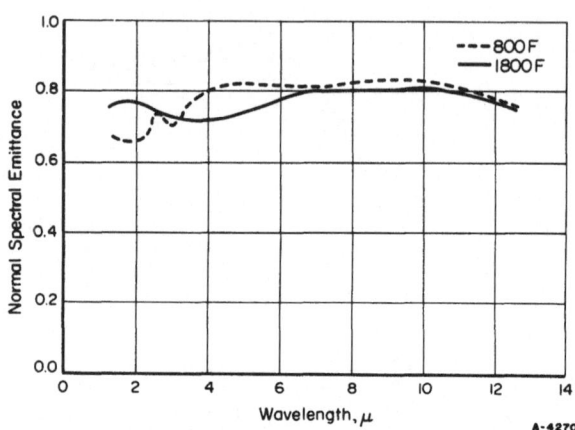

NORMAL SPECTRAL EMITTANCE OF SYLVESTER FCT-10 ON INCONEL X

NORMAL SPECTRAL EMITTANCE OF SYLVESTER FCT-10 ON INCONEL X--REFERENCE INFORMATION

Reference	Investigator	Symbol	Composition and Surface Condition	Test Method	Remarks
13	Gravina and Katz		Sylvester ceramic coating FCT-10, a light gray titanium dioxide. Flame sprayed on degreased, sand blasted, preheated Inconel X. Surface roughness approximately 180 to 200 micro-inches. Coating thickness not given. Measured at: 800 F 1800 F	Normal spectral emittance. Resistance-heated strip specimen. Thermistor-bolometer detector. Monochromator. Reference blackbody. Temperatures measured with thermocouples.	Measured in air. Data taken from curves.

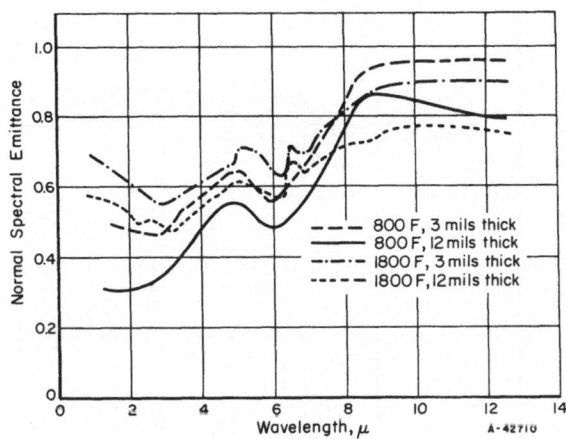

NORMAL SPECTRAL EMITTANCE OF SYLVESTER FCR-11 ON INCONEL X

NORMAL SPECTRAL EMITTANCE OF SYLVESTER FCR-11 ON INCONEL X--REFERENCE INFORMATION

Reference	Investigator	Symbol	Composition and Surface Condition	Test Method	Remarks
13	Gravina and Katz		Sylvester ceramic coating FCR-11, a dark gray rare-earth oxide mixture, (50 per cent cerium). Flame sprayed on degreased, sand blasted, preheated Inconel X. Surface roughness approximately 180 to 200 micro-inches. Measured at: 800 F, 3 mils thick 800 F, 12 mils thick 1800 F, 3 mils thick 1800 F, 12 mils thick	Normal spectral emittance. Resistance-heated strip specimen. Thermistor-bolometer detector. Monochromator. Reference blackbody. Temperatures measured with thermocouples.	Measured in air. Data taken from curves.

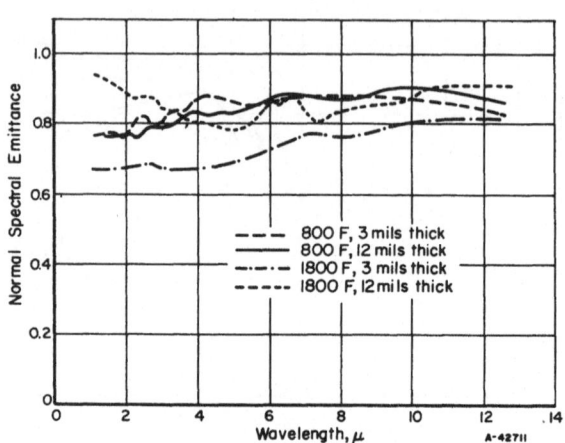

NORMAL SPECTRAL EMITTANCE OF SYLVESTER FCT-11 ON INCONEL X

SPECTRAL EMITTANCE OF SYLVESTER FCT-11 CERAMIC COATING ON INCONEL X—REFERENCE INFORMATION

Reference	Investigator	Symbol	Composition and Surface Condition	Test Method	Remarks
13	Gravina and Katz		Sylvester ceramic coating FCT-11, a dark gray sillimanite. Flame sprayed on degreased, sand blasted, preheated Inconel X. Surface roughness approximately 180 to 200 micro-inches. Measured at: 800 F, 3 mils thick 800 F, 12 mils thick 1800 F, 3 mils thick 1800 F, 12 mils thick	Normal spectral emittance. Resistance-heated strip specimen. Thermistor-bolometer detector. Monochromator. Reference blackbody. Temperatures measured with thermocouples.	Measured in air. Data taken from curves.

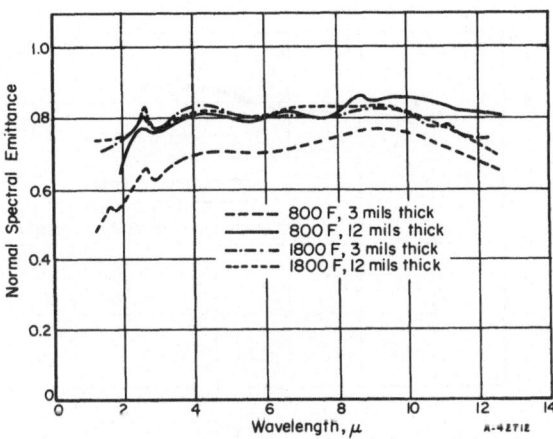

NORMAL SPECTRAL EMITTANCE OF SYLVESTER FCT-12 ON A-286 STEEL

NORMAL SPECTRAL EMITTANCE OF SYLVESTER FCT-12 ON A-286 STEEL—REFERENCE INFORMATION

Reference	Investigator	Symbol	Composition and Surface Condition	Test Method	Remarks
13	Gravina and Katz		Sylvester ceramic coating FCT-12, a black sillimanite. Flame sprayed on degreased, sand blasted, preheated material. Surface roughness approximately 180 to 200 micro-inches. Measured at: 800 F, 3 mils thick 800 F, 12 mils thick 1800 F, 3 mils thick 1800 F, 12 mils thick	Normal spectral emittance. Resistance-heated strip specimen. Thermistor-bolometer detector. Monochromator. Reference blackbody. Temperatures measured with thermocouples.	Measured in air. Data taken from curves.

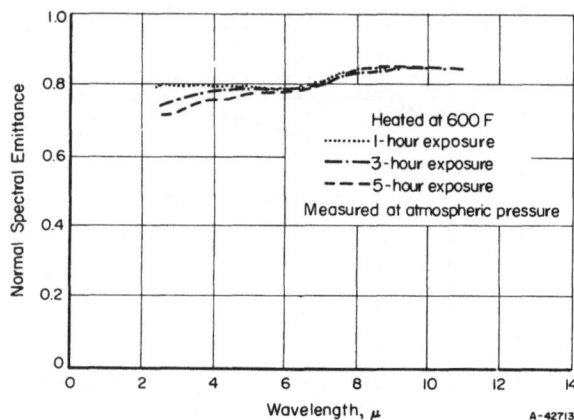

NORMAL SPECTRAL EMITTANCE OF SYLVESTER FCT-12 ON A-286 STEEL

NORMAL SPECTRAL EMITTANCE OF SYLVESTER FCT-12 ON A-286 STEEL AT 600 F--REFERENCE INFORMATION

Reference	Investigator	Symbol	Composition and Surface Condition	Test Method	Remarks
13	Gravina and Katz		Sylvester ceramic coating FCT-12, a black sillimanite. Flame sprayed on de-greased, sand blasted, preheated material. Surface roughness approximately 180 to 200 micro-inches. Heated at 600 F: 1 hour 3 hours 5 hours	Normal spectral emittance. Resistance-heated strip specimen. Thermistor-bolometer detector. Monochromator. Reference blackbody. Temperatures measured with thermocouples.	Measured in air. Data taken from curves.

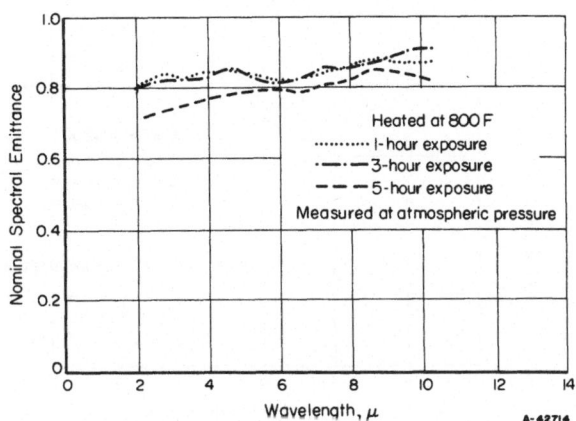

NORMAL SPECTRAL EMITTANCE OF SYLVESTER FCT-12 ON A-286 STEEL

NORMAL SPECTRAL EMITTANCE OF SYLVESTER FCT-12 CERAMIC COATING ON A-286 STEEL AT 800 F--REFERENCE INFORMATION

Reference	Investigator	Symbol	Composition and Surface Condition	Test Method	Remarks
13	Gravina and Katz		Sylvester ceramic coating FCT-12, a black sillimanite. Flame sprayed on degreased, sand blasted, preheated material. Surface roughness approximately 180 to 200 microinches. Heated at 800 F: 1 hour 3 hours 5 hours	Normal spectral emittance. Resistance-heated strip specimen. Thermistor-bolometer detector. Monochromator. Reference blackbody. Temperatures measured with thermocouples.	Measured in air. Data taken from curves.

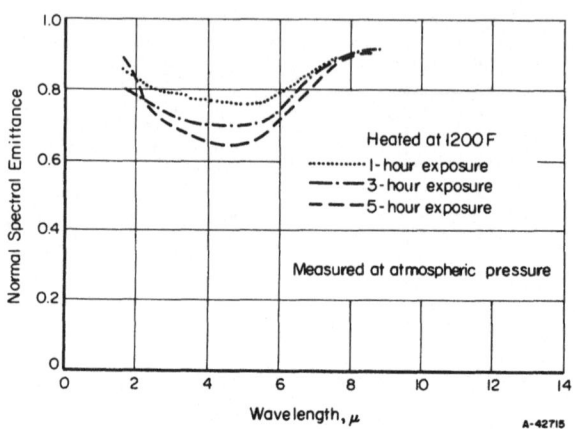

NORMAL SPECTRAL EMITTANCE OF SYLVESTER FCT-12 ON A-286 STEEL

NORMAL SPECTRAL EMITTANCE OF SYLVESTER FCT-12 ON A-286 STEEL AT 1200 F—REFERENCE INFORMATION

Reference	Investigator	Symbol	Composition and Surface Condition	Test Method	Remarks
13	Gravina and Katz		Sylvester ceramic coating FCT-12, a black sillimanite. Flame sprayed on degreased, sand blasted, preheated material. Surface roughness approximately 180 to 200 microinches. Heated at 1200 F: 1 hour 3 hours 5 hours	Normal spectral emittance. Resistance-heated strip specimen. Thermistor-bolometer detector. Monochromator. Reference blackbody. Temperatures measured with thermocouples.	Measured in air. Data taken from curves.

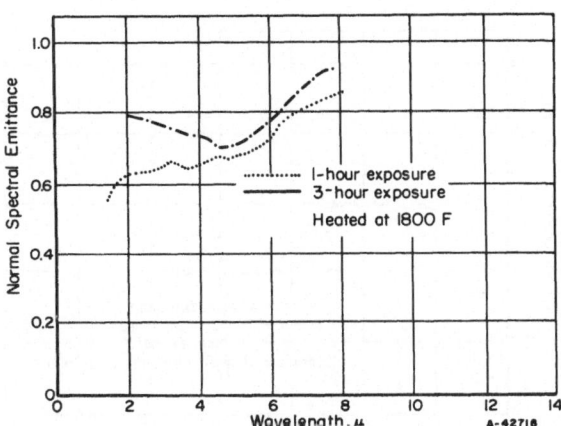

NORMAL SPECTRAL EMITTANCE OF SYLVESTER FCT-12 ON A-286 STEEL

NORMAL SPECTRAL EMITTANCE OF SYLVESTER FCT-12 CERAMIC COATING ON A-286 STEEL AT 1800 F—REFERENCE INFORMATION

Reference	Investigator	Symbol	Composition and Surface Condition	Test Method	Remarks
13	Gravina and Katz		Sylvester ceramic coating FCT-12, a black sillimanite. Flame sprayed on degreased, sand blasted, preheated material. Surface roughness approximately 180 to 200 microinches. Heated at 1800 F: 1 hour 3 hours	Normal spectral emittance. Resistance-heated strip specimen. Thermistor-bolometer detector. Monochromator. Reference blackbody. Temperatures measured with thermocouples.	Measured in air. Data taken from curves.

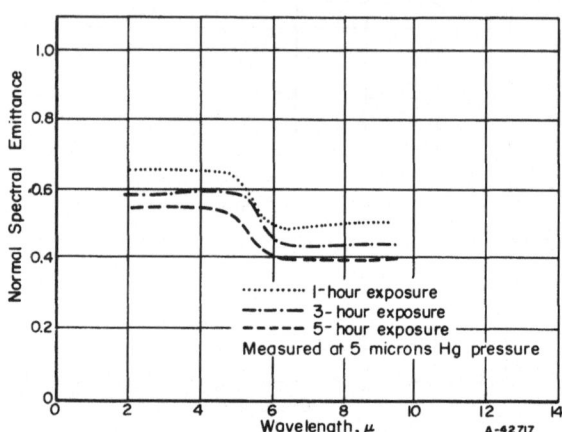

NORMAL SPECTRAL EMITTANCE OF SYLVESTER FCT-12 ON A-286 STEEL

NORMAL SPECTRAL EMITTANCE OF SYLVESTER FCT-12 ON A-286 STEEL AT 600 F--REFERENCE INFORMATION

Reference	Investigator	Symbol	Composition and Surface Condition	Test Method	Remarks
13	Gravina and Katz		Sylvester ceramic coating FCT-12, a black sillimanite. Flame sprayed on degreased, sand blasted, preheated material. Surface roughness approximately 180 to 200 microinches. Heated at 600 F: 1 hour 3 hours 5 hours	Normal spectral emittance. Resistance-heated strip specimen. Thermistor-bolometer detector. Monochromator. Reference blackbody. Temperatures measured with thermocouples.	Measured in 5 microns Hg pressure. Data taken from curves.

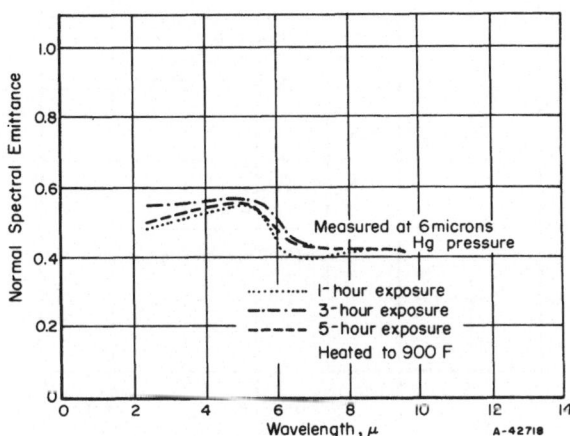

A-42718

NORMAL SPECTRAL EMITTANCE OF SYLVESTER FCT-12 ON A-286 STEEL

NORMAL SPECTRAL EMITTANCE OF SYLVESTER FCT-12 ON A-286 STEEL AT 800 F—REFERENCE INFORMATION

Reference	Investigator	Symbol	Composition and Surface Condition	Test Method	Remarks
13	Gravina and Katz		Sylvester ceramic coating FC-12, a black sillimanite. Flame sprayed on degreased, sand blasted, preheated material. Surface roughness approximately 180 to 200 micro-inches. Heated at 800 F: 1 hour 3 hours 5 hours	Normal spectral emittance. Resistance-heated strip specimen. Thermistor-bolometer detector. Monochromator. Reference blackbody. Temperatures measured with thermocouples.	Measured in 6 microns Hg pressure Data taken from curves.

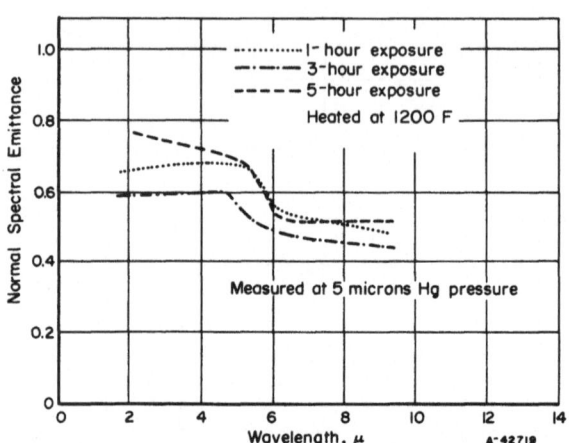

NORMAL SPECTRAL EMITTANCE OF SYLVESTER FCT-12 ON A-286 STEEL

NORMAL SPECTRAL EMITTANCE OF SYLVESTER FCT-12 ON A-286 STEEL AT 1200 F--REFERENCE INFORMATION

Reference	Investigator	Symbol	Composition and Surface Condition	Test Method	Remarks
13	Gravina and Katz		Sylvester ceramic coating FCT-12, a black sillimanite. Flame sprayed on degreased, sand blasted, preheated material. Surface roughness approximately 180 to 200 micro-inches. Heated at 1200 F: 1 hour 3 hours 5 hours	Normal spectral emittance. Resistance-heated strip specimen. Thermistor-bolometer detector. Monochromator. Reference blackbody. Temperatures measured with thermocouples.	Measured in 5 microns Hg pressure. Data taken from curves.

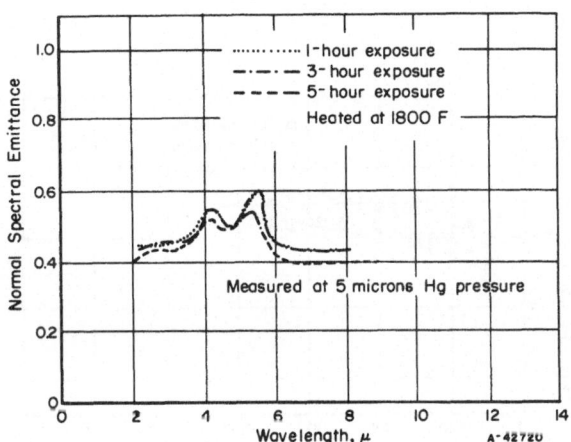

NORMAL SPECTRAL EMITTANCE OF SYLVESTER FCT-12 ON A-286 STEEL

NORMAL SPECTRAL EMITTANCE OF SYLVESTER FCT-12 ON A-286 STEEL AT 1800 F--REFERENCE INFORMATION

Reference	Investigator	Symbol	Composition and Surface Condition	Test Method	Remarks
13	Gravina and Katz		Sylvester ceramic coating FCT-12, a black sillimanite. Flame sprayed on degreased, sand blasted, preheated material. Surface roughness approximately 180 to 200 micro-inches. Heated at 1800 F: 1 hour 3 hours 5 hours	Normal spectral emittance. Resistance-heated strip specimen. Thermistor-bolometer detector. Monochromator. Reference blackbody. Temperatures measured with thermocouples.	Measured in 5 microns Hg pressure. Data taken from curves.

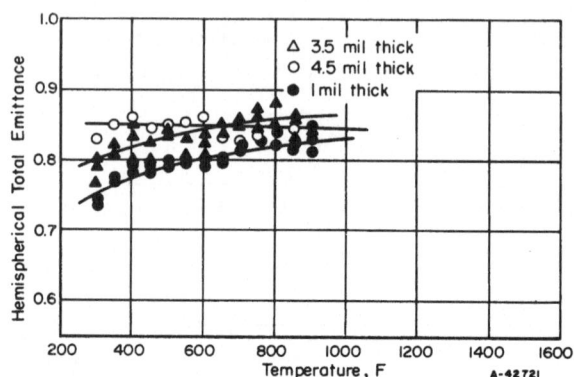

HEMISPHERICAL TOTAL EMITTANCE OF TITANIUM OXIDE ON ALUMINUM

HEMISPHERICAL TOTAL EMITTANCE OF TITANIUM OXIDE ON ALUMINUM--REFERENCE INFORMATION.

Reference	Investigator	Symbol	Composition and Surface Condition	Test Method	Remarks
16	Pratt & Whitney Aircraft		Metco plasma flame spray powder XP-1114. Flame sprayed on aluminum strip. Coating thickness: 1 mil 4.5 mils 3.5 mils	Hemispherical total emittance. Resistance-heated strip specimen. Power dissipated in measured area. Temperatures measured with thermocouples.	Measured in vacuum. Data taken from curves.

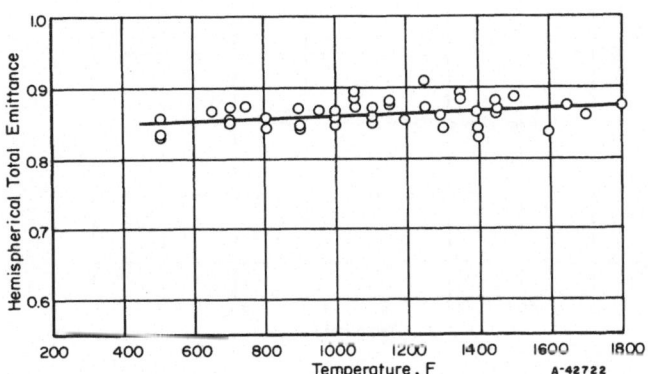

HEMISPHERICAL TOTAL EMITTANCE OF "TITANIA BASE" POWDER ON TYPE 310 STAINLESS STEEL

HEMISPHERICAL TOTAL EMITTANCE OF TITANIA BASE POWDER ON TYPE 310 STAINLESS STEEL--REFERENCE INFORMATION

Reference	Investigator	Symbol	Composition and Surface Condition	Test Method	Remarks
16	Pratt & Whitney Aircraft		Plasmadyne powder. Flame sprayed on Type 310 stainless steel	Hemispherical total emittance. Resistance-heated strip specimen. Power dissipated in measured area. Temperatures measured with thermocouples.	Measured in vacuum. Data taken from curve.

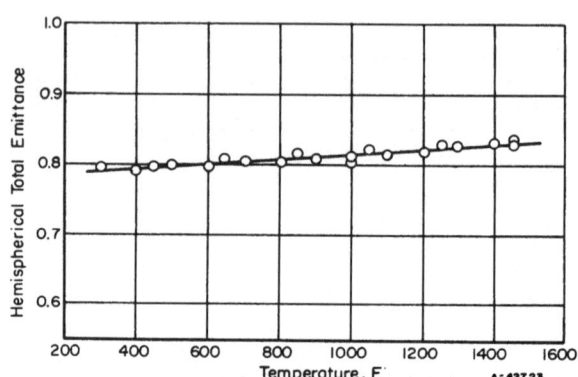

HEMISPHERICAL TOTAL EMITTANCE OF 50% TITANIUM OXIDE-50% ALUMINUM OXIDE ON TYPE 310 STAINLESS STEEL

HEMISPHERICAL TOTAL EMITTANCE OF 50% TITANIUM OXIDE-50% ALUMINUM OXIDE ON
TYPE 310 STAINLESS STEEL--REFERENCE INFORMATION

Reference	Investigator	Symbol	Composition and Surface Condition	Test Method	Remarks
16	Pratt & Whitney Aircraft		Metco plasma flame spray powder XP-1121.	Hemispherical total emittance. Resistance–heated strip specimen. Power dissipated in measured area. Temperatures measured with thermocouples.	Measured in vacuum. Data taken from curve.

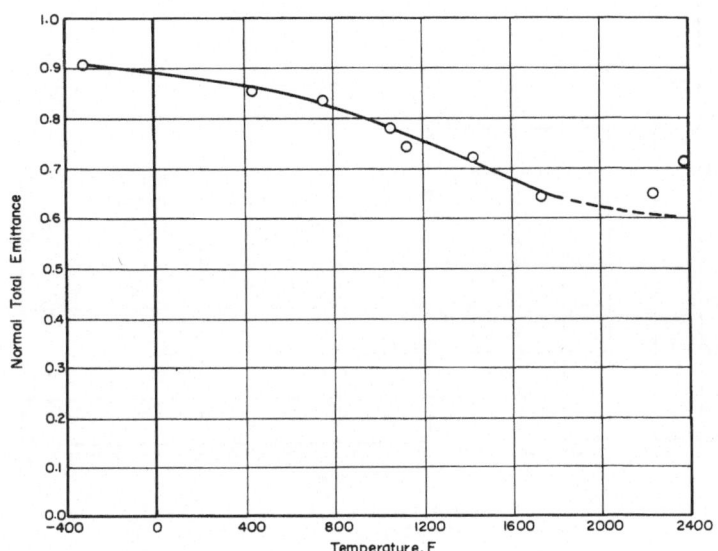

NORMAL TOTAL EMITTANCE OF ZIRCONIUM OXIDE ON INCONEL

NORMAL TOTAL EMITTANCE OF ZIRCONIUM OXIDE ON INCONEL--REFERENCE INFORMATION

Reference	Investigator	Symbol	Composition and Surface Condition	Test Method	Remarks
11	Olson and Morris	O	Zirconium oxide on Inconel. Thickness or surface condition not given. (Coating burned off-- probably near 2000 F.)	Normal total emittance. Comparison blackbody. Furnace heated specimens. Temperatures measured with thermocouples. Thermistor-bolometer detector.	Measured in air. Data taken from curve.

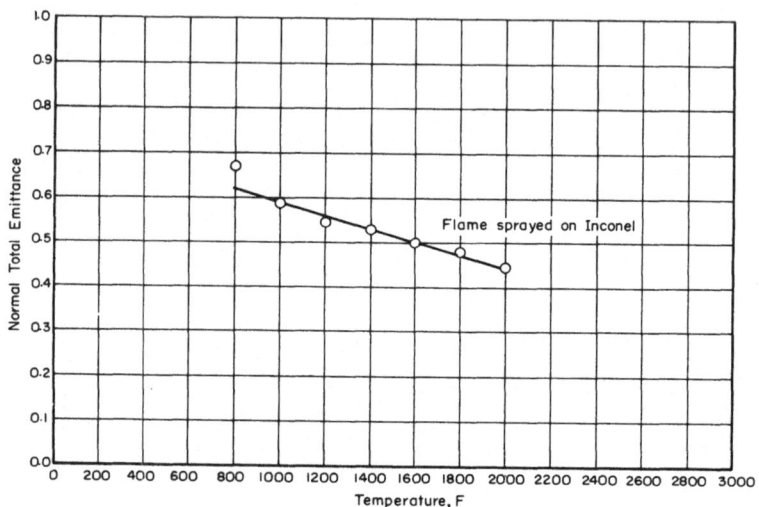

NORMAL TOTAL EMITTANCE OF ZIRCONIUM OXIDE ON INCONEL

NORMAL TOTAL EMITTANCE OF ZIRCONIUM OXIDE ON INCONEL--REFERENCE INFORMATION

Reference	Investigator	Symbol	Composition and Surface Condition	Test Method	Remarks
10	Wade, W. R.	○	Flame-sprayed zirconia on Inconel heater strip. Coating thickness not given.	Normal total emittance. Thermopile detector. Resistance-heated Inconel strip with test material flame sprayed to "opaque" thickness. Comparison blackbody. Temperatures measured with thermocouples.	Measured in air. Temperatures given are those of Inconel heater strip. Data taken from curve.

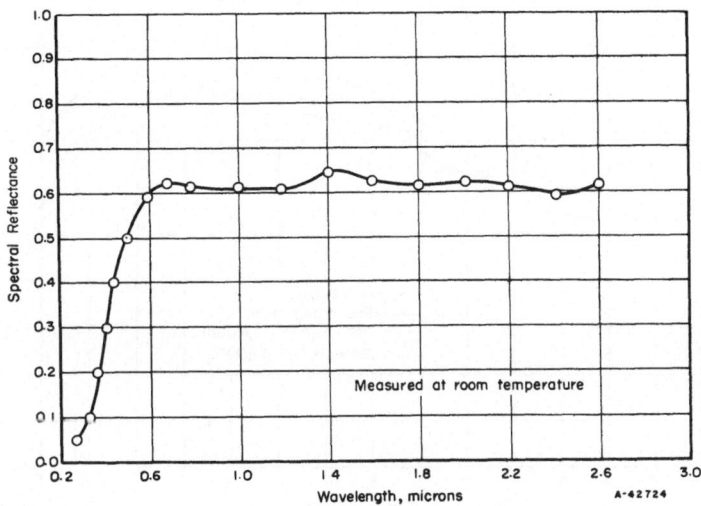

SPECTRAL REFLECTANCE OF ZIRCONIUM OXIDE ON INCONEL

SPECTRAL REFLECTANCE OF ZIRCONIUM OXIDE ON INCONEL--REFERENCE INFORMATION

Reference	Investigator	Symbol	Composition and Surface Condition	Test Method	Remarks
11	Olson and Morris	O	Zirconium oxide flame sprayed on Inconel. Thickness or surface condition not given.	Spectral reflectance at 9 degrees from normal (incident radiation). Recording spectrophotometer. Integrating sphere. Lead sulphide detector. (Normal illumination--hemispherical viewing.)	Measured in air at room temperature. Data taken from curve.

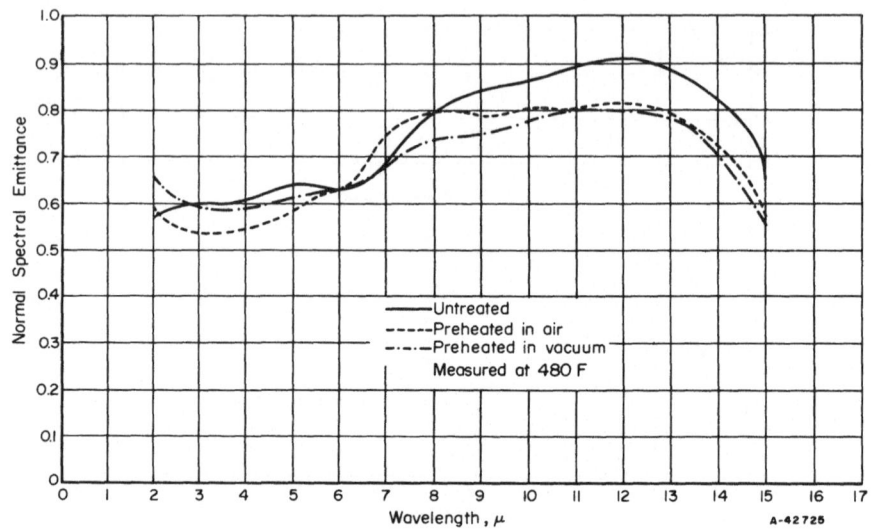

NORMAL SPECTRAL EMITTANCE OF ZIRCONIA ON INCONEL X AT 480 F

NORMAL SPECTRAL EMITTANCE OF ZIRCONIA ON INCONEL X AT 480 F--REFERENCE INFORMATION

Reference	Investigator	Symbol	Composition and Surface Condition	Test Method	Remarks
14	Adams, J. G.		Flame sprayed on Inconel X (untreated). Heated 30 minutes in air at 1500 F. Heated 30 minutes in 6.2 x 10⁻⁵ mm Hg pressure at 1500 F.	Normal spectral emittance. Furnace-heated disk specimen. Comparison blackbody (Hohlraun). Spectrometer-mono-chromator with photo-multiplier, lead sulphide, and thermo-couple detectors. Temperatures measured with thermocouples.	Measured in air.

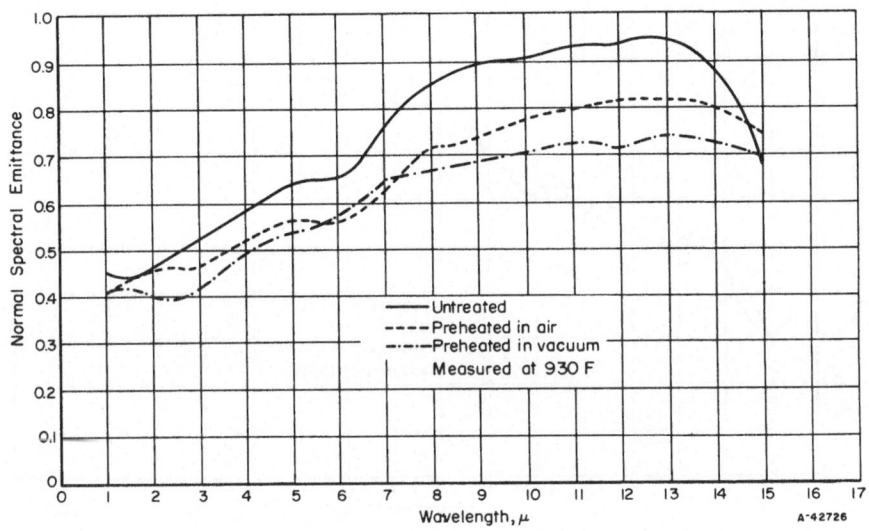

NORMAL SPECTRAL EMITTANCE OF ZIRCONIA ON INCONEL X AT 930 F

NORMAL SPECTRAL EMITTANCE OF ZIRCONIA ON INCONEL X AT 930 F--REFERENCE INFORMATION

Reference	Investigator	Symbol	Composition and Surface Condition	Test Method	Remarks
14	Adams, J. G.		Flame sprayed on Inconel X. Untreated Heated 30 minutes in air at 1500 F Heated 30 minutes in 6.2 x 10⁻⁵ mm Hg pressure at 1500 F	Normal spectral emittance. Furnace-heated disk specimen. Comparison blackbody (Hohlraun). Spectrometer-mono- chromator with photo- multiplier, lead sulphide, and thermo- couple detectors. Temperatures measured with thermocouples.	Measured in air.

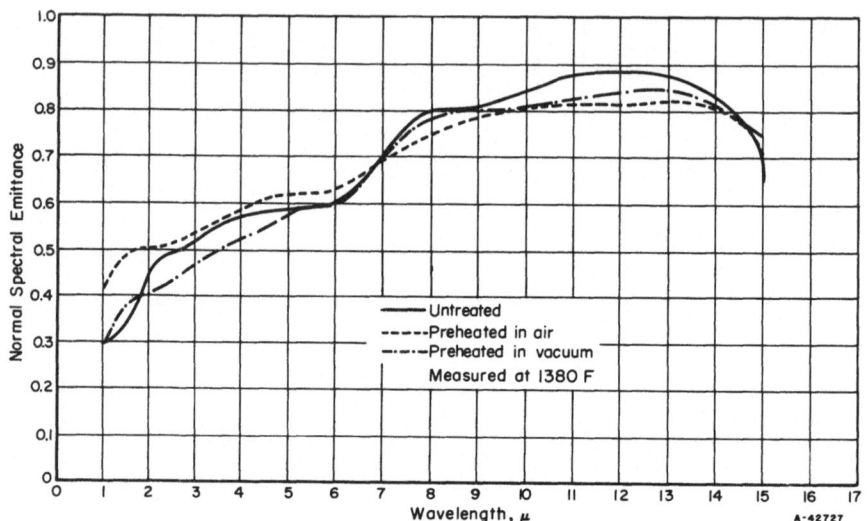

NORMAL SPECTRAL EMITTANCE OF ZIRCONIA ON INCONEL X AT 1380 F

NORMAL SPECTRAL EMITTANCE OF ZIRCONIA ON INCONEL X AT 1380 F--REFERENCE INFORMATION

Reference	Investigator	Symbol	Composition and Surface Condition	Test Method	Remarks
14	Adams, J. G.		Flame sprayed on Inconel X. Untreated Heated 30 minutes in air at 1500 F Heated 30 minutes in 6.2 x 10⁻⁵ mm Hg pressure at 1500 F	Normal spectral emittance. Furnace-heated disk specimen. Comparison blackbody (Hohlraun). Spectrometer-mono-chromator with photo-multiplier, lead sulphide, and thermo-couple detectors. Temperatures measured with thermocouples.	Measured in air.

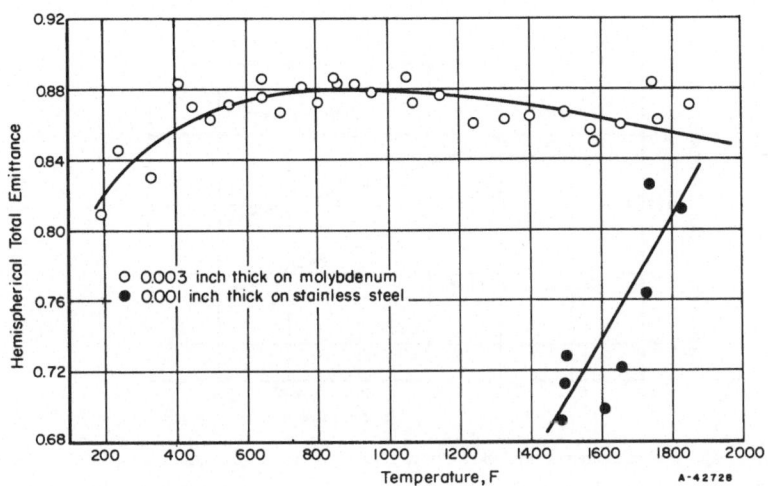

A-42728

HEMISPHERICAL TOTAL EMITTANCE OF TWO ZIRCONIUM OXIDE COATED SPECIMENS

HEMISPHERICAL TOTAL EMITTANCE OF ZIRCONIUM OXIDE ON MOLYBDENUM AND
TYPE 310 STAINLESS STEEL--REFERENCE INFORMATION

Reference	Investigator	Symbol	Composition and Surface Condition	Test Method	Remarks
15	Pratt & Whitney Aircraft		3-mil-thick coating applied by Linde Plasmarc process to molybdenum strip. 1-mil-thick coating flame sprayed by Metco process on Type 310 stainless steel strip.	Hemispherical total emittance. Resistance-heated strip specimen. Power dissipated in measured area. Temperatures measured with thermocouples.	Measured in vacuum. Data taken from curves.

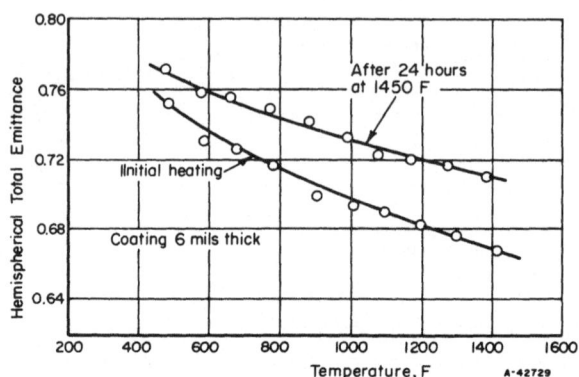

HEMISPHERICAL TOTAL EMITTANCE OF ZIRCONIA ON TYPE 310 STAINLESS STEEL

HEMISPHERICAL TOTAL EMITTANCE OF ZIRCONIA ON TYPE 310 STAINLESS STEEL--REFERENCE INFORMATION

Reference	Investigator	Symbol	Composition and Surface Condition	Test Method	Remarks
18	Pratt & Whitney Aircraft		6-mil-thick coating applied by Metco plasma flame spray process on Type 310 stainless steel tube. Initial heating After 24 hours at 1450 F (cooling and heating show change to be permanent)	Hemispherical total emittance. Resistance-heated tube specimen. Power dissipated in measured area. Temperatures measured with thermocouples.	Measured in vacuum. Data taken from curves.

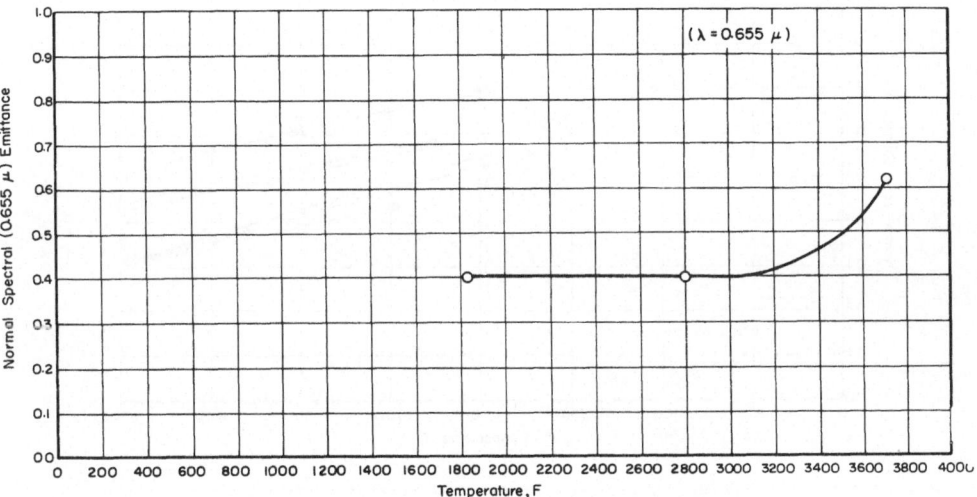

NORMAL SPECTRAL EMITTANCE OF THORIUM OXIDE ON TUNGSTEN AND MOLYBDENUM

NORMAL SPECTRAL EMITTANCE OF THORIUM OXIDE ON TUNGSTEN AND MOLYBDENUM--REFERENCE INFORMATION

Reference	Investigator	Symbol	Composition and Surface Condition	Test Method	Remarks
6	Morgan, F. H.	o	Purity or coating thickness not given. Thoria cataphoretically coated on tungsten (or molybdenum) ribbon previously flashed in hydrogen.	Resistance-heated, coated ribbon. Temperatures measured with thermocouples. Brightness temperature measured with optical pyrometer.	Measured in vacuum. Coatings on tungsten and molybdenum gave identical results. Data taken from curve and discussion.

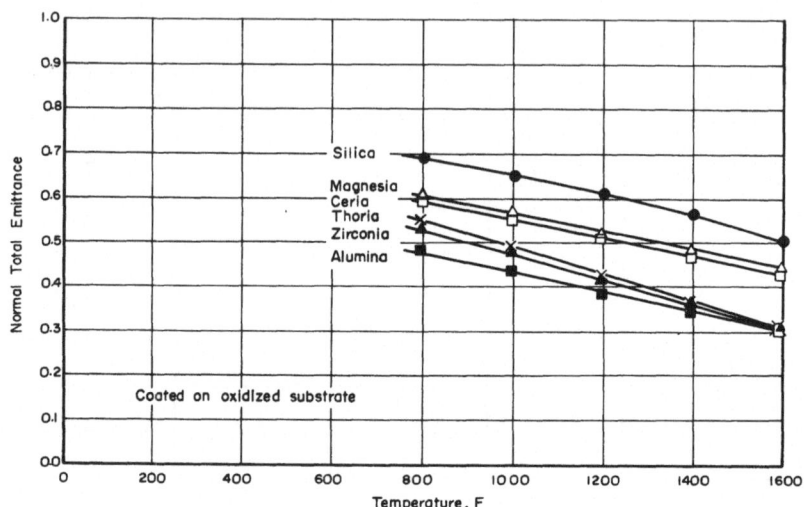

NORMAL TOTAL EMITTANCE OF VARIOUS REFRACTORY OXIDES ON NIMONIC 75

NORMAL TOTAL EMITTANCE OF VARIOUS REFRACTORY OXIDES ON NIMONIC 75--REFERENCE INFORMATION

Reference	Investigator	Symbol	Composition and Surface Condition	Test Method	Remarks
12	Sully, Brandes, and Waterhouse		Purest commercially available materials. Applied to oxidized Nimonic 75 strip as water suspension. Coating thickness not given.	Normal total emittance. Resistance-heated metal-strip specimens with ceramic coated surface. Comparison blackbody. Temperatures measured with thermocouples. Thermopile detector.	Measured in air. Data taken from curves. Hemispherical total emittance found to equal normal total emittance for the alumina coated specimen. (Should hold true for the others also.)
		●	Silica		
		△	Magnesia		
		□	Ceria		
		×	Thoria		
		▲	Zirconia		
		■	Alumina		

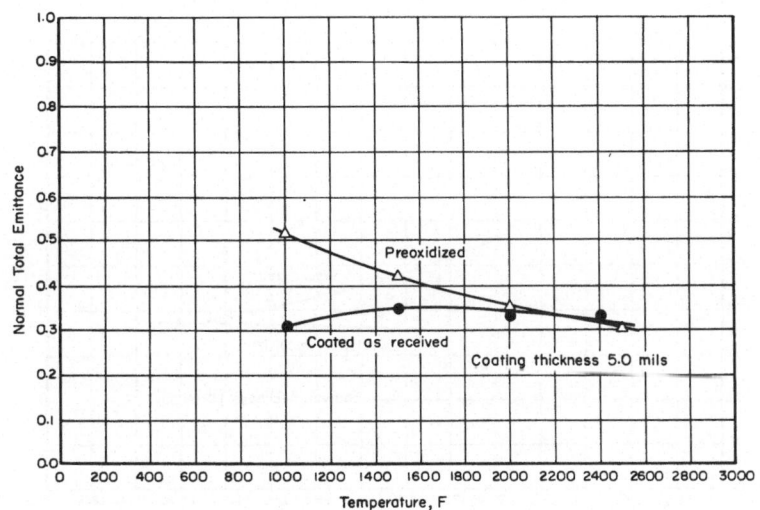

NORMAL TOTAL EMITTANCE OF Al-Si ON MOLYBDENUM

NORMAL TOTAL EMITTANCE OF Al-Si ON MOLYBDENUM--REFERENCE INFORMATION

Reference	Investigator	Symbol	Composition and Surface Condition	Test Method	Remarks
7	Anthony and Pearl	● △	N.R.C. Al-Si on molybdenum. Coated as received. Preoxidized, then coated. Coating thickness 5 mils.	Normal total emittance. Induction heated specimen. Thermopile detector. Comparison blackbody. Temperatures measured with thermocouples and optical pyrometer.	Measured in continuous purge of helium gas.

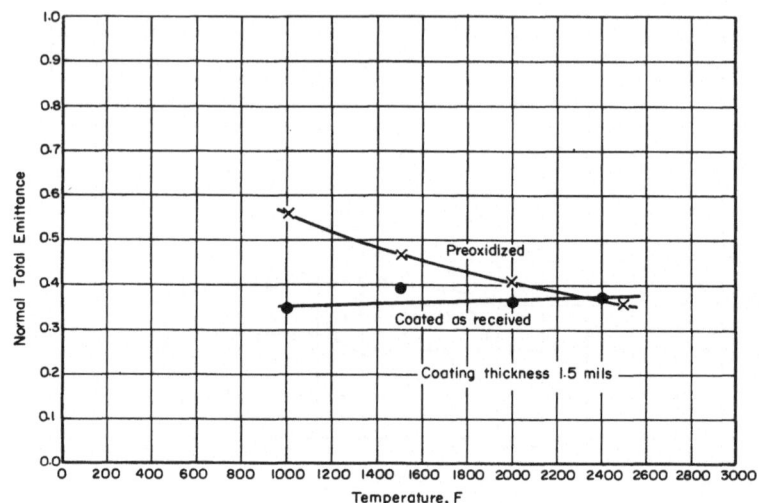

NORMAL TOTAL EMITTANCE OF DURAK-MG ON MOLYBDENUM

NORMAL TOTAL EMITTANCE OF DURAK-MG ON MOLYBDENUM--REFERENCE INFORMATION

Reference	Investigator	Symbol	Composition and Surface Condition	Test Method	Remarks
7	Anthony and Pearl	● ✕	Durak-MG coating on molybdenum. Coated as received. Preoxidized, then coated. Coating thickness 1.5 mils, nominal.	Normal total emittance. Induction-heated specimen. Thermopile detector. Comparison blackbody. Temperatures measured with thermocouples and optical pyrometer.	Measured in continuous purge of helium gas. Data taken from tables.

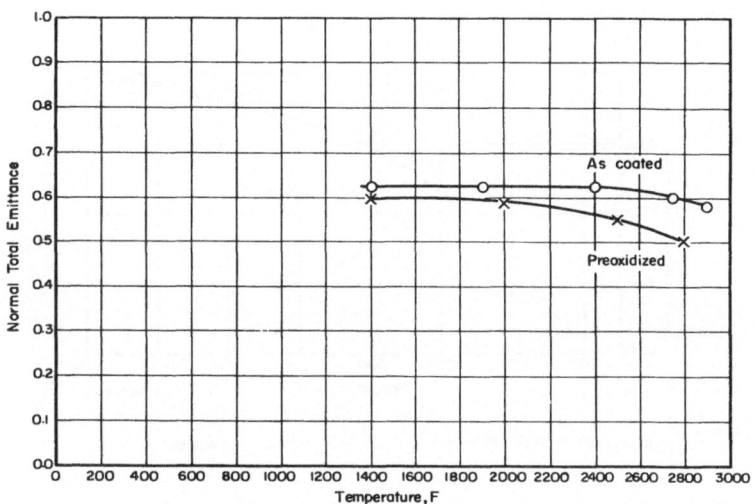

NORMAL TOTAL EMITTANCE OF DURAK-MG ON MOLYBDENUM-0.5 TITANIUM ALLOY

NORMAL TOTAL EMITTANCE OF DURAK-MG ON MOLYBDENUM-0.5 TITANIUM ALLOY—REFERENCE INFORMATION

Reference	Investigator	Symbol	Composition and Surface Condition	Test Method	Remarks
8	Fieldhouse, Lang, and Blau		Durak-MG coating on molybdenum-0.5 per cent titanium alloy.	Normal and angular total emittance. Induction heated specimen.	Measured in 90 per cent argon, 10 per cent H_2 gas. Measurements made at
		o	As coated.	Spectrometer with prism replaced by	angles of 0, 30, 45, and 60 degrees
		x	Preoxidized at 2000 F for 1 hour. Specimens "flat and smooth" (coating thickness not given).	plane mirror. Thermocouple detector. Blackbody hole in specimen. Temperature calibration with blackbody and optical pyrometer.	with normal to specimen surface. Normal total emittance equals hemispherical total emittance within reported experimental error of ± 5 per cent. Data taken from curves.

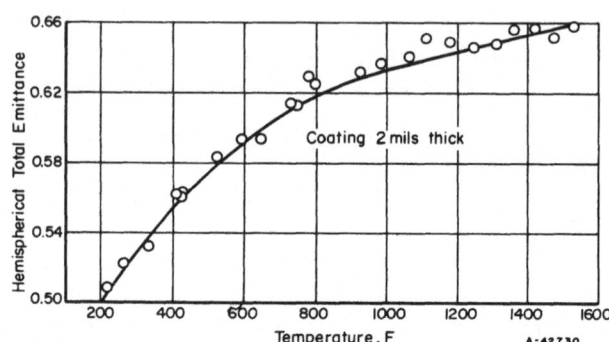

HEMISPHERICAL TOTAL EMITTANCE OF MOLYBDENUM DISILICIDE ON MOLYBDENUM

HEMISPHERICAL TOTAL EMITTANCE OF MOLYBDENUM DISILICIDE ON MOLYBDENUM--REFERENCE INFORMATION

Reference	Investigator	Symbol	Composition and Surface Condition.	Test Method	Remarks
15	Pratt and Whitney Aircraft		2-mil-thick coating of' $MoSi_2$ applied by the Linde Plasmarc process to both sides of a molybdenum strip.	Hemispherical total emittance. Resistance-heated strip specimen. Power dissipated in measured area. Temperatures measured with thermocouples.	Measured in vacuum. Data taken from curve.

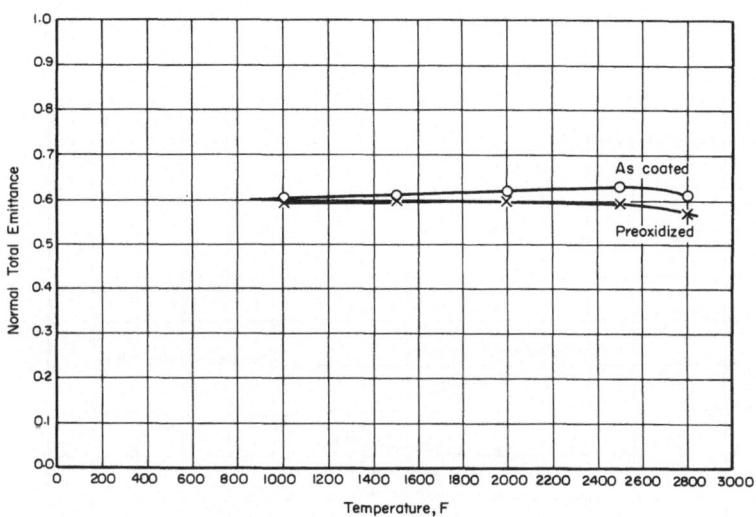

NORMAL TOTAL EMITTANCE OF W-2 ON MOLYBDENUM-0.5 TITANIUM ALLOY

NORMAL TOTAL EMITTANCE OF W-2 ON MOLYBDENUM-0.5 TITANIUM ALLOY--REFERENCE INFORMATION

Reference	Investigator	Symbol	Composition and Surface Condition	Test Method	Remarks
8	Fieldhouse, Lang, and Blau		Chromalloy W-2 coating on molybdenum-0.5 per cent titanium alloy.	Normal and angular total emittance. Induction heated specimen.	Measured in 90 per cent argon, 10 per cent H_2 atmosphere. Measurements made
		o	As coated.	Spectrometer with prism replaced	at angles of 0, 30,
		x	Preoxidized, at 2000 F for 1 hour. Coating thickness not given. Specimen "flat and smooth". (W-2 coating thought to be molybdenum disilicide.)	by plane mirror. Thermocouple detector. Blackbody hole in specimen. Temperature calibration with blackbody and optical pyrometer.	45, and 60 degrees with the normal to the specimen surface. Normal total emittance equals hemispherical total emittance within reported experimental error of ± 5 per cent. Data taken from curves.

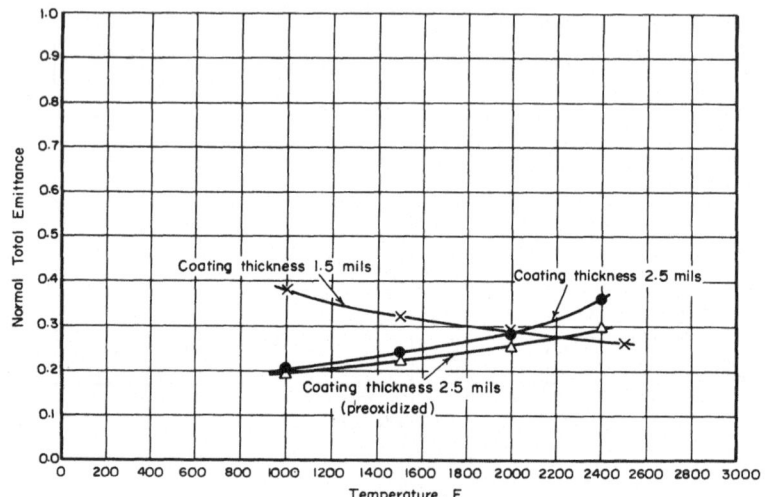

NORMAL TOTAL EMITTANCE OF W-2 ON MOLYBDENUM-0.5 TITANIUM ALLOY

NORMAL TOTAL EMITTANCE OF W-2 ON MOLYBDENUM-0.5 TITANIUM ALLOY--REFERENCE INFORMATION

Reference	Investigator	Symbol	Composition and Surface Condition	Test Method	Remarks
7	Anthony and Pearl		Molybdenum-0.5 per cent titanium alloy coated with W-2.	Normal total emittance. Induction-heated specimen.	Measured in continuous purge of helium gas.
		Δ	Coating thickness 1.5 mils.	Thermopile detector.	
		●	Coating thickness 2.5 mils.	Comparison blackbody.	
		×	Preoxidized, then 2.5-mil thick coating applied. (Alloy not defined.) (W-2 coating thought to be molybdenum disilicide.)	Temperatures measured with thermocouples and optical pyrometer.	

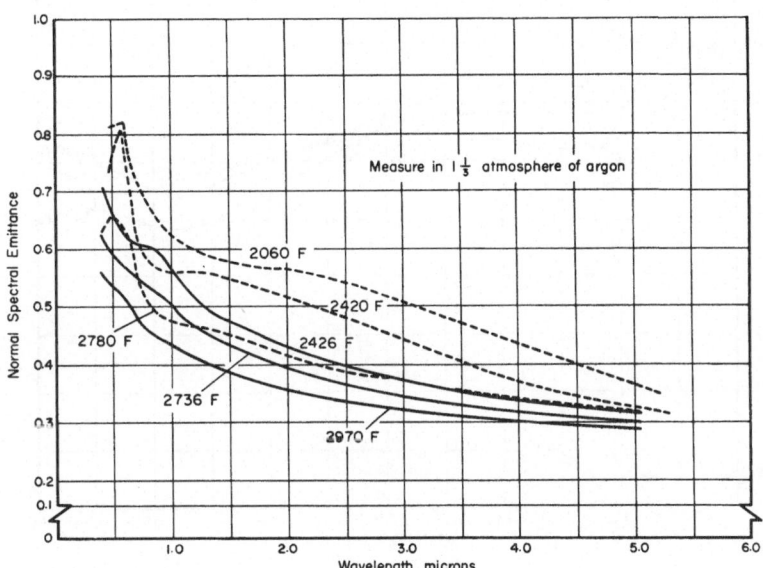

NORMAL SPECTRAL EMITTANCE OF W-2 ON MOLYBDENUM-0.5 TITANIUM ALLOY

NORMAL SPECTRAL EMITTANCE OF W-2 ON MOLYBDENUM-0.5 TITANIUM ALLOY--REFERENCE INFORMATION

Reference	Investigator	Symbol	Composition and Surface Condition	Test Method	Remarks
9	Coffman, Kibler, and Riethof		Surface condition: as received. Coating thickness not given. Specimen No. 4 - measured at 2426, 2736, and 2970 F. Specimen No. 5 - measured at 2060, 2420, and 2780 F. (W-2 coating thought to be molybdenum disilicide.)	Normal spectral emittance. Induction-heated specimen. Spectrometer-mono-chromator. Comparison blackbody.	Measured in 1-1/3 atmosphere of argon. Results not repro-ducible at lower temperatures after heating to higher temperature.

372

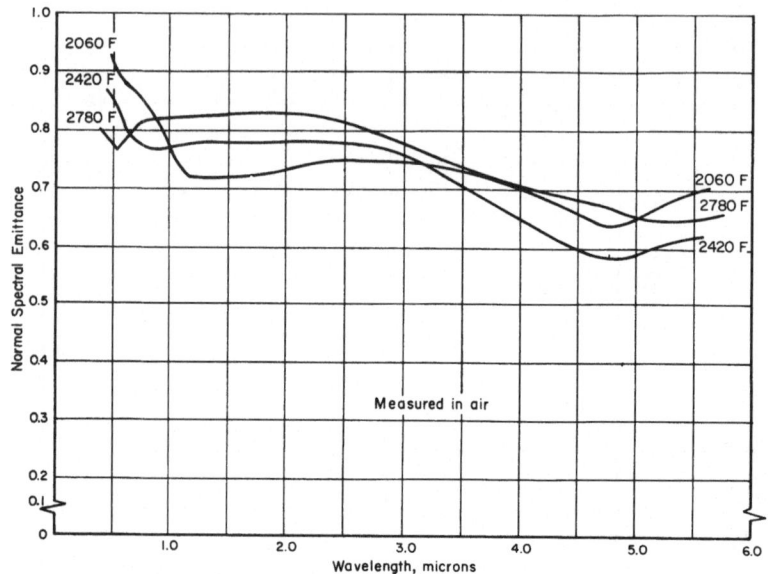

NORMAL SPECTRAL EMITTANCE OF W-2 ON MOLYBDENUM-0.5 TITANIUM ALLOY

NORMAL SPECTRAL EMITTANCE OF W-2 ON MOLYBDENUM-0.5 TITANIUM ALLOY--REFERENCE INFORMATION

Reference	Investigator	Symbol	Composition and Surface Condition	Test Method	Remarks
9	Coffman, Kibler, and Riethof		Surface condition--as received. Coating thickness not given. (W-2 coating thought to be molybdenum disilicide.)	Normal spectral emittance. Induction-heated specimens. Spectrometer-monochromator. Comparison blackbody.	Measured in air. Data taken from curves.

Note: Specimens run at 2060 F, held for 2 hours and rerun; run at 2420 F, held for 2 hours and rerun; run at 2780 F, held for 2 hours and rerun. First run at each temperature is shown. The final run at 2780 F showed no further change. |

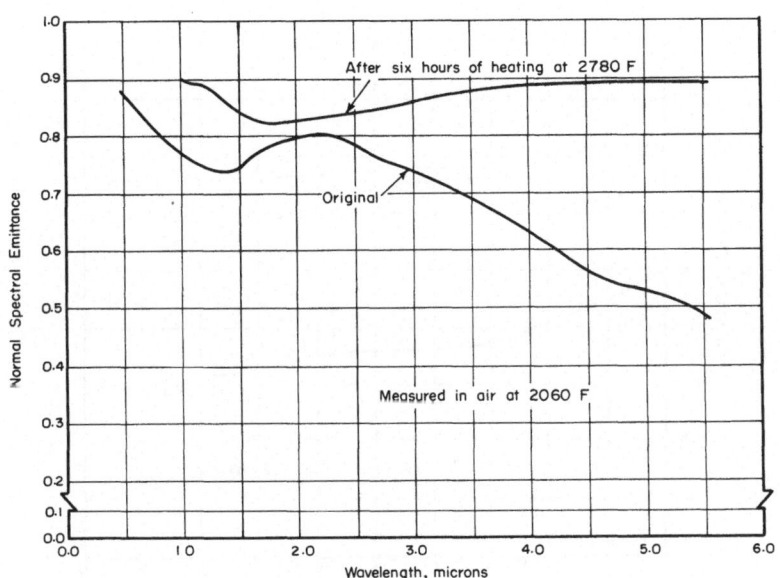

NORMAL SPECTRAL EMITTANCE OF W-2 ON MOLYBDENUM-0.5 TITANIUM ALLOY

NORMAL SPECTRAL EMITTANCE OF W-2 ON MOLYBDENUM-0.5 TITANIUM ALLOY--REFERENCE INFORMATION

Reference	Investigator	Symbol	Composition and Surface Condition	Test Method	Remarks
9	Coffman, Kibler, and Riethof		Surface condition--as received. Coating thickness not given. (W-2 coating thought to be molybdenum disilicide.)	Normal spectral emittance. Inductively heated specimen. Spectrometer-mono-chromator. Comparison blackbody.	Measured in air. Variation with thermal treatment. Data taken from curves.

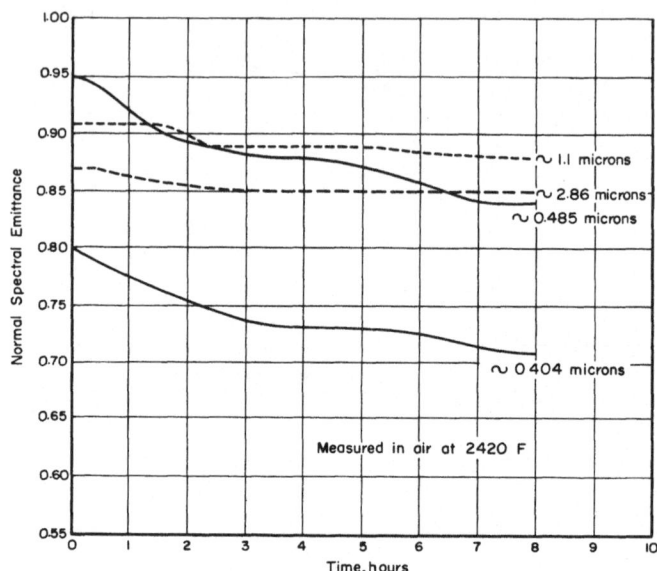

VARIATION OF NORMAL SPECTRAL EMITTANCE OF W-2 ON MOLYBDENUM-0.5 TITANIUM ALLOY WITH HEATING TIME IN AIR

VARIATION OF NORMAL SPECTRAL EMITTANCE OF W-2 ON MOLYBDENUM-0.5 TITANIUM
ALLOY WITH HEATING TIME IN AIR--REFERENCE INFORMATION

Reference	Investigator	Symbol	Composition and Surface Condition	Test Method	Remarks
9	Coffman, Kibler, and Riethof		Surface condition--as received. Coating thickness not given. (W-2 coating thought to be molybdenum disilicide.)	Normal spectral emittance. Inductively heated specimen. Spectrometer-monochromator. Comparison blackbody.	Measured in air at 2420 F. Data taken from curves. Measured at wavelengths of 0.404, 0.485, 1.1 and 2.86 microns.

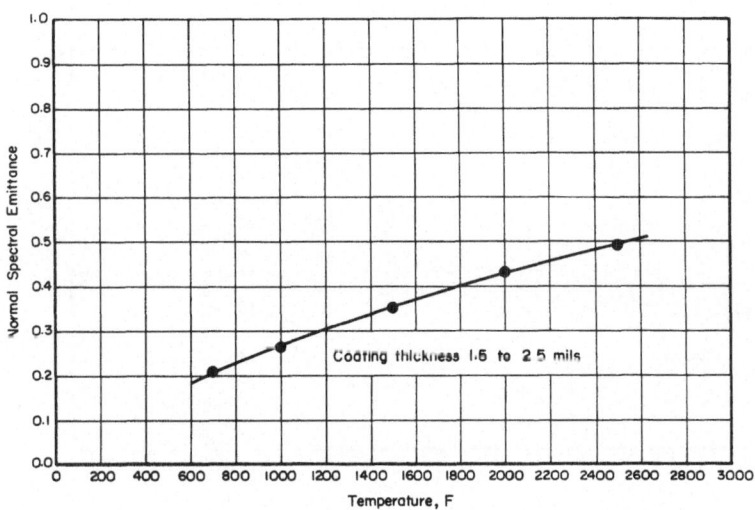

NORMAL TOTAL EMITTANCE OF MODIFIED W-2 ON COLUMBIUM ALLOY

NORMAL TOTAL EMITTANCE OF MODIFIED W-2 ON COLUMBIUM ALLOY--REFERENCE INFORMATION

Reference	Investigator	Symbol	Composition and Surface Condition	Test Method	Remarks
7	Anthony and Pearl	●	Columbium-10Ti - 10Mo. As received. Coating: T-1 modified W-2, 1.5 to 2.5 mils nominal thickness.	Normal total emittance. Induction-heated speci- men. Thermopile detector. Comparison blackbody. Temperatures measured with thermocouples and optical pyrometer.	Measured in continuous purge of helium gas.

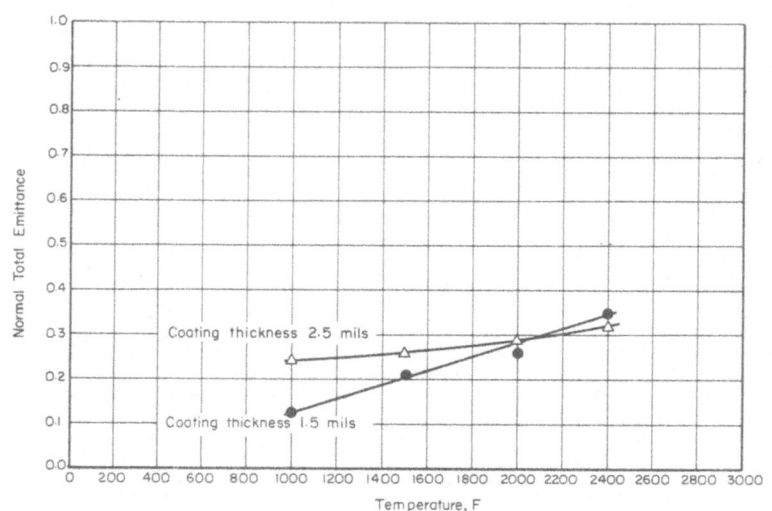

NORMAL TOTAL EMITTANCE OF MODIFIED W-2 ON TUNGSTEN

NORMAL TOTAL EMITTANCE OF MODIFIED W-2 ON TUNGSTEN--REFERENCE INFORMATION

Reference	Investigator	Symbol	Composition and Surface Condition	Test Method	Remarks
7	Anthony and Pearl	● Δ	Modified W-2 coating on tungsten. Coating thickness 1.5 mils. Coating thickness 2.5 mils.	Normal total emittance. Induction-heated speci- men. Thermopile detector. Comparison blackbody. Temperatures measured with thermocouples and optical pyrometer.	Measured in continuous purge of helium gas.

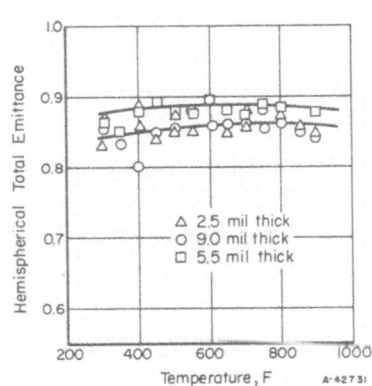

HEMISPHERICAL TOTAL EMITTANCE OF ALUMINUM PHOSPHATE BONDED COATING ON ALUMINUM

HEMISPHERICAL TOTAL EMITTANCE OF ALUMINUM PHOSPHATE BONDED COATING ON ALUMINUM--REFERENCE INFORMATION

Reference	Investigator	Symbol	Composition and Surface Condition	Test Method	Remarks
16	Pratt & Whitney Aircraft		Silicon carbide and silicon dioxide filler. 2.5 mils thick 9.0 mils thick 5.5 mils thick	Hemispherical total emittance. Resistance-heated strip specimen. Power dissipated in measured area. Temperatures measured with thermocouples.	Measured in vacuum. Data taken from curves.

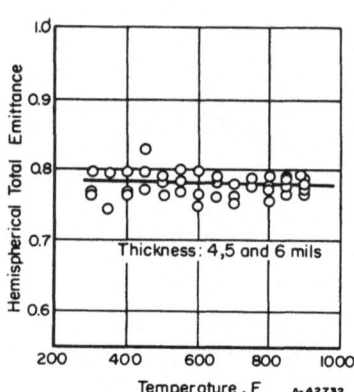

HEMISPHERICAL TOTAL EMITTANCE OF ALUMINUM PHOSPHATE BONDED COATING ON ALUMINUM

HEMISPHERICAL TOTAL EMITTANCE OF ALUMINUM PHOSPHATE COATING ON ALUMINUM--REFERENCE INFORMATION

Reference	Investigator	Symbol	Composition and Surface Condition	Test Method	Remarks
16	Pratt & Whitney Aircraft		Boron and silicon dioxide filler applied to aluminum strip.	Hemispherical total emittance. Resistance-heated strip specimen. Power dissipated in measured area. Temperatures measured with thermocouples.	Measured in vacuum. Data taken from curves.

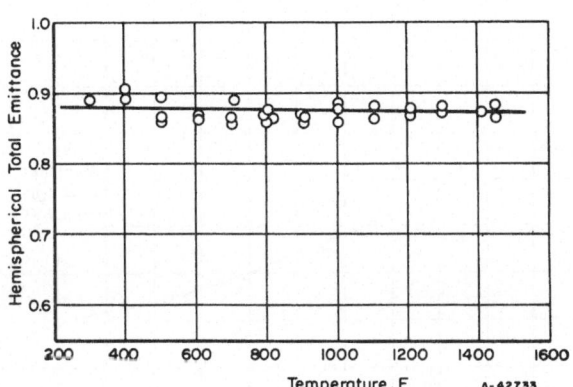

HEMISPHERICAL TOTAL EMITTANCE OF ALUMINUM PHOSPHATE COATING ON TYPE 310 STAINLESS STEEL

HEMISPHERICAL TOTAL EMITTANCE OF ALUMINUM PHOSPHATE COATING ON TYPE 310 STAINLESS STEEL—REFERENCE INFORMATION

Reference	Investigator	Symbol	Composition and Surface Condition	Test Method	Remarks
17	Pratt & Whitney Aircraft		Aluminum phosphate with nickel chrome spinel and silicon dioxide filler. Coated both sides.	Hemispherical total emittance. Resistance-heated strip and tube specimens. Power dissipated in measured area. Temperatures measured with thermocouples.	Measured in vacuum. Data taken from curve.

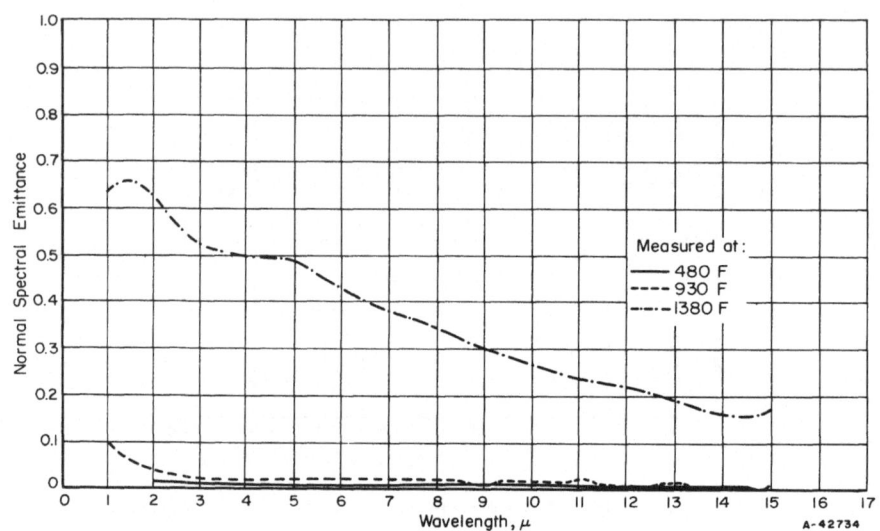

NORMAL SPECTRAL EMITTANCE OF CERAMIC GOLD ON TITANIUM (SHINY FINISH)

NORMAL SPECTRAL EMITTANCE OF CERAMIC GOLD ON TITANIUM--REFERENCE INFORMATION

Reference	Investigator	Symbol	Composition and Surface Condition	Test Method	Remarks
14	Adams, J. G.		As received - shiny finish. Engelhard Industries Bright Gold No. 6854. Applied by spray and fired at 600 C for 5 minutes. Measured at: 480 F 930 F 1380 F	Normal spectral emittance. Furnace-heated disk specimen. Comparison blackbody (Hohlraun). Spectrometer-mono-chromator with photo-multiplier, lead sulphide, and thermo-couple detectors. Temperatures measured with thermocouples.	Measured in air.

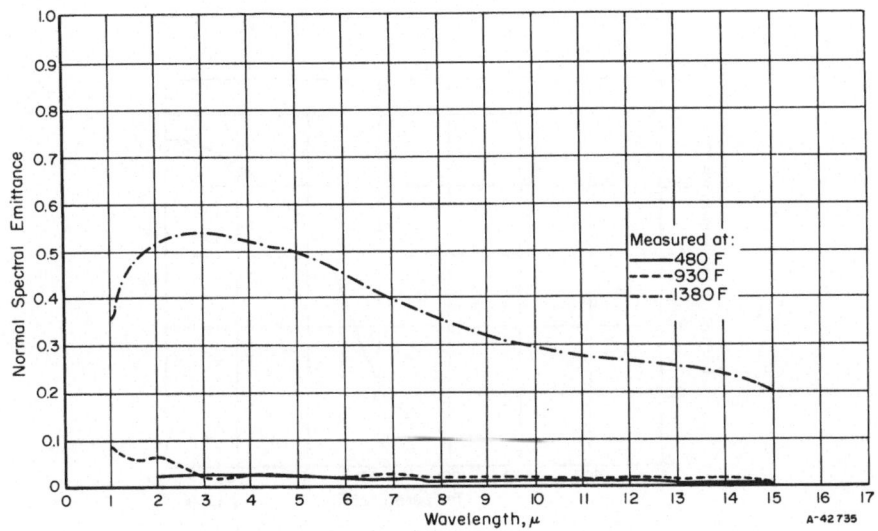

NORMAL SPECTRAL EMITTANCE OF CERAMIC GOLD ON TITANIUM (MATTE FINISH)

NORMAL SPECTRAL EMITTANCE OF CERAMIC GOLD ON TITANIUM (MATTE FINISH)--REFERENCE INFORMATION

Reference	Investigator	Symbol	Composition and Surface Condition	Test Method	Remarks
14	Adams, J. G.		As received - matte finish. Engelhard Industries Bright Gold No. 6854. Applied by spray and fired at 600 C for 5 minutes. Measured at: 480 F 930 F 1380 F	Normal spectral emittance. Furnace-heated disk specimen. Comparison blackbody (Hohlraun). Spectrometer-mono-chromator with photo-multiplier, lead sulphide, and thermo-couple detectors. Temperatures measured with thermocouples.	Measured in air.

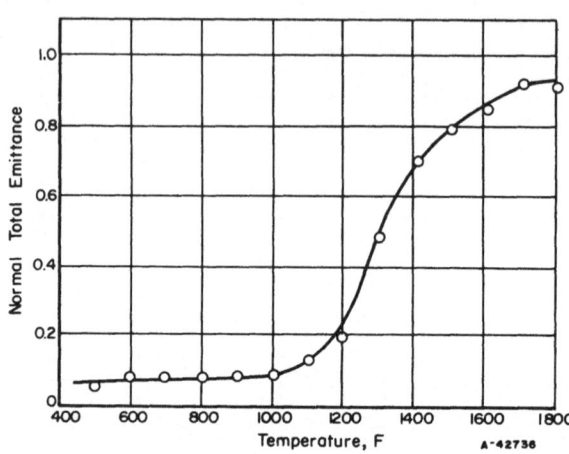

NORMAL TOTAL EMITTANCE OF HANOVIA LIQUID GOLD NO. 6896 ON A-286 STEEL

NORMAL TOTAL EMITTANCE OF HANOVIA LIQUID GOLD NO. 6896 ON A-286 STEEL--REFERENCE INFORMATION

Reference	Investigator	Symbol	Composition and Surface Condition	Test Method	Remarks
13	Gravina and Katz		Hanovia Liquid Gold No. 6896, resinous gold compound dissolved in essential oils. Coating thickness not given.	Normal total emittance. Resistance-heated strip specimen. Thermistor-bolometer detector. Reference blackbody. Temperatures measured with thermocouples.	Measured in air. Data taken from curves.

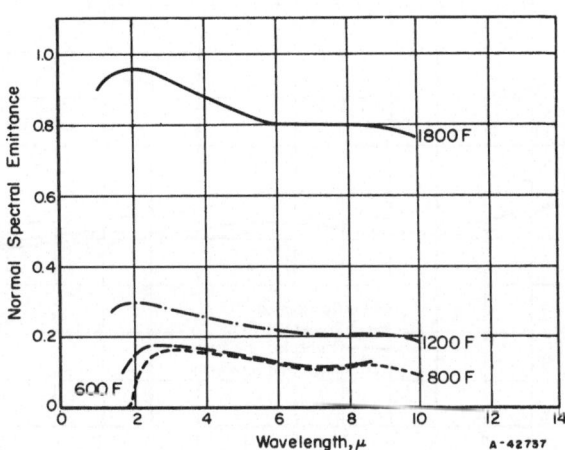

NORMAL SPECTRAL EMITTANCE OF HANOVIA LIQUID GOLD NO. 6896 ON A-286 STEEL AT 600, 800, 1200 AND 1800

NORMAL SPECTRAL EMITTANCE OF HANOVIA LIQUID GOLD NO. 6896 ON A-286 STEEL--REFERENCE INFORMATION

Reference	Investigator	Symbol	Composition and Surface Condition	Test Method	Remarks
13	Gravina and Katz		Hanovia Liquid Bright Gold No. 6896, a resinous gold compound dissolved in essential oils. Gold content 8 to 20 per cent. Coating thickness not given. Measured at: 600 F 800 F 1200 F 1800 F	Normal spectral emittance. Resistance—heated strip specimen. Thermistor—bolometer detector. Monochromator. Reference blackbody. Temperatures measured with thermocouples.	Measured in air. Data taken from curves.

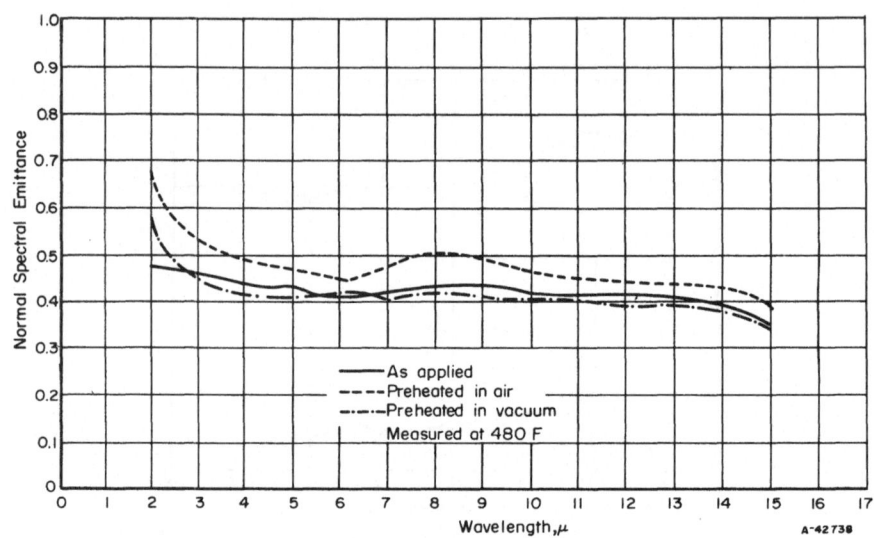

NORMAL SPECTRAL EMITTANCE OF CHROMIUM NICKEL ON INCONEL X AT 480 F

NORMAL SPECTRAL EMITTANCE OF CHROMIUM-NICKEL ON INCONEL X AT 480 F--REFERENCE INFORMATION

Reference	Investigator	Symbol	Composition and Surface Condition	Test Method	Remarks
14	Adams, J. G.		20 per cent chromium – 80 per cent nickel. Flame sprayed on Inconel X. As applied – untreated Heated 30 minutes in air at 1500 F Heated 30 minutes in 6.8×10^{-5} mm Hg pressure at 1500 F	Normal spectral emittance. Furnace-heated disk specimen. Comparison blackbody (Hohlraun). Spectrometer-monochromator with photo-multiplier, lead sulphide, and thermo-couple detectors. Temperatures measured with thermocouples.	Measured in air.

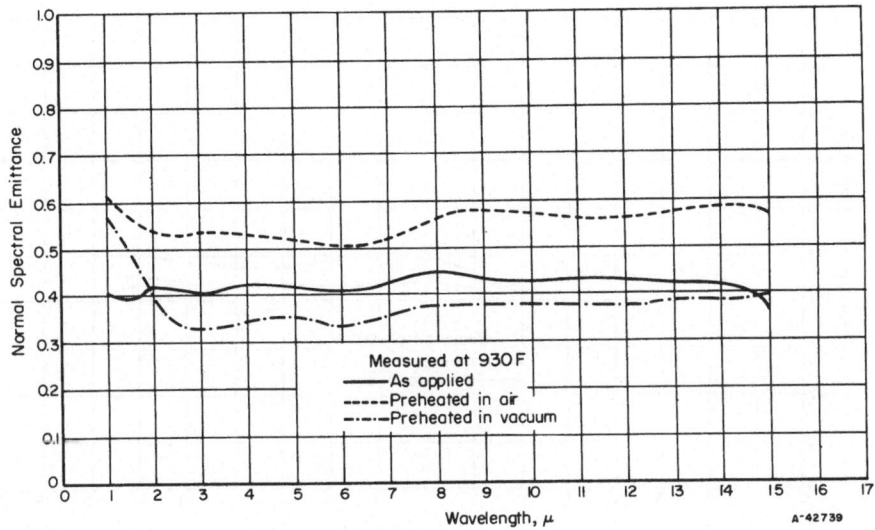

NORMAL SPECTRAL EMITTANCE OF CHROMIUM-NICKEL AT 930 F

NORMAL SPECTRAL EMITTANCE OF CHROMIUM-NICKEL ON INCONEL X AT 930 F—REFERENCE INFORMATION

Reference	Investigator	Symbol	Composition and Surface Condition	Test Method	Remarks
14	Adams, J. G.		20 per cent chromium – 80 per cent nickel. Flame sprayed on Inconel X. As applied – untreated Heated 30 minutes in air at 1500 F Heated 30 minutes in 6.8 x 10⁻⁵ mm Hg pressure at 1500 F	Normal spectral emittance. Furnace-heated disk specimen. Comparison blackbody (Hohlraun). Spectrometer-mono- chromator with photo- multiplier, lead sulphide, and thermo- couple detectors. Temperatures measured with thermocouples.	Measured in air.

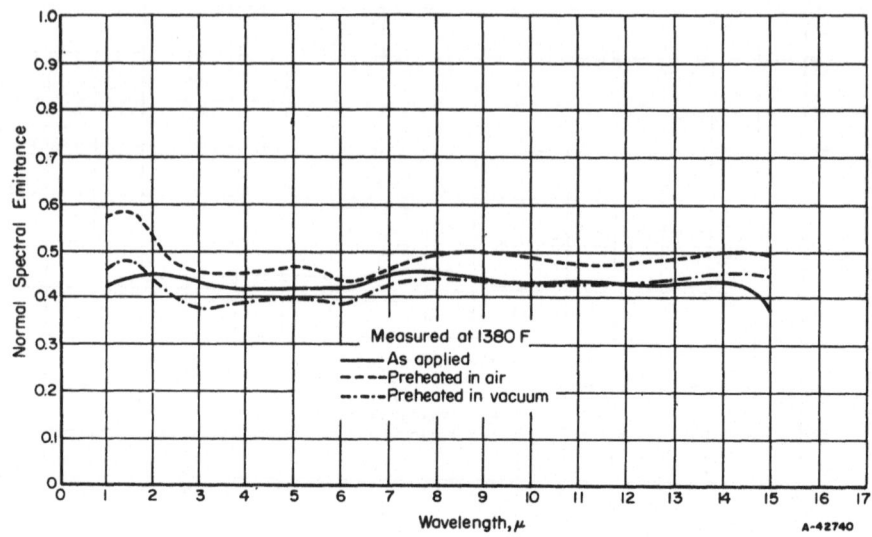

NORMAL SPECTRAL EMITTANCE OF CHROMIUM-NICKEL ON INCONEL X AT 1380 F

NORMAL SPECTRAL EMITTANCE OF CHROMIUM-NICKEL ON INCONEL X AT 1380 F--REFERENCE INFORMATION

Reference	Investigator	Symbol	Composition and Surface Condition	Test Method	Remarks
14	Adams, J. G.		20 per cent chromium - 80 per cent nickel. Flame sprayed on Inconel X. As applied - untreated Heated 30 minutes in air at 1500 F Heated 30 minutes in 6.8 x 10⁻⁵ mm Hg pressure at 1500 F	Normal spectral emittance. Furnace-heated disk specimen. Comparison blackbody (Hohlraun). Spectrometer-mono-chromator with photo-multiplier, lead sulphide, and thermo-couple detectors. Temperatures measured with thermocouples.	Measured in air.

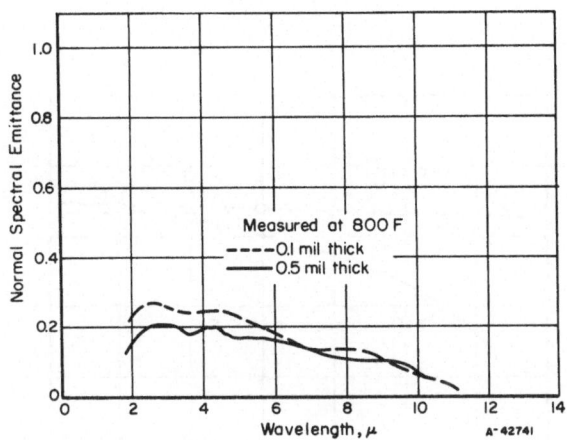

NORMAL SPECTRAL EMITTANCE OF KANIGEN NICKEL COATING ON A-286 STEEL

NORMAL SPECTRAL EMITTANCE OF KANIGEN NICKEL COATING ON A-286 STEEL--REFERENCE INFORMATION

Reference	Investigator	Symbol	Composition and Surface Condition	Test Method	Remarks
13	Gravina and Katz		Chemically deposited nickel alloy. Composition given below. Coating thickness: 0.1 mil 0.5 mil	Normal spectral emittance. Resistance-heated strip specimen. Thermistor-bolometer detector. Monochromator. Reference blackbody. Temperatures measured with thermocouples.	Measured in air. Data taken from curves.

Composition, per cent

Ni 90-92
P   8-10
C   .0400
O_2 .0023
N_2 .0047
H_2 .0016

Trace impurities of:
  Co, Al, Cu, Mn, Fe, Pb, and Si.

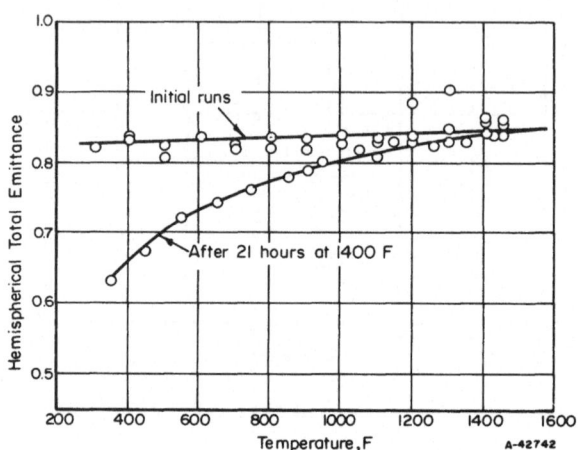

HEMISPHERICAL TOTAL EMITTANCE OF SINTERED NICKEL "C" ON TYPE 310 STAINLESS STEEL

HEMISPHERICAL TOTAL EMITTANCE OF SINTERED NICKEL "C" ON TYPE 310 STAINLESS STEEL--REFERENCE INFORMATION

Reference	Investigator	Symbol	Composition and Surface Condition	Test Method	Remarks
17	Pratt & Whitney Aircraft		Sintered Nickel "C", lithiated and oxidized. Nickel "C" slurry sprayed on Type 310 stainless steel, sintered in H_2, lithiated, and oxidized. Initial runs After 21 hours at 1450 F	Hemispherical total emittance. Resistance-heated strip specimen. Power dissipated in measured area. Temperatures measured with thermocouples.	Measured in vacuum. Data taken from curves.

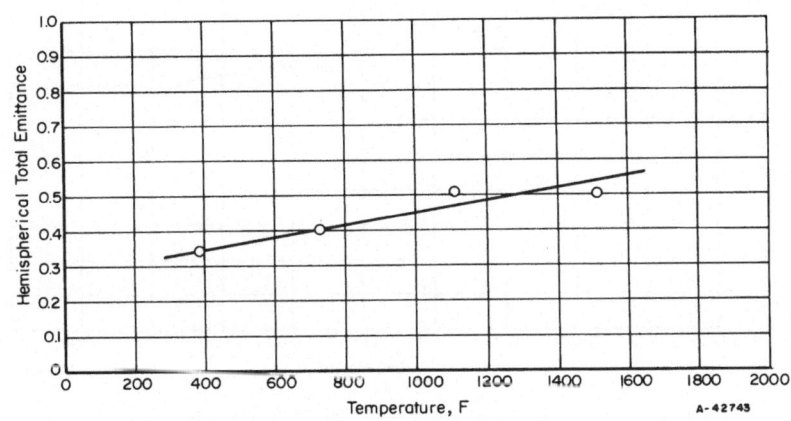

HEMISPHERICAL TOTAL EMITTANCE OF TUNGSTEN ON ARMCO IRON

HEMISPHERICAL TOTAL EMITTANCE OF TUNGSTEN ON ARMCO IRON--REFERENCE INFORMATION

Reference	Investigator	Symbol	Composition and Surface Condition	Test Method	Remarks
6	Butler, Jenkins, Rudkin, and Laughridge		Metco XP-1106 crystalline tungsten (-200 mesh + 30 micron) plasma flame sprayed on Armco iron. Surface uniformity judged by eye only. Coating thickness not given.	Hemispherical total emittance. Disk specimen. Temperature measured with thermocouples. Emittance calculated from mass, specific heat, and rate of change of temperature of the specimen.	Measured in vacuum. Data taken from curve.

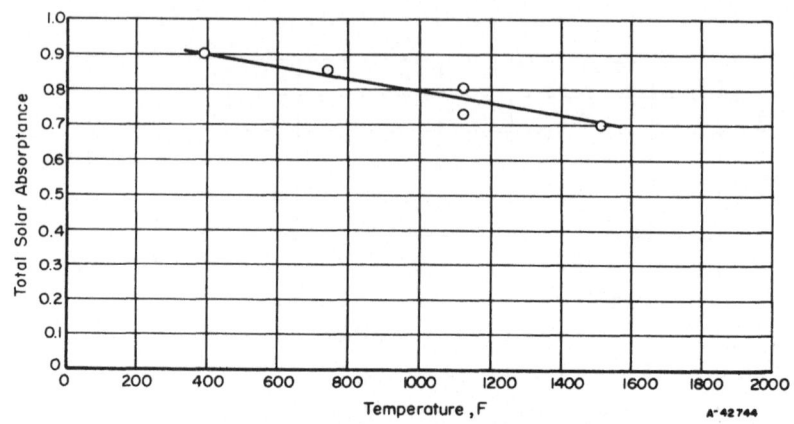

TOTAL SOLAR ABSORPTANCE OF TUNGSTEN ON ARMCO IRON

TOTAL SOLAR ABSORPTANCE OF TUNGSTEN ON ARMCO IRON--REFERENCE INFORMATION

Reference	Investigator	Symbol	Composition and Surface Condition	Test Method	Remarks
6	Butler, Jenkins, Rudkin, and Laughridge		Metco XP-1106 crystalline tungsten (-200 mesh + 30 micron). Plasma flame sprayed on Armco iron. Surface uniformity judged by eye only. Coating thickness not given.	Total solar absorptance. Carbon-arc-image furnace. Disk specimen. Temperatures measured with thermocouples. Absorptance calculated from mass, specific heat, rate of change of temperature, and known irradiance of the surface. (Solar spectrum simulated by carbon arc)	Measured in vacuum. Data taken from curves.

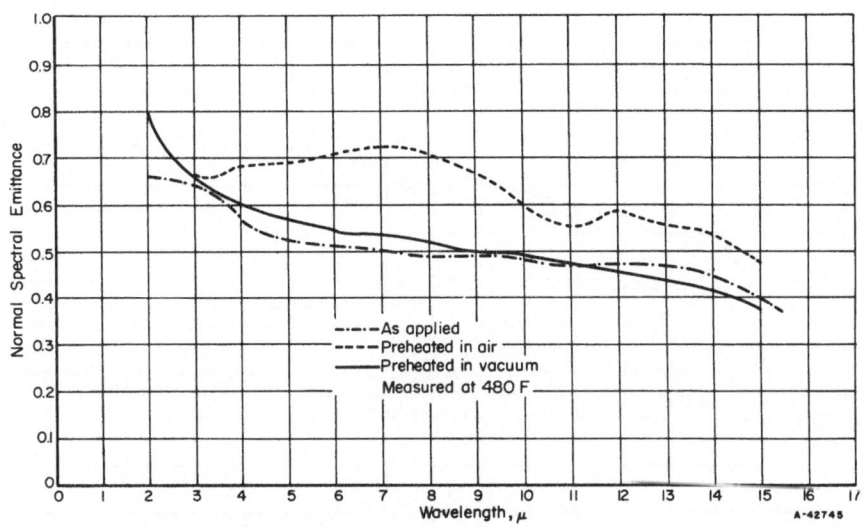

NORMAL SPECTRAL EMITTANCE OF TUNGSTEN ON INCONEL X AT 480 F

NORMAL SPECTRAL EMITTANCE OF TUNGSTEN ON INCONEL X AT 480 F--REFERENCE INFORMATION

Reference	Investigator	Symbol	Composition and Surface Condition	Test Method	Remarks
14	Adams, J. G.		Flame sprayed on Inconel X. Heated 30 minutes in 6.8×10^{-5} mm Hg pressure at 1500 F Heated 30 minutes in air at 1500 F As applied - untreated	Normal spectral emittance. Furnace-heated disk specimen. Comparison blackbody (Hohlraun). Spectrometer-monochromator with photomultiplier, lead sulphide, and thermocouple detectors. Temperatures measured with thermocouples.	Measured in air.

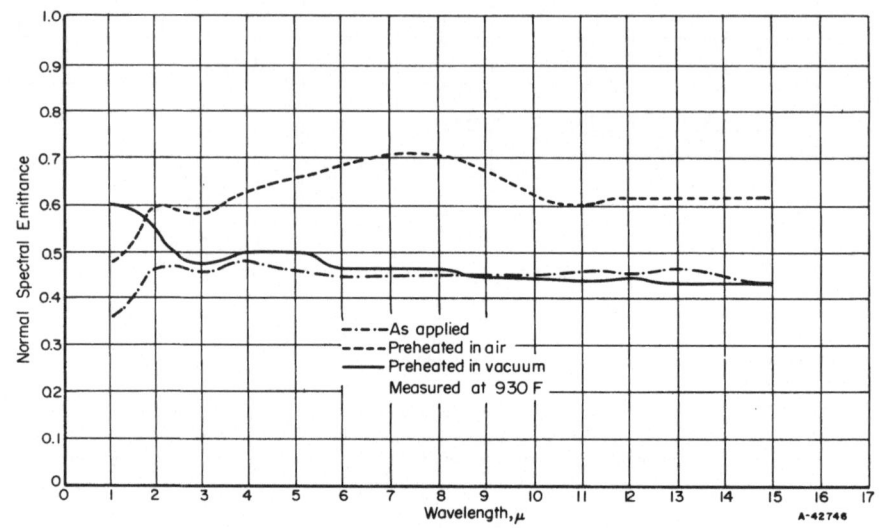

NORMAL SPECTRAL EMITTANCE OF TUNGSTEN ON INCONEL X AT 930 F

NORMAL SPECTRAL EMITTANCE OF TUNGSTEN ON INCONEL X AT 930 F--REFERENCE INFORMATION

Reference	Investigator	Symbol	Composition and Surface Condition	Test Method	Remarks
14	Adams, J. G.		Flame sprayed on Inconel X. Heated 30 minutes in air at 1500 F Heated 30 minutes in 6.8 x 10⁻⁵ mm Hg pressure at 1500 F As applied - untreated	Normal spectral emittance. Furnace-heated disk specimen. Comparison blackbody (Hohlraun). Spectrometer-monochromator with photomultiplier, lead sulphide, and thermocouple detectors. Temperatures measured with thermocouples.	Measured in air.

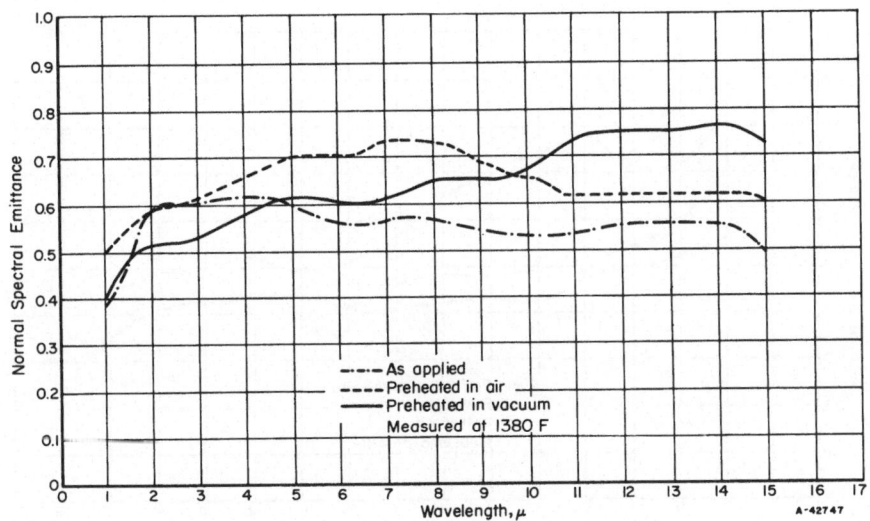

NORMAL SPECTRAL EMITTANCE OF TUNGSTEN ON INCONEL X AT 1380 F

NORMAL SPECTRAL EMITTANCE OF TUNGSTEN ON INCONEL X AT 1380 F--REFERENCE INFORMATION

Reference	Investigator	Symbol	Composition and Surface Condition	Test Method	Remarks
14	Adams, J. G.		Flame sprayed on Inconel X. As applied - untreated Heated 30 minutes in air at 1500 F Heated 30 minutes in 6.8 x 10⁻⁵ mm Hg pressure at 1500 F	Normal spectral emittance. Furnace-heated disk specimen. Comparison blackbody (Hohlraun). Spectrometer-mono-chromator with photo-multiplier, lead sulphide, and thermo-couple detectors. Temperatures measured with thermocouples.	Measured in air.

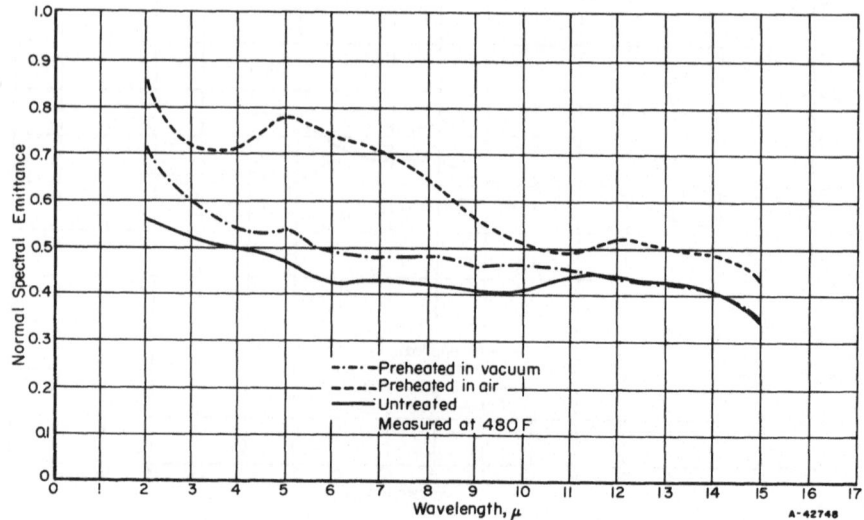

NORMAL SPECTRAL EMITTANCE OF TUNGSTEN-50 PER CENT COBALT ON INCONEL X AT 480 F

NORMAL SPECTRAL EMITTANCE OF TUNGSTEN - 50 PER CENT COBALT ON INCONEL X AT 480 F--REFERENCE INFORMATION

Reference	Investigator	Symbol	Composition and Surface Condition	Test Method	Remarks
14	Adams, J. G.		50 per cent tungsten - 50 per cent cobalt. Flame sprayed on Inconel X. Untreated Heated 30 minutes in air at 1500 F Heated 30 minutes in 6.8 x 10⁻5 mm Hg pressure at 1500 F	Normal spectral emittance. Furnace-heated disk specimen. Comparison blackbody (Hohlraun). Spectrometer-mono-chromator with photo-multiplier, lead sulphide, and thermo-couple detectors. Temperatures measured with thermocouples.	Measured in air.

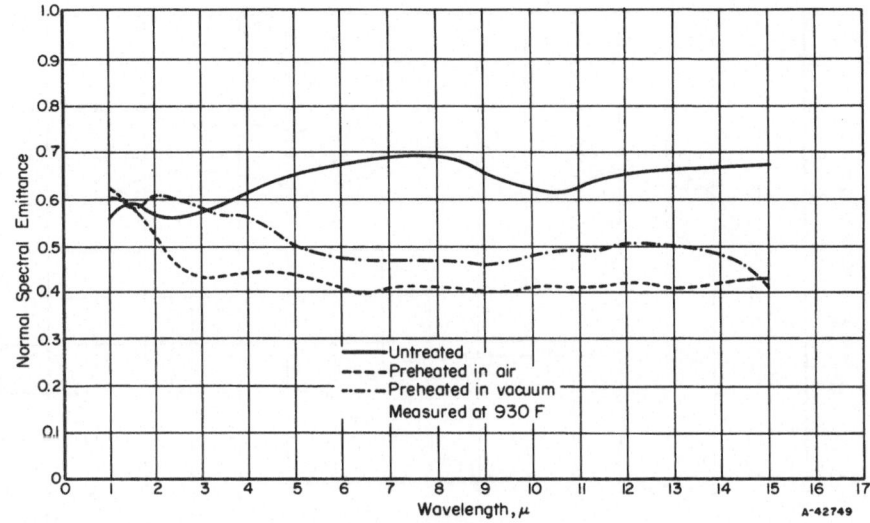

NORMAL SPECTRAL EMITTANCE OF TUNGSTEN-50% COBALT ON INCONEL X AT 930 F

NORMAL SPECTRAL EMITTANCE OF TUNGSTEN - 50 PER CENT COBALT ON INCONEL X AT 930 F—REFERENCE INFORMATION

Reference	Investigator	Symbol	Composition and Surface Condition	Test Method	Remarks
14	Adams, J. G.		50 per cent tungsten – 50 per cent cobalt. Flame sprayed on Inconel X. Untreated – as sprayed Heated 30 minutes in air at 1500 F Heated 30 minutes in 6.8×10^{-5} mm Hg pressure at 1500 F	Normal spectral emittance. Furnace-heated disk specimen. Comparison blackbody (Hohlraun). Spectrometer-mono-chromator with photo-multiplier, lead sulphide, and thermo-couple detectors. Temperatures measured with thermocouples.	Measured in air.

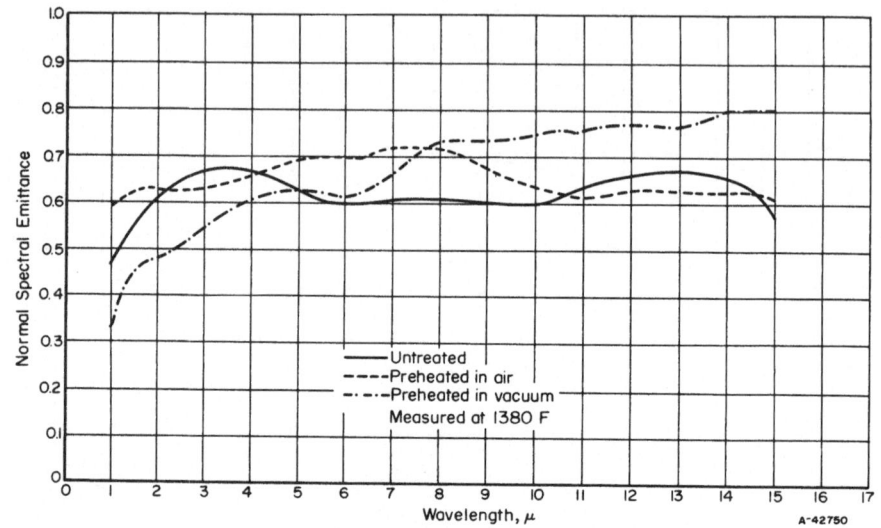

NORMAL SPECTRAL EMITTANCE OF TUNGSTEN-50% COBALT ON INCONEL X AT 1380 F

NORMAL SPECTRAL EMITTANCE OF TUNGSTEN – 50 PER CENT COBALT ON INCONEL X AT 1380 F--REFERENCE INFORMATION

Reference	Investigator	Symbol	Composition and Surface Condition	Test Method	Remarks
14	Adams, J. G.		50 per cent tungsten – 50 per cent cobalt. Flame sprayed on Inconel X. As sprayed – untreated Heated 30 minutes in air at 1500 F Heated 30 minutes in 6.8 x 10⁻⁵ mm Hg pressure at 1500 F	Normal spectral emittance. Furnace-heated disk specimen. Comparison blackbody (Hohlraun). Spectrometer-monochromator with photomultiplier, lead sulphide, and thermocouple detectors. Temperatures measured with thermocouples.	Measured in air.

NORMAL TOTAL EMITTANCE OF HIGH EMITTANCE COATINGS AT 1200 AND 1500 F

Material	Coating	Coating Thickness, mils	Normal Total Emittance	
			1200 F	1500 F
25-52 base with 200-mesh overspray coatings of:	Ferrosilicon	2	0.93	0.95
	Chrome oxide	1 1/2	0.95	0.95
	Mild steel scale	2 1/2	0.97	0.98
	Chromite ore No. 1	1 1/2	0.86	0.88
	Chromite ore No. 2 (high Cr_2O_3)	1 1/2	0.85	0.89
	Manganese dioxide	3/4	0.92	0.94
	Iron manganate spinel	1 1/2	0.88	0.95
	Nickel oxide	1 1/2	0.92	0.95
25-52 base with 200-mesh Chromite No. 1 as a blend containing:	20% Chromite No. 1	1	0.91	0.93
	30% Chromite No. 1	2	0.93	0.95
	30% Chromite No. 1	1 1/4	0.97	0.96
	40% Chromite No. 1	3	0.95	0.96
	40% Chromite No. 1	1 1/4	0.95	0.98
	50% Chromite No. 1	1 3/4	0.93	0.93
Overspray of nickel oxide on mild steel 25-52 base with overspray of:	Minus 100 and 200-mesh Chromite No. 1	2 1/2	0.85	0.87
	Minus 200 and 325-mesh Chromite No. 1	2 1/2	0.89	0.91
	Minus 325 mesh Chromite No. 1	2 1/2	0.92	0.93
25-52 base with 325-mesh:	Chromite No. 1 overspray	1/2	0.90	0.94
	Chromite No. 1 overspray	3/4	0.96	0.96
	Chromite No. 1 overspray	1 1/2	0.86	0.88
	Chromite No. 1 overspray	2	0.73	0.79

NORMAL TOTAL EMITTANCE OF HIGH EMITTANCE COATINGS--REFERENCE INFORMATION

Reference	Investigator	Symbol	Composition and Surface Condition	Test Method	Remarks
5	Douglass, E. A.			Normal total emittance. Total radiation pyrometer. Coatings on rotating steel cylinder containing blackbody hole. Temperatures measured with thermocouples.	Measured in air. Data taken from table.

NORMAL TOTAL EMITTANCE OF VARIOUS COATINGS AT 1200 AND 1500 F

	Thickness, mils				Normal Total Emittance	
Top-Coat Oxide	Base[1]	Cover[2]	Top	Total	1200 F	1500 F
*Feldspar	5.0	13.0	2.0	20.0	0.32	0.27
*Treopax	5.0	11.0	3.0	19.0	0.27	0.23
*Quartz	4.0	15.0	2.0	21.0	0.49	0.34
*Zirconium spinel	4.5	13.0	3.0	20.5	0.23	0.22
*Alumina	5.0	12.5	2.0	19.5	0.42	0.35
Black Label clay	5.0	14.0	3.0	22.0	0.67	0.60
*Uverite	5.0	12.5	3.0	20.5	0.45	0.33
Zircon	4.0	15.0	2.0	21.0	0.61	0.52
Antimony oxide	5.0	11.0	4.0	20.0	0.62	0.57
Calcium carbonate	4.0	13.0	2.0	19.0	0.62	0.68
*Fused magnesia	5.0	11.5	2.0	18.5	0.57	0.63
*Zinc oxide	4.5	15.5	2.0	22.0	0.51	0.60
*Tin oxide	4.5	13.0	1.5	19.0	0.34	0.35
*Zirconia	5.0	13.0	1.5	19.5	0.40	0.34
*Diaspore clay	4.5	11.5	2.0	18.0	0.49	0.42
*Cerium oxide	5.0	13.0	2.0	20.0	0.35	0.37
*Calcium metaphosphate	4.5	13.5	2.0	20.0	0.42	0.65
*Vanadium pentoxide	4.5	13.5	2.0	20.0	0.74	0.68
Chrome oxide	4.0	13.0	1.5	18.5	0.79	0.79
XM-1	5.0	15.0	-	20.0	0.69	0.69

*Indicates oxides with 5 per cent water glass added as a binder.

NORMAL TOTAL EMITTANCE OF VARIOUS COATINGS--REFERENCE INFORMATION

Reference	Investigator	Symbol	Composition and Surface Condition	Test Method	Remarks
5	Douglass, E. A.			Normal total emittance. Total radiation pyrometer. Coatings on rotating steel cylinder containing blackbody hole. Temperatures measured with thermocouples.	Measured in air. Data taken from table.

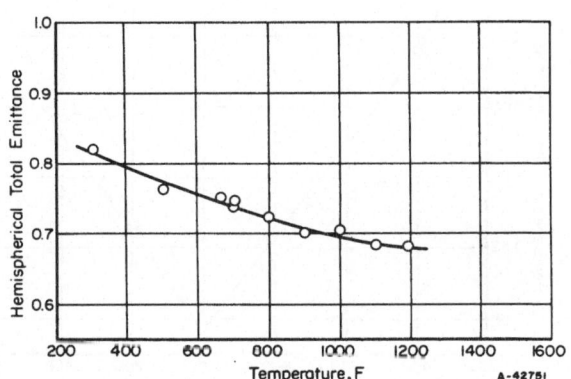

HEMISPHERICAL TOTAL EMITTANCE OF BORON NITRIDE ON TANTALUM

HEMISPHERICAL TOTAL EMITTANCE OF BORON NITRIDE ON TANTALUM--REFERENCE INFORMATION

Reference	Investigator	Symbol	Composition and Surface Condition	Test Method	Remarks
17	Pratt & Whitney Aircraft		Boron nitride with Synar binder. Coated on both sides of tantalum strip	Hemispherical total emittance. Resistance-heated strip specimen. Power dissipated in measured area. Temperatures measured with thermocouples.	Measured in vacuum. Data taken from curve.

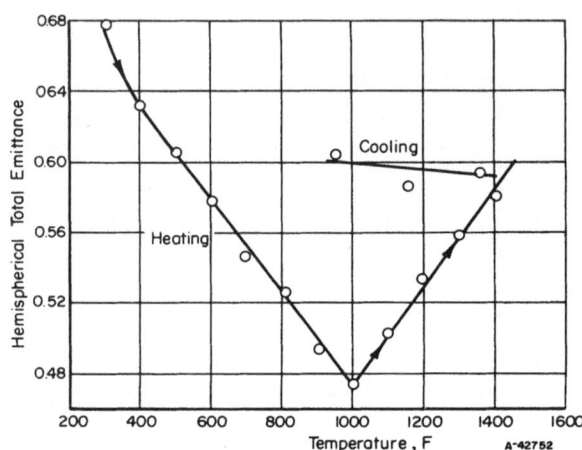

HEMISPHERICAL TOTAL EMITTANCE OF CALCIUM FLUORIDE ON TYPE 310 STAINLESS STEEL

HEMISPHERICAL TOTAL EMITTANCE OF CALCIUM FLUORIDE ON TYPE 310 STAINLESS STEEL--REFERENCE INFORMATION

Reference	Investigator	Symbol	Composition and Surface Condition	Test Method	Remarks
18	Pratt & Whitney Aircraft		A dispersion of calcium fluoride (Acheson Colloid Co. DAG EC 1789). Coated on both sides of a Type 310 stainless strip. Note: cooling change shown after 20 hours at 1450 F.	Hemispherical total emittance. Resistance-heated strip specimen. Power dissipated in measured area. Temperatures measured with thermocouples.	Measured in vacuum. Data taken from curves.

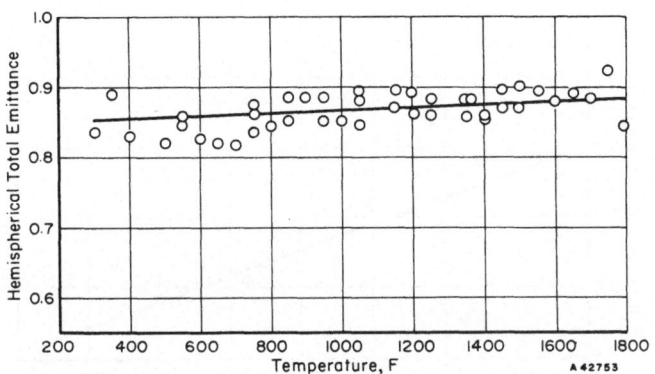

HEMISPHERICAL TOTAL EMITTANCE OF STRONTIUM TITANATE ON TYPE 310 STAINLESS STEEL

HEMISPHERICAL TOTAL EMITTANCE OF STRONTIUM TITANATE ON TYPE 310 STAINLESS STEEL--REFERENCE INFORMATION

Reference	Investigator	Symbol	Composition and Surface Condition	Test Method	Remarks
16	Pratt & Whitney Aircraft		Metco plasma flame spray powder.	Hemispherical total emittance. Resistance-heated strip specimen. Power dissipated in measured area. Temperatures measured with thermocouples.	Measured in vacuum. Data taken from curve.

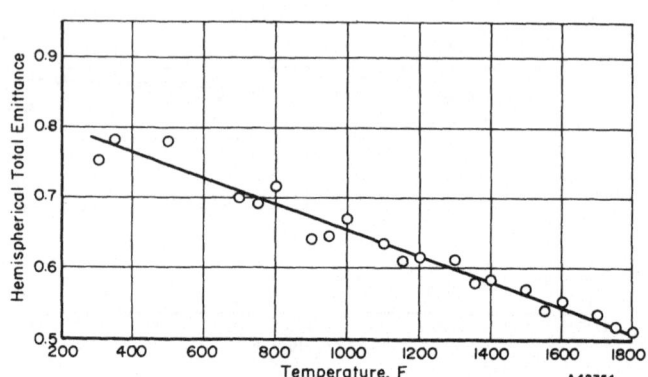

HEMISPHERICAL TOTAL EMITTANCE OF ZIRCONIUM SILICATE ON TYPE 310 STAINLESS STEEL

HEMISPHERICAL TOTAL EMITTANCE OF ZIRCONIUM SILICATE ON TYPE 310 STAINLESS STEEL--REFERENCE INFORMATION

Reference	Investigator	Symbol	Composition and Surface Condition	Test Method	Remarks
16	Pratt & Whitney Aircraft		Metco plasma flame spray powder XP-1116. Flame sprayed on Type 310 stainless steel strip.	Hemispherical total emittance. Resistance-heated strip specimen. Power dissipated in measured area. Temperatures measured with thermocouples.	Measured in vacuum. Data taken from curve.

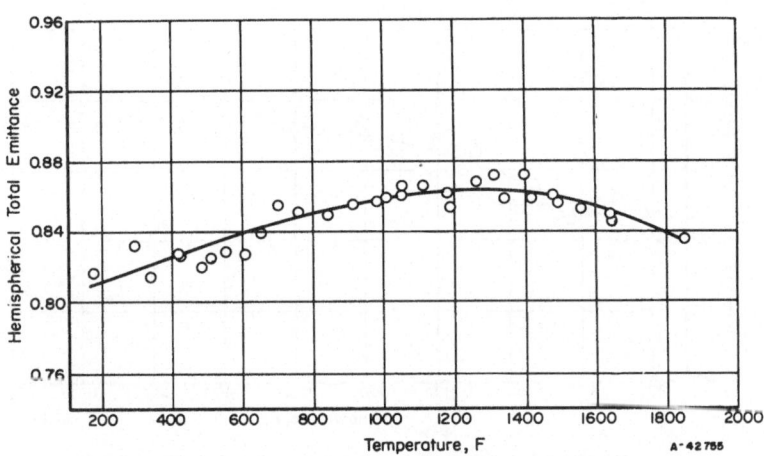

HEMISPHERICAL TOTAL EMITTANCE OF ACETYLENE BLACK AND XYLOL ON TYPE 310 STAINLESS STEEL

HEMISPHERICAL TOTAL EMITTANCE OF ACETYLENE BLACK AND XYLOL ON TYPE 310 STAINLESS STEEL--REFERENCE INFORMATION

Reference	Investigator	Symbol	Composition and Surface Condition	Test Method	Remarks
15	Pratt & Whitney Aircraft		Colloidal suspension of acetylene black in xylol (Acheson Colloid Co. DAG EC 1652) sprayed on Type 310 stainless steel strip.	Hemispherical total emittance. Resistance-heated strip specimen. Power dissipated in measured area. Temperatures measured with thermocouples.	Measured in vacuum. Data taken from curve.

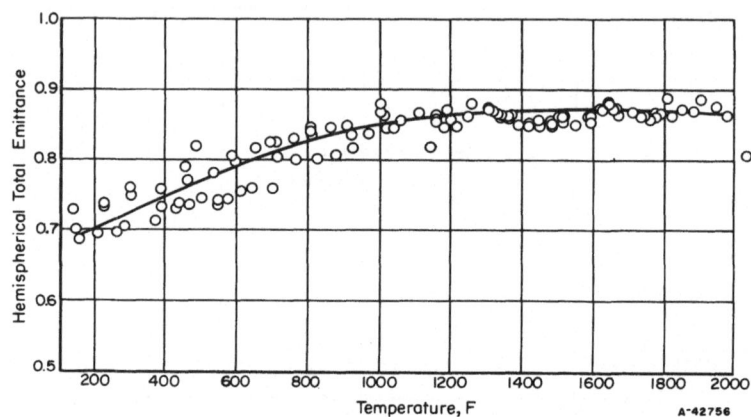

HEMISPHERICAL TOTAL EMITTANCE OF CHROMIUM BLACK ON TYPE 310 STAINLESS STEEL

HEMISPHERICAL TOTAL EMITTANCE OF CHROMIUM BLACK ON TYPE 310 STAINLESS STEEL—REFERENCE INFORMATION

Reference	Investigator	Symbol	Composition and Surface Condition	Test Method	Remarks
15	Pratt & Whitney Aircraft		Chromium black deposited by a variation of the Solvay process on Type 310 stainless steel strip.	Hemispherical total emittance. Resistance-heated strip specimen. Power dissipated in measured area. Temperatures measured with thermocouples.	Measured in vacuum. Data taken from curve.

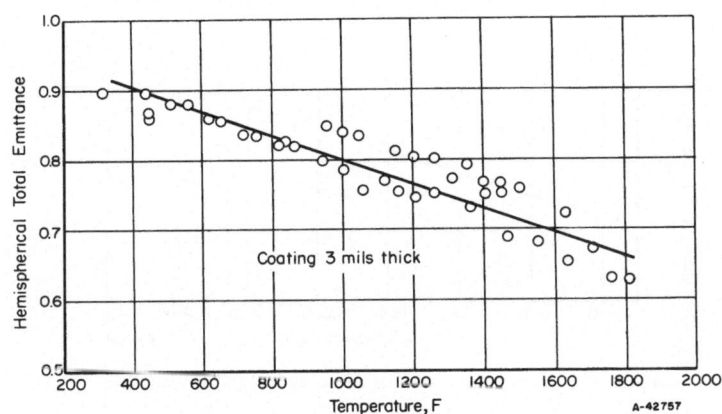

Coating 3 mils thick

A-42757

HEMISPHERICAL TOTAL EMITTANCE OF KRYLON BLACK ON TYPE 310 STAINLESS STEEL

HEMISPHERICAL TOTAL EMITTANCE OF KRYLON BLACK ON TYPE 310 STAINLESS STEEL—REFERENCE INFORMATION

Reference	Investigator	Symbol	Composition and Surface Condition	Test Method	Remarks
18	Pratt & Whitney Aircraft		Commercial Krylon Black, a mixture of carbon black and silicates in a lacquer carrier.	Hemispherical total emittance. Resistance-heated strip specimen. Power dissipated in measured area. Temperatures measured with thermocouples.	Measured in vacuum. Data taken from curve.

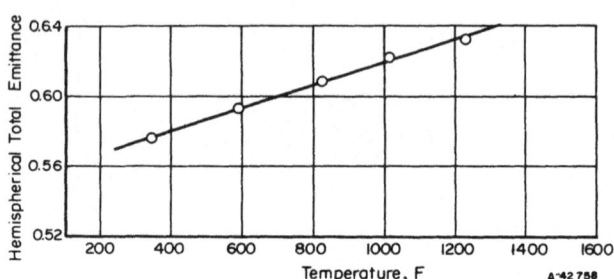

HEMISPHERICAL TOTAL EMITTANCE OF GRAPHITE VARNISH ON TYPE 310 STAINLESS STEEL

HEMISPHERICAL TOTAL EMITTANCE OF GRAPHITE VARNISH ON TYPE 310 STAINLESS STEEL--REFERENCE INFORMATION

Reference	Investigator	Symbol	Composition and Surface Condition	Test Method	Remarks
15	Pratt & Whitney Aircraft		Spray coated graphite varnish on Type 310 stainless steel. Note: coating flaked off near 1500 F.	Hemispherical total emittance. Resistance-heated wedge specimen. Power dissipated in measured area. Temperatures measured with thermocouples.	Measured in vacuum. Data taken from curve.

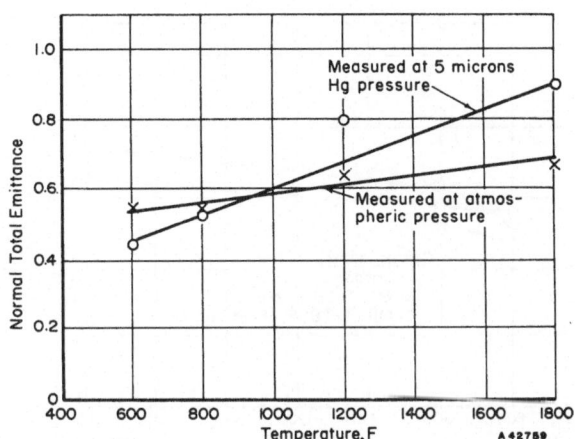

NORMAL TOTAL EMITTANCE OF OXIDIZED A-286 STEEL (CONTAMINATED)

NORMAL TOTAL EMITTANCE OF OXIDIZED A-286 STEEL (CONTAMINATED)--REFERENCE INFORMATION

Reference	Investigator	Symbol	Composition and Surface Condition	Test Method	Remarks
13	Gravina and Katz		Surface oxidized and contaminated with JP-4 fuel. At 5 microns (Hg) pressure At atmospheric pressure	Normal total emittance. Resistance-heated strip specimen. Thermistor-bolometer detector. Reference blackbody. Temperatures measured with thermocouples.	Measured in air and vacuum. Data taken from curves.

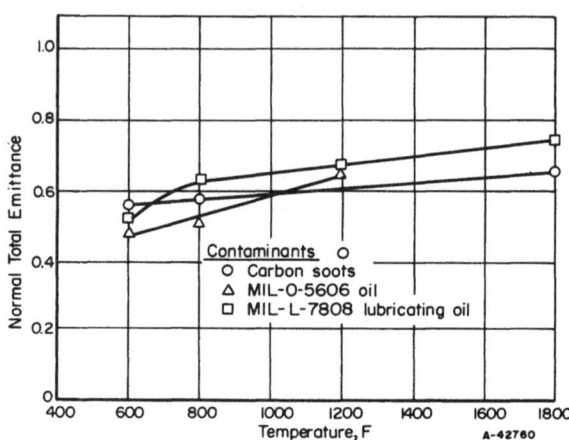

NORMAL TOTAL EMITTANCE OF OXIDIZED A-286 STEEL CONTAMINATED

NORMAL TOTAL EMITTANCE OF OXIDIZED TYPE A-286 STEEL (CONTAMINATED)--REFERENCE INFORMATION

Reference	Investigator	Symbol	Composition and Surface Condition	Test Method	Remarks
13	Gravina and Katz		Type A-286 steel, oxidized and contaminated with: carbon soot MIL-O-5606 oil MIL-L-7808 lubricating oil	Normal total emittance. Resistance-heated strip specimen. Thermistor-bolometer detector. Reference blackbody. Temperatures measured with thermocouples.	Measured at atmospheric pressure. Data taken from curves.

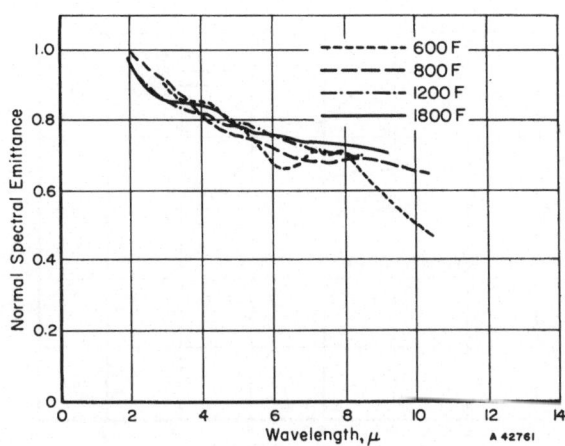

NORMAL SPECTRAL EMITTANCE OF OXIDIZED A-286 STEEL CONTAMINATED WITH JP-4 FUEL

NORMAL SPECTRAL EMITTANCE OF OXIDIZED A-286 STEEL (CONTAMINATED)--REFERENCE INFORMATION

Reference	Investigator	Symbol	Composition and Surface Condition	Test Method	Remarks
13	Gravina and Katz		Air oxidized A-286 steel contaminated with JP-4 fuel. Measured at: 600 F 800 F 1200 F 1800 F	Normal spectral emittance. Resistance-heated strip specimen. Thermistor-bolometer detector. Monochromator. Reference blackbody. Temperatures measured with thermocouples.	Measured in air. Data taken from curves.

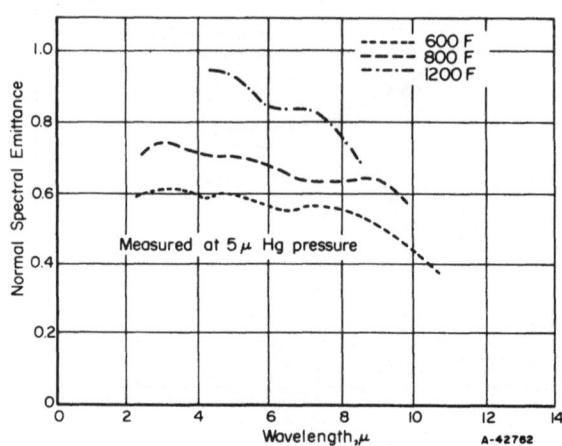

NORMAL SPECTRAL EMITTANCE OF OXIDIZED A-286 STEEL CONTAMINATED WITH JP-4 FUEL

NORMAL SPECTRAL EMITTANCE OF OXIDIZED A-286 STEEL (CONTAMINATED)--REFERENCE INFORMATION

Reference	Investigator	Symbol	Composition and Surface Condition	Test Method	Remarks
13	Gravina and Katz		Air oxidized type A-286 steel contaminated with JP-4 fuel. Measured at: 600 F 800 F 1200 F	Normal spectral emittance. Resistance-heated strip specimen. Thermistor-bolometer detector. Monochromator. Reference blackbody. Temperatures measured with thermocouples.	Measured in 5 micron Hg pressure. Data taken from curves.

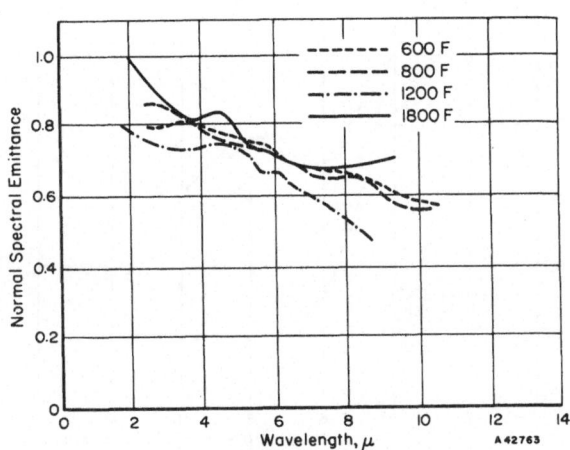

NORMAL SPECTRAL EMITTANCE OF OXIDIZED A-286 STEEL CONTAMINATED WITH CARBON DEPOSITS

NORMAL SPECTRAL EMITTANCE OF OXIDIZED A-286 STEEL (CONTAMINATED)--REFERENCE INFORMATION

Reference	Investigator	Symbol	Composition and Surface Condition	Test Method	Remarks
13	Gravina and Katz		Air oxidized type A-286 steel contaminated with carbon deposits. Measured at: 600 F 800 F 1200 F 1800 F	Normal spectral emittance. Resistance-heated strip specimen. Thermistor-bolometer detector. Monochromator. Reference blackbody. Temperatures measured with thermocouples.	Measured in air. Data taken from curves.

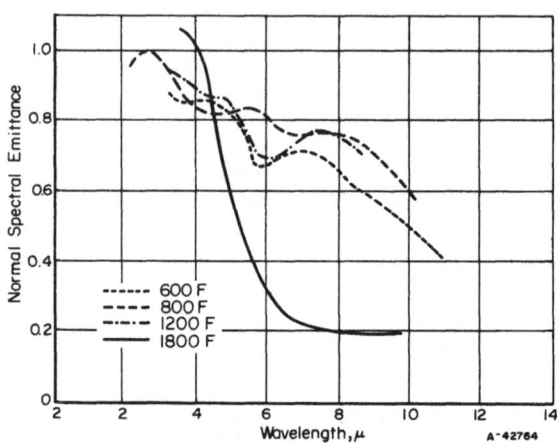

NORMAL SPECTRAL EMITTANCE OF OXIDIZED A-286 STEEL CONTAMINATED WITH MIL-L-7808

NORMAL SPECTRAL EMITTANCE OF OXIDIZED A-286 STEEL (CONTAMINATED)--REFERENCE INFORMATION

Reference	Investigator	Symbol	Composition and Surface Condition	Test Method	Remarks
13	Gravina and Katz		Air oxidized type A-286 steel contaminated with MIL-L-7808. Measured at: 600 F 800 F 1200 F 1800 F	Normal spectral emittance. Resistance-heated strip specimen. Thermistor-bolometer detector. Monochromator. Reference blackbody. Temperatures measured with thermocouples.	Measured in air. Data taken from curves.

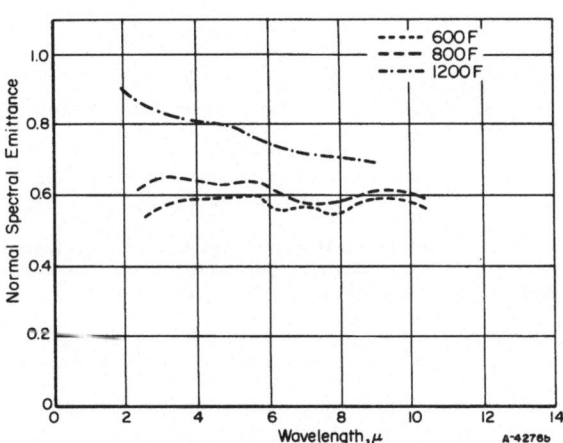

NORMAL SPECTRAL EMITTANCE OF OXIDIZED A-286 STEEL CONTAMINATED WITH MIL-O-5606

NORMAL SPECTRAL EMITTANCE OF OXIDIZED A-286 STEEL (CONTAMINATED)--REFERENCE INFORMATION

Reference	Investigator	Symbol	Composition and Surface Condition	Test Method	Remarks
13	Gravina and Katz		Air oxidized type A-286 steel contaminated with MIL-O-5606. Tested at: 600 F 800 F 1200 F	Normal spectral emittance. Resistance-heated strip specimen. Thermistor-bolometer detector. Monochromator. Reference blackbody. Temperatures measured with thermocouples.	Measured in air. Data taken from curves.

References

(1)   Burgess, D. G., Jasperse, J. R., Marcus, L., Martin, W. S., and Flint, E. P., "Research on Ceramic Coatings With Controlled Reflective and Emissive Properties", WADC TR 60-317 (July, 1960).

(2)   Richmond, J. C., and Stewart, J. E., "Spectral Emittance of Uncoated and Ceramic-Coated Inconel and Type 321 Stainless Steel", NASA Memorandum 4-9-59 W.

(3)   Bevans, J. T., Gier, J. T., and Dunkle, R. V., "Comparison of Total Emittances With Values Computed From Spectral Measurements", Trans. ASME, 80 (2), 1405-1416 (October, 1958).

(4)   Dull, R. L., "Resistance Heating of Titanium", Republic Aviation Corp., Part I, Contract No. AF 33(600)-38042 (April 15 to July 15, 1959).

(5)   Douglass, E. A., "Investigation Directed Toward the Development of Ceramic Coatings With High Reflectivities and Emissivities for Use in Aircraft Power Plants", WADC TR 56-110 (February, 1956), Contract No. AF 33(616)-2376.

(6)   Morgan, F. H., "Spectral Emissivity of Coatings of Thoria and Other Refractories as a Function of Temperature", Jour. Appl. Phys., 22, 108-109 (1951).

(7)   Anthony, F. M., and Pearl, H. A., "Investigation of Feasibility of Utilizing Available Heat-Resistant Materials for Hypersonic Leading Edge Applications", WADC TR 59-744, Vol III (July, 1960).

(8)   Fieldhouse, I. B., Lang, J. I., and Blau, H. H., Jr., "Investigation of Feasibility of Utilizing Available Heat Resistant Materials for Hypersonic Leading Edge Applications", WADC TR 59-744, Vol IV (October, 1960).

(9)   Coffman, J. A., Kibler, G. M., and Riethof, T. R., "Carbonization of Plastics and Refractory Materials Research", Third Quarterly Progress Report, ASTIA No. AD-245223, Contract No. AF 33(616)-6841 (September 30, 1960).

(10)   Wade, W. R., "Measurements of Total Hemispherical Emissivity of Several Stably Oxidized Metals and Some Refractory Oxide Coatings", NASA Memo 1-20-59L (January, 1959).

(11)   Olson, O. H., and Morris, J. C., "Determination of Emissivity and Reflectivity Data on Aircraft Structural Materials", WADC TR 56-222, Part III (April, 1960).

(12)   Sully, A. H., Brandes, E. A., and Waterhouse, R. B., "Some Measurements of the Total Emissivity of Metals and Pure Refractory Oxides and the Variation of Emissivity With Temperature", Brit. J. Appl. Physics, 3, 97-101 (March, 1952).

(13)   Gravina, A., and Katz, M., "Investigation of High Emittance Coatings to Extend the Mach Number Range of Application of Structural Materials", WADD TR 60-102 (December, 1960).

(14) Adams, J. G., "The Determination of Spectral Emissivities, Reflectivities, and Absorptivities of Materials and Coatings", Northrop Corporation Report No. NOR-61-189 (Aug. 3, 1961).

(15) Pratt and Whitney Aircraft, "Measurement of Spectral and Total Emittance of Materials and Surfaces Under Simulated Space Conditions", Report No. PWA-1863 (July 1, 1959 through June 30, 1960) Contract No. NASW-4.

(16) Pratt and Whitney Aircraft, "Determination of the Emissivity of Materials", Report PWA-2043.

(17) Pratt and Whitney Aircraft, "Determination of the Emissivity of Materials", Report PWA-1994.

(18) Pratt and Whitney Aircraft, "Determination of the Emissivity of Materials", Report PWA-1966.

RADIATIVE PROPERTY DATA

Ceramics and Graphite

TABLE OF CONTENTS

TABLE OF CONTENTS
(Continued)

TABLE OF CONTENTS
(Continued)

TABLE OF CONTENTS
(Continued)

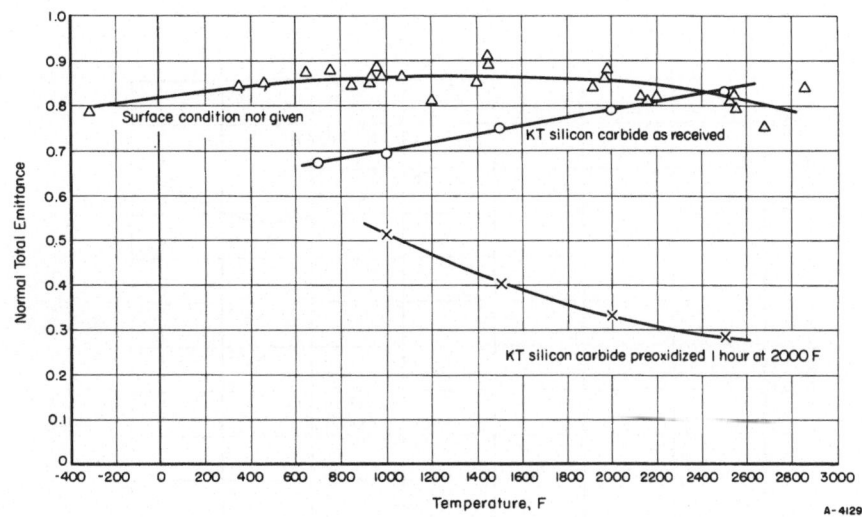

NORMAL TOTAL EMITTANCE OF SILICON CARBIDE

NORMAL TOTAL EMITTANCE OF SILICON CARBIDE--REFERENCE INFORMATION

Reference	Investigator	Symbol	Composition and Surface Condition	Test Method	Remarks
1	Anthony and Pearl		KT Silicon carbide	Normal total emittance. Induction-heated specimen. Comparison blackbody. Thermopile detector. Temperatures measured with thermocouples.	Measured in purge of dry helium gas. Data taken from table.
		O	As received		
		X	Pre-oxidized in air 1 hour at 2000 F		
2	Olson and Morris	△	Silicon carbide Surface condition not given	Normal total emittance. Furnace-heated specimen. Comparison blackbody. Thermistor detector. Temperatures measured with thermocouples.	Measured in air. Data taken from curves.

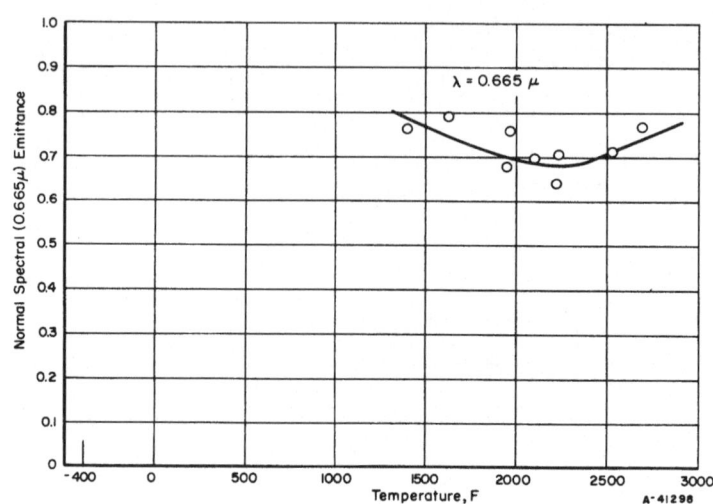

NORMAL SPECTRAL EMITTANCE OF SILICON CARBIDE

NORMAL SPECTRAL EMITTANCE OF SILICON CARBIDE--REFERENCE INFORMATION

Reference	Investigator	Symbol	Composition and Surface Condition	Test Method	Remarks
2	Olson and Morris	O	Silicon carbide Surface condition not given	Normal spectral emittance. Furnace-heated specimen. Comparison blackbody. Commercial detector and filter system for peak response at 0.665μ. Temperatures measured with thermocouples.	Measured in air. Data taken from curves. (λ = 0.665μ)

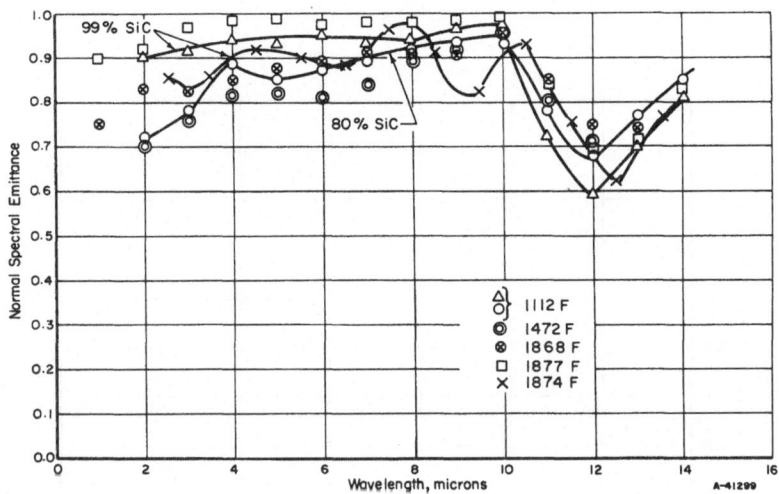

NORMAL SPECTRAL EMITTANCE OF SILICON CARBIDE

NORMAL SPECTRAL EMITTANCE OF SILICON CARBIDE--REFERENCE INFORMATION

Reference	Investigator	Symbol	Composition and Surface Condition	Test Method	Remarks
3	Blau, Marsh, Martin, Jasperse, and Chaffee		Silicon carbide Diamond wheel finish as supplied by manufacturer	Normal spectral emittance. Specimen mounted in wall of cylindrical Globar (SiC) heater.	Measured in air. Data taken from curves. (Curves are drawn through the 1112 F points only.)
			Crystolon R (Norton) 99% + pure	Comparison blackbody hole also in heater wall.	
		△	Measured at 1112 F	Temperatures measured with thermocouples.	
		□	Measured at 1877 F	Monochromator and thermocouple detector.	
			RC4237 (Norton) 80% pure		
		O	Measured at 1112 F		
		⊙	Measured at 1472 F		
		⊗	Measured at 1868 F		
4	Blau, Chaffee, Jasperse, and Martin		99 per cent silicon carbide (Norton Crystalon R)	Normal spectral emittance. Induction-heated specimen.	Measured in 90% argon, 10% hydrogen atmosphere.
		×	Flat smooth surface from diamond wheel cutting.	Comparison blackbody. Monochromator and thermocouple detector. Temperatures measured with micro-optical pyrometer.	Data taken from curve.
			The minima at about 9 and 12 microns are attributed to a thin SiO$_2$ surface film. Measured at 1874 F		

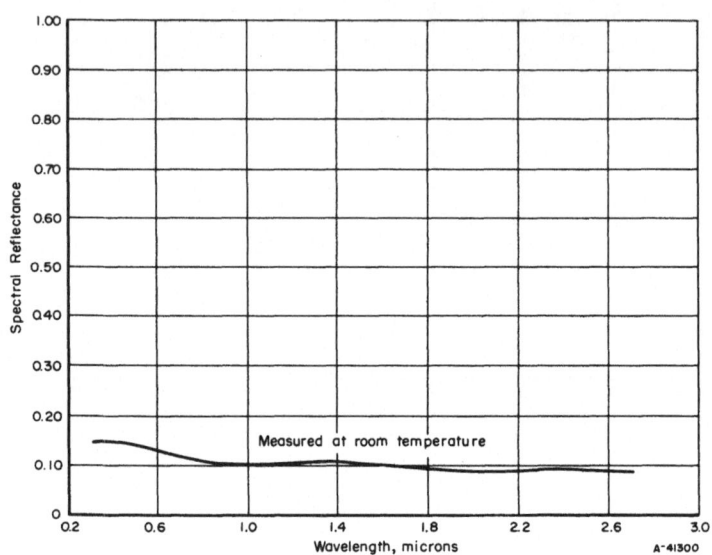

SPECTRAL REFLECTANCE OF SILICON CARBIDE

SPECTRAL REFLECTANCE OF SILICON CARBIDE—REFERENCE INFORMATION

Reference	Investigator	Symbol	Composition and Surface Condition	Test Method	Remarks
2	Olson and Morris		Silicon carbide, purity and surface condition not given	Spectral reflectance. Incident radiation 9 degrees from normal to specimen surface. Integrating sphere reflectometer. Monochromator and lead sulphide detector. Normal (9 degrees) illumination Diffuse reflection.	Measured in air at room temperature. Data taken from curves.

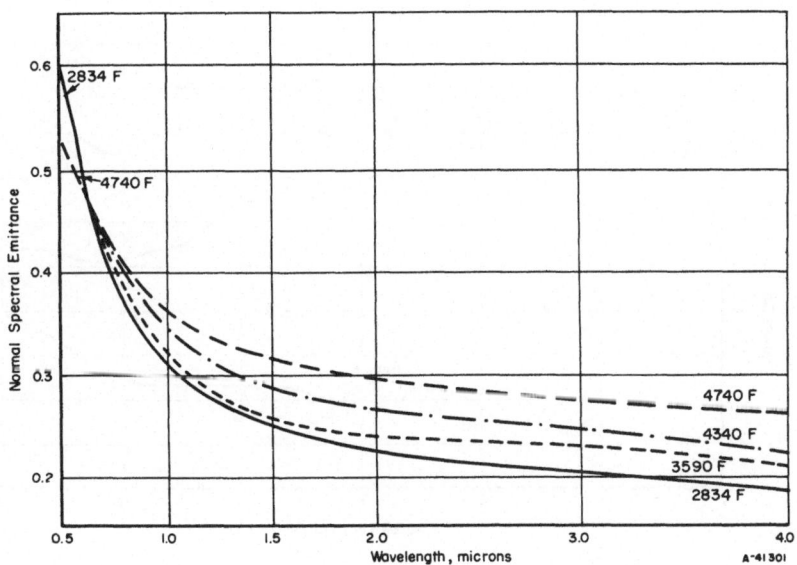

NORMAL SPECTRAL EMITTANCE OF TANTALUM CARBIDE (0.5 TO 4 MICRONS)

NORMAL SPECTRAL EMITTANCE OF TANTALUM CARBIDE (0.5 TO 4 MICRONS)--REFERENCE INFORMATION

Reference	Investigator	Symbol	Composition and Surface Condition	Test Method	Remarks
6	Riethof		Tantalum carbide Composition or surface condition not given Measured at 2834, 3590, 4340, and 4740 F	Normal spectral emittance. Induction-heated specimen. Blackbody hole in specimen surface. Thermocouple detector. Monochromator. Temperatures measured with optical pyrometer.	Measured in argon. Data taken from curves.

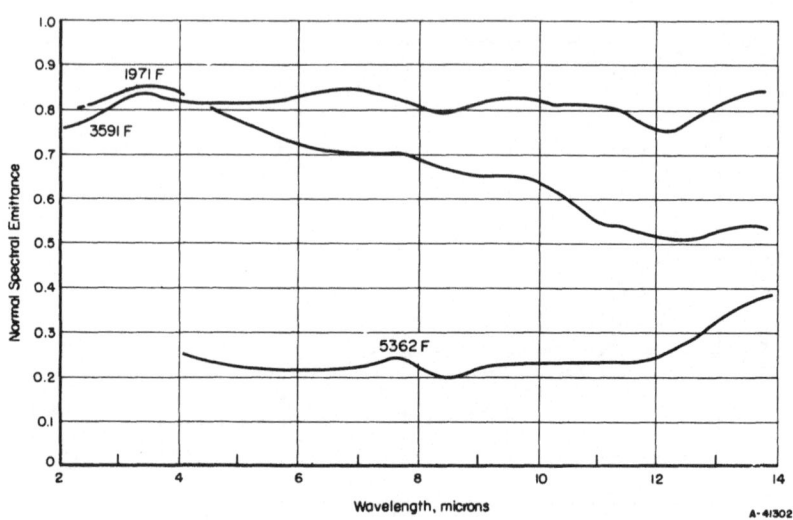

NORMAL SPECTRAL EMITTANCE OF TANTALUM CARBIDE (2 TO 14 MICRONS)

NORMAL SPECTRAL EMITTANCE OF TANTALUM CARBIDE (2 TO 14 MICRONS)--REFERENCE INFORMATION

Reference	Investigator	Symbol	Composition and Surface Condition	Test Method	Remarks
4	Blau, Chaffee, Jasperse, and Martin		Tantalum carbide Purity not given Surface flat and smooth but not polished (Note: Surface analysis after 3234 K (5362 F) run showed thin tantalum oxide film)	Normal spectral emittance. Induction-heated specimen. Comparison blackbody. Monochromator and thermo-couple detector. Temperatures measured with optical pyrometer.	Measured in 90% argon 10% hydrogen atmosphere. Data taken from curves.

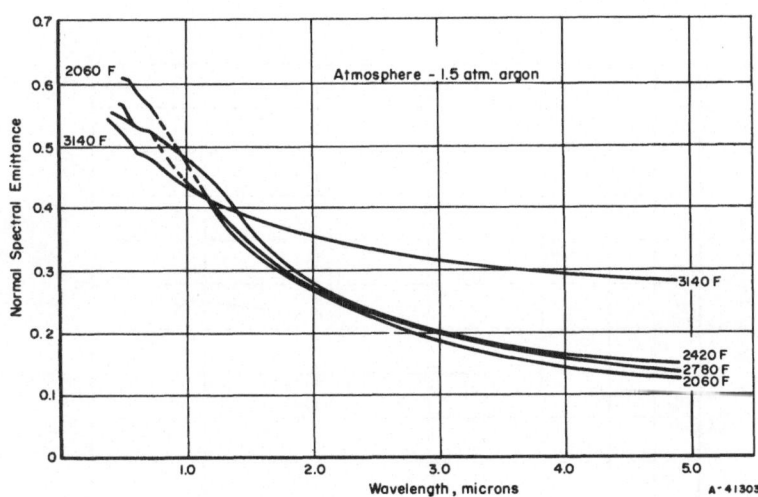

NORMAL SPECTRAL EMITTANCE OF TUNGSTEN CARBIDE

NORMAL SPECTRAL EMITTANCE OF TUNGSTEN CARBIDE--REFERENCE INFORMATION

Reference	Investigator	Symbol	Composition and Surface Condition	Test Method	Remarks
5	Coffman, Coulson, and Kibler		Tungsten carbide (WC) Surface condition or purity not given Note: Surface transformation from WC to W_2C at 3140 F Measured at 2060, 2780, 2420, and 3140 F	Normal spectral emittance. Induction-heated specimen. Blackbody hole in specimen surface. Thermocouple detector. Monochromator. Temperatures measured with optical pyrometer.	Measured in 1.5 atmosphere of argon. Data taken from curves.

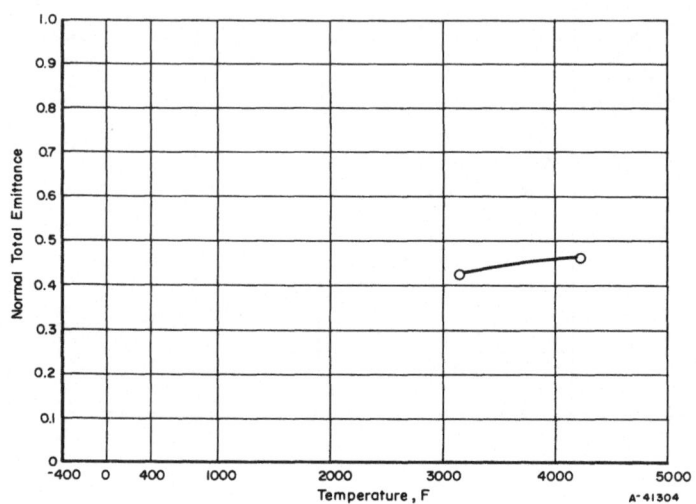

NORMAL TOTAL EMITTANCE OF ZIRCONIUM CARBIDE

NORMAL TOTAL EMITTANCE OF ZIRCONIUM CARBIDE--REFERENCE INFORMATION

Reference	Investigator	Symbol	Composition and Surface Condition	Test Method	Remarks
5	Coffman, Coulson, and Kibler	O	Formed into "toadstool" shaped specimen Composition and surface condition not given	Normal total emittance. Induction-heated specimen. Comparison /blackbody. Temperatures measured with optical pyrometer.	Measured in 1.5 atmosphere of dry, pure, argon. Data taken from curve.

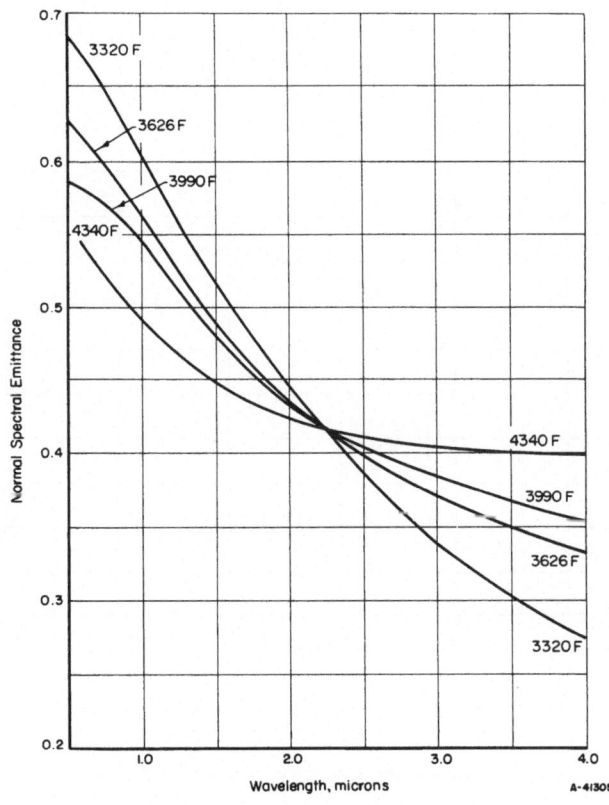

NORMAL SPECTRAL EMITTANCE OF ZIRCONIUM CARBIDE

NORMAL SPECTRAL EMITTANCE OF ZIRCONIUM CARBIDE--REFERENCE INFORMATION

Reference	Investigator	Symbol	Composition and Surface Condition	Test Method	Remarks
6	Riethof		Zirconium carbide Composition or surface condition not given Measured at 3320, 3626, 3990, and 4340 F	Normal spectral emittance. Induction-heated specimen. Blackbody hole in specimen surface. Thermocouple detector. Monochromator. Temperatures measured with optical pyrometer.	Measured in argon. Data taken from curves.

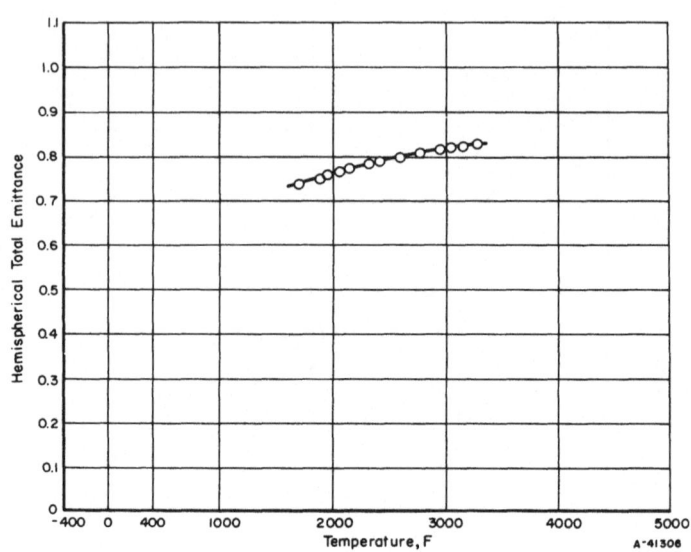

HEMISPHERICAL TOTAL EMITTANCE OF ACHESON GRAPHITE

HEMISPHERICAL TOTAL EMITTANCE OF ACHESON GRAPHITE--REFERENCE INFORMATION

Reference	Investigator	Symbol	Composition and Surface Condition	Test Method	Remarks
9	Jain and Krishnan	O	Acheson graphite Sample held at 2000 K for 1 hour in vacuum, until emittance became steady and reproducible	Hemispherical total emittance. Hole-in-tube method. Correction of inside blackbody temperature to surface temperature made using known thermal conductivity and wall thickness. Blackbody temperature measured with optical pyrometer.	Measured in vacuum. Data taken from curves.

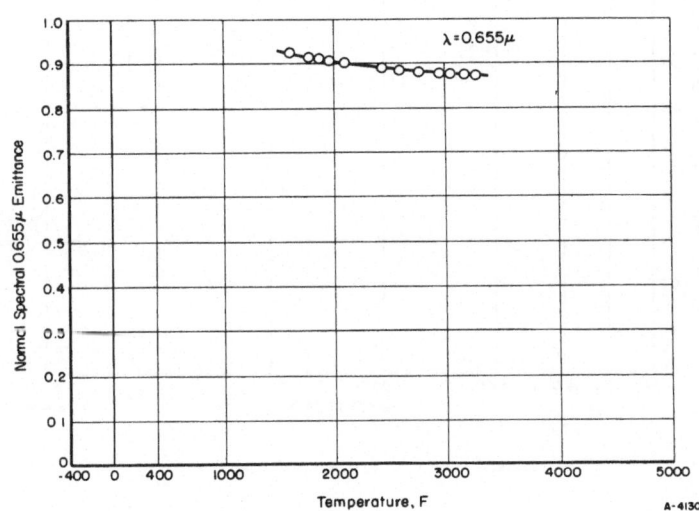

NORMAL SPECTRAL EMITTANCE OF ACHESON GRAPHITE

NORMAL SPECTRAL EMITTANCE OF ACHESON GRAPHITE--REFERENCE INFORMATION

Reference	Investigator	Symbol	Composition and Surface Condition	Test Method	Remarks
9	Jain and Krishnan	O	Acheson graphite Specimen held at 2000 K for 1 hour in vacuum until emittance became steady and reproducible	Normal spectral emittance. Hole-in-tube method. Temperatures measured with optical pyrometer.	Measured in vacuum. Data taken from curves. $(\lambda = 0.665\mu)$

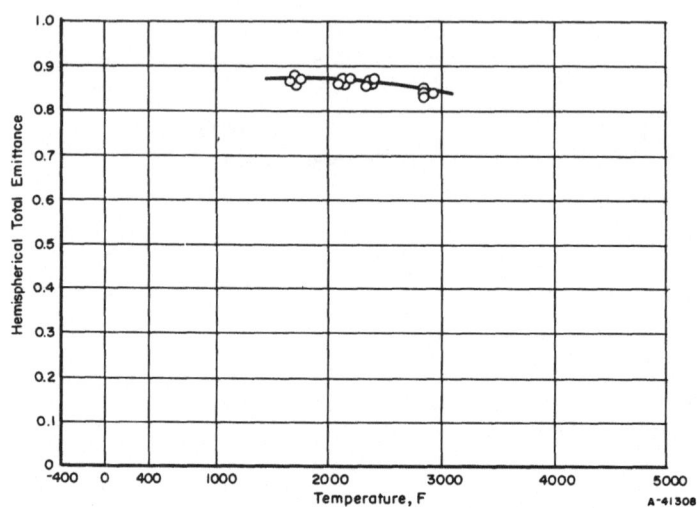

HEMISPHERICAL TOTAL EMITTANCE OF ATJ GRAPHITE

HEMISPHERICAL TOTAL EMITTANCE OF ATJ GRAPHITE--REFERENCE INFORMATION

Reference	Investigator	Symbol	Composition and Surface Condition	Test Method	Remarks
4	Blau, Chaffee, Jasperse, and Martin	O	ATJ graphite Surface condition not given	Normal total emittance. (Hemispherical emittance equals normal emittance for this specimen.) Induction-heated specimen. Monochromator with prism replaced by plane mirror. Thermocouple detector. Blackbody hole drilled in specimen surface. Temperatures measured with micro-optical pyrometer.	Measured in 90% argon – 10% hydrogen atmosphere. Data taken from curves.

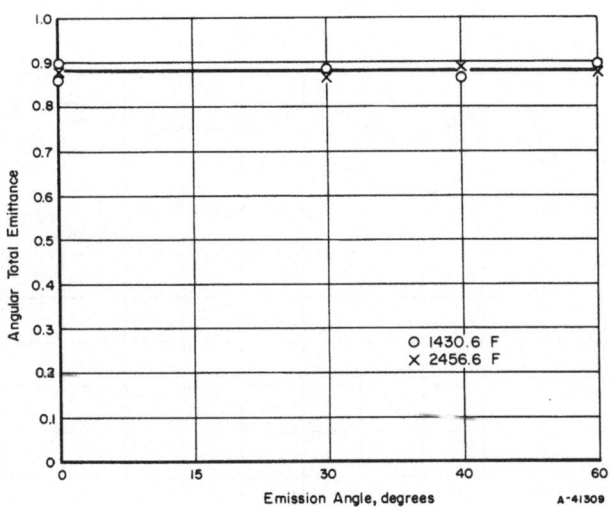

TOTAL EMITTANCE VERSUS EMISSION ANGLE OF ATJ GRAPHITE

TOTAL EMITTANCE VERSUS EMISSION ANGLE OF ATJ GRAPHITE--REFERENCE INFORMATION

Reference	Investigator	Symbol	Composition and Surface Condition	Test Method	Remarks
4	Blau, Chaffee, Jasperse, and Martin		ATJ graphite Surface smooth and flat, but not polished.	Total emittance measured normally and at 30, 45, and 60 degrees from the normal. Induction-heated specimen. Monochromator with prism replaced by plane mirror. Thermocouple detector. Blackbody hole drilled in specimen surface. Temperatures measured with micro-optical pyrometer.	Measured in 90% argon – 10% hydrogen atmosphere. Data taken from curves. Normal emittance equals hemispherical emittance for this specimen.
		O	Measured at 1431 F		
		X	Measured at 2457 F		

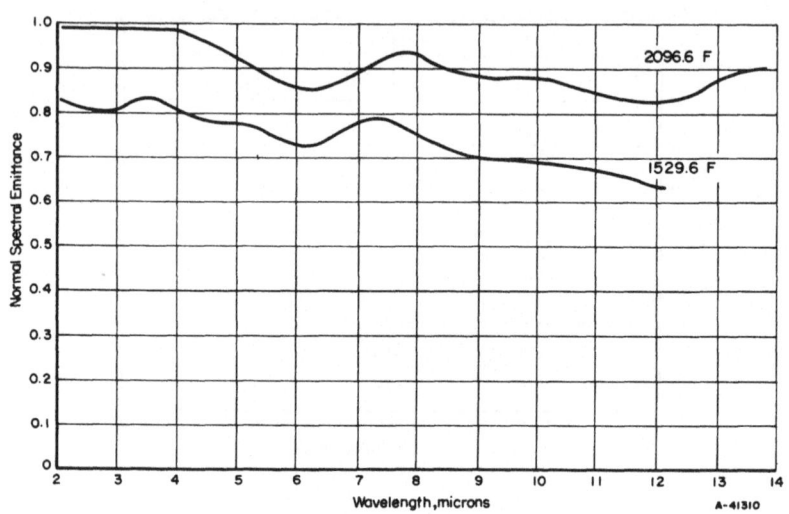

NORMAL SPECTRAL EMITTANCE OF ATJ GRAPHITE

NORMAL SPECTRAL EMITTANCE OF ATJ GRAPHITE--REFERENCE INFORMATION

Reference	Investigator	Symbol	Composition and Surface Condition	Test Method	Remarks
4	Blau, Chaffee, Jasperse, and Martin		ATJ graphite Surface smooth and flat but not polished	Normal spectral emittance. Induction-heated specimen. Monochromator and thermo-couple detector. Blackbody hole drilled in specimen surface. Temperatures measured with micro-optical pyrometer.	Measured in 90% argon - 10% hydrogen atmos-phere. Data taken from curves.

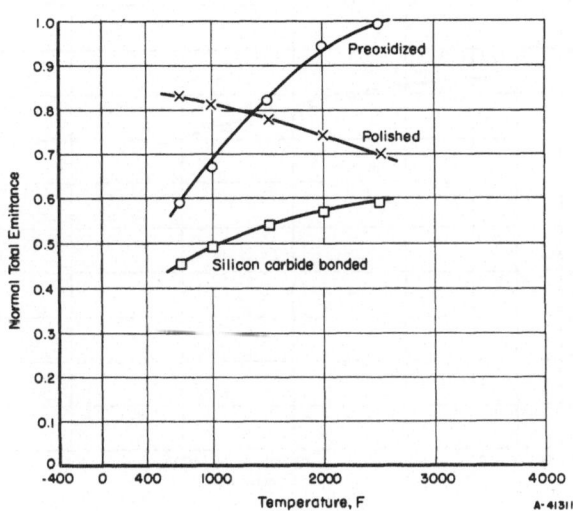

NORMAL TOTAL EMITTANCE OF ELECTRODE GRAPHITE

NORMAL TOTAL EMITTANCE OF ELECTRODE GRAPHITE--REFERENCE INFORMATION

Reference	Investigator	Symbol	Composition and Surface Condition	Test Method	Remarks
1	Anthony and Pearl		Electrode graphite	Normal total emittance. Induction-heated specimen. Comparison blackbody. Thermopile detector. Temperatures measured with thermocouples.	Measured in purge of helium gas. Data taken from table.
		O	Preoxidized		
		X	Polished		
		□	Silicon carbide bonded		

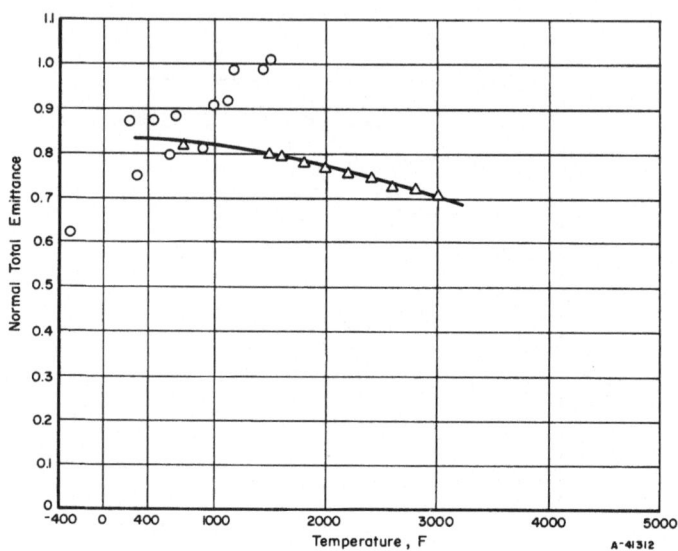

NORMAL TOTAL EMITTANCE OF GBE GRAPHITE

NORMAL TOTAL EMITTANCE OF GBE GRAPHITE--REFERENCE INFORMATION

Reference	Investigator	Symbol	Composition and Surface Condition	Test Method	Remarks
8	Olson and Morris	O	National GBE graphite Surface condition not given	Normal total emittance. Resistance—heated strip specimen. Comparison blackbody. Temperatures measured with thermocouples. Thermistor detector.	Measured in vacuum. Data taken from curves.
7	Betz, Olson, Schurin, and Morris	△	Same as above	Same as above.	Same as above.

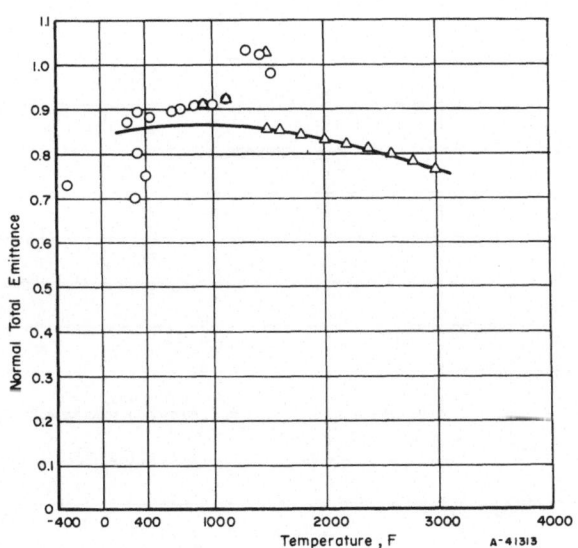

NORMAL TOTAL EMITTANCE OF TYPE GBH GRAPHITE

NORMAL TOTAL EMITTANCE OF TYPE GBH GRAPHITE--REFERENCE INFORMATION

Reference	Investigator	Symbol	Composition and Surface Condition	Test Method	Remarks
8	Olson and Morris	O	National GBH graphite Surface condition not given Note: Changed with cycling	Normal total emittance. Resistance-heated strip specimen. Comparison blackbody. Thermistor detector. Temperatures measured with thermocouples.	Measured in vacuum. Data taken from curves.
7	Betz, Olson, Schurin, and Morris	Δ	Surface condition not given	Same as above.	Measured in vacuum. Data taken from table.

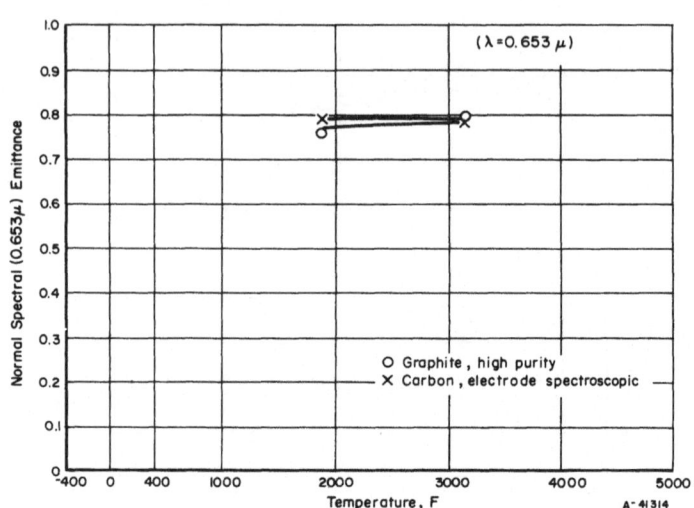

NORMAL SPECTRAL EMITTANCE OF GRAPHITE AND CARBON

NORMAL SPECTRAL EMITTANCE OF GRAPHITE AND CARBON--REFERENCE INFORMATION

Reference	Investigator	Symbol	Composition and Surface Condition	Test Method	Remarks
10	Thorn and Simpson	O	High-purity, medium-density graphite	Normal spectral emittance. Modified hole-in-tube method.	Measured in vacuum. Data taken from curves.
		X	Spectroscopic electrode carbon	Temperatures measured with calibrated optical pyrometer.	$(\lambda = 0.653\mu)$
			Surface condition, polished and then heated to 1800 K in vacuum for 3 hours		

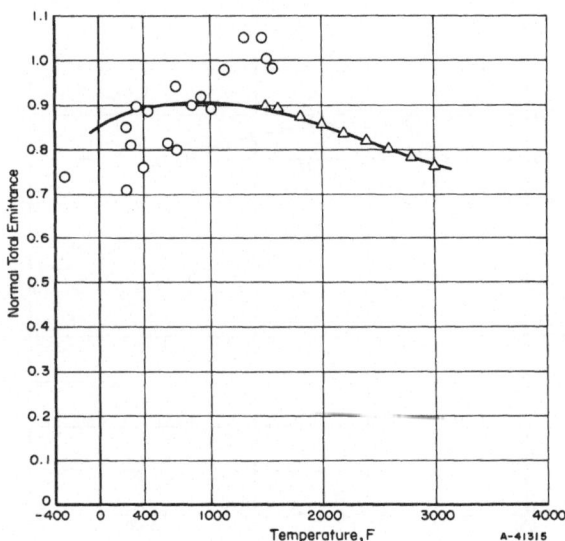

NORMAL TOTAL EMITTANCE OF TYPE 3474D GRAPHITE

NORMAL TOTAL EMITTANCE OF TYPE 3474D GRAPHITE--REFERENCE INFORMATION

Reference	Investigator	Symbol	Composition and Surface Condition	Test Method	Remarks
8	Olson and Morris	O	Speer 3474D graphite Surface condition not given Note: Changed with cycling	Normal total emittance. Resistance-heated strip specimen. Comparison blackbody. Thermistor detector. Temperatures measured with thermocouples.	Measured in vacuum. Data taken from curves.
7	Betz, Olson, Schurin, and Morris	Δ	Surface condition not given	Same as above.	Measured in vacuum. Data taken from table.

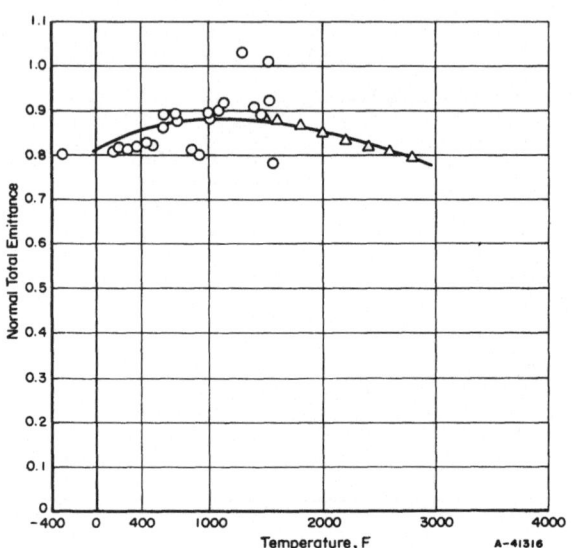

NORMAL TOTAL EMITTANCE OF TYPE 7087 GRAPHITE

NORMAL TOTAL EMITTANCE OF TYPE 7087 GRAPHITE--REFERENCE INFORMATION

Reference	Investigator	Symbol	Composition and Surface Condition	Test Method	Remarks
8	Olson and Morris	O	Speer 7087 graphite Surface condition not given Note: Changed with cycling	Normal total emittance. Resistance-heated strip specimen. Comparison blackbody. Thermistor detector. Temperatures measured with thermocouples.	Measured in vacuum. Data taken from curves.
7	Betz, Olson, Schurin, and Morris	△	Surface condition not given	Same as above.	Measured in vacuum. Data taken from table.

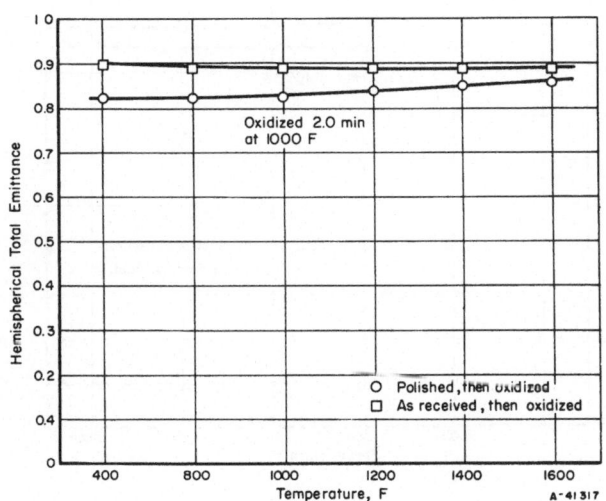

HEMISPHERICAL TOTAL EMITTANCE OF OXIDIZED K150A Ni-TiC HARD METAL

HEMISPHERICAL TOTAL EMITTANCE OF OXIDIZED K150A Ni-TiC HARD METAL--REFERENCE INFORMATION

Reference	Investigator	Symbol	Composition and Surface Condition	Test Method	Remarks
11	Wade and Casey		Composition: 10Ni, 80TiC, 10CbC	Hemispherical total emittance. (Total emittance measured normally and at various angles. Normal emittance equals hemispherical emittance.) Thermopile total ratiation detector. Resistance-heated specimen. Comparison blackbody. Temperatures measured with thermocouples.	Measured in air. Data taken from curves.
		□	As received, then oxidized		
		○	Polished: Hand lapped with 3 micron and 1 micron diamond paste, then oxidized		

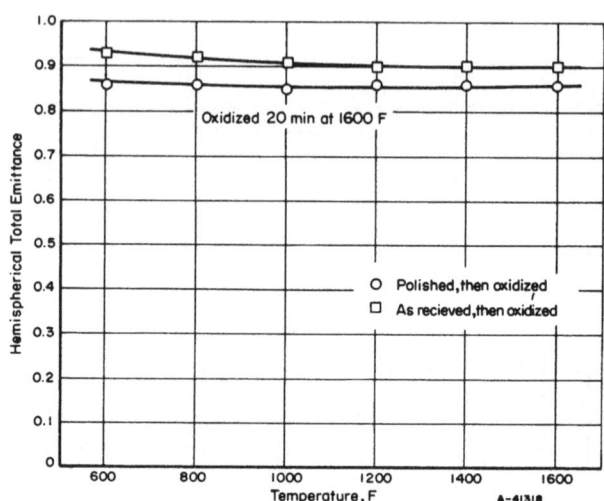

HEMISPHERICAL TOTAL EMITTANCE OF OXIDIZED K151A Ni-TiC HARD METAL

HEMISPHERICAL TOTAL EMITTANCE OF OXIDIZED K151A Ni-TiC HARD METAL—REFERENCE INFORMATION

Reference	Investigator	Symbol	Composition and Surface Condition	Test Method	Remarks
11	Wade and Casey		Composition: 20Ni, 70TiC, 10CbC	Hemispherical total emittance. (Total emittance measured normally and at various angles. Normal emittance equals hemispherical emittance.) Thermopile total radiation detector. Resistance-heated specimen. Comparison blackbody. Temperatures measured with thermocouples.	Measured in air. Data taken from curves.
		□	As received, then oxidized		
		○	Polished; hand lapped with 3-micron and 1-micron diamond paste, then oxidized		
	Composition: 20Ni, 70TiC, 10CbC				

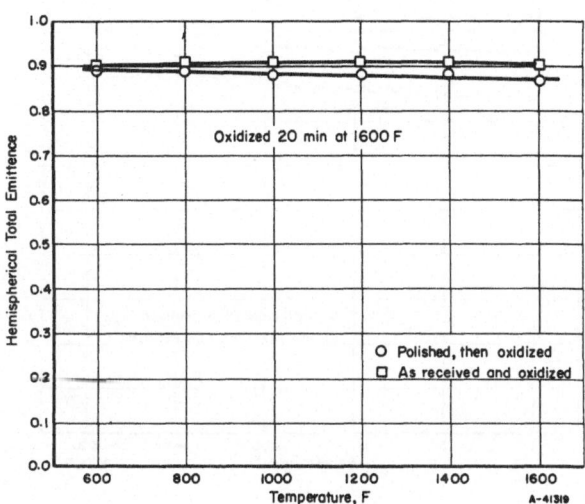

HEMISPHERICAL TOTAL EMITTANCE OF OXIDIZED K152B Ni-TiC HARD METAL

HEMISPHERICAL TOTAL EMITTANCE OF OXIDIZED K152B Ni-TiC HARD METAL--REFERENCE INFORMATION

Reference	Investigator	Symbol	Composition and Surface Condition	Test Method	Remarks
11	Wade and Casey		Composition: 30Ni, 65TiC, 5CbC	Hemispherical total emittance. (Total emittance measured normally and at various angles. Normal emittance equals hemispherical emittance.) Thermopile total radiation detector. Resistance-heated specimen. Comparison blackbody. Temperatures measured with thermocouples.	Measured in air. Data taken from curves.
		□ ○	As received, then oxidized Polished; hand lapped with 3-micron and 1-micron diamond paste, then oxidized		

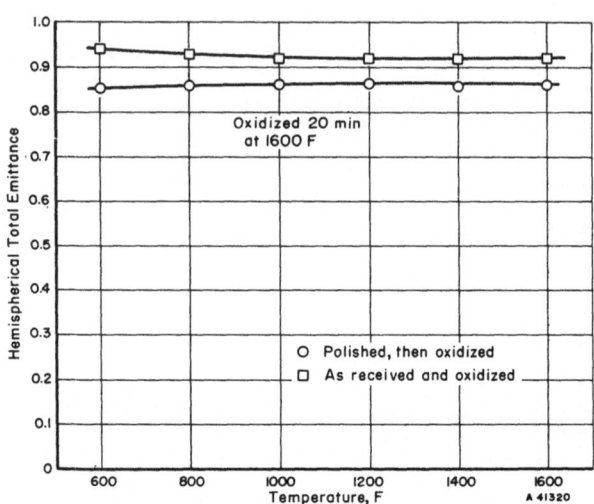

HEMISPHERICAL TOTAL EMITTANCE OF OXIDIZED K153B Ni-TiC HARD METAL

HEMISPHERICAL TOTAL EMITTANCE OF OXIDIZED K153B Ni-TiC HARD METAL--REFERENCE INFORMATION

Reference	Investigator	Symbol	Composition and Surface Condition	Test Method	Remarks
11	Wade and Casey		Composition: 40Ni, 54TiC, 6CbC	Hemispherical total emittance. (Total emittance measured normally and at various angles. Normal emittance equals hemispherical emittance.) Thermopile total radiation detector. Resistance-heated specimen. Comparison blackbody. Temperatures measured with thermocouples.	Measured in air. Data taken from curves.
		□	As received, then oxidized 20 minutes at 1600 F		
		O	Polished; lapped with 3-micron and 1-micron diamond paste, then oxidized 20 minutes at 1600 F		

445

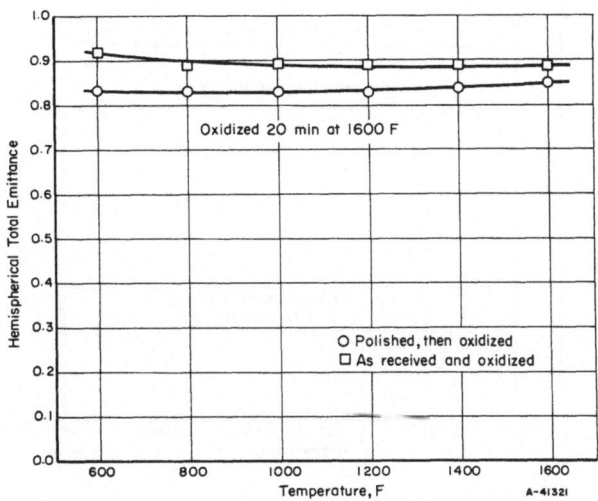

HEMISPHERICAL TOTAL EMITTANCE OF OXIDIZED K163B1 Ni-TiC HARD METAL

HEMISPHERICAL TOTAL EMITTANCE OF OXIDIZED K163B1 Ni-TiC HARD METAL--REFERENCE INFORMATION

Reference	Investigator	Symbol	Composition and Surface Condition	Test Method	Remarks
11	Wade and Casey		Composition: 33.3Ni, 54TiC, 6.7Mo, 6CbC	Hemispherical total emittance. (Total emittance measured normally and at various angles. Normal emittance equals hemispherical emittance.) Thermopile total radiation detector. Resistance-heated specimen. Comparison blackbody. Temperatures measured with thermocouples.	Measured in air. Data taken from curves.
		□	As received, then oxidized 20 minutes at 1600 F		
		O	Polished; lapped with 3-micron and 1-micron diamond paste, then oxidized 20 minutes at 1600 F		

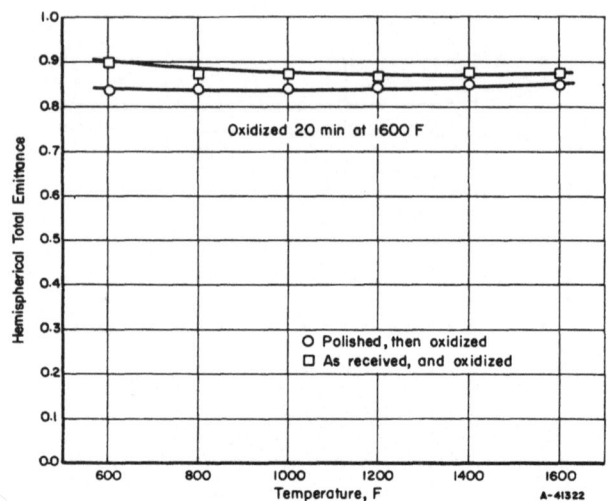

HEMISPHERICAL TOTAL EMITTANCE OF OXIDIZED K184B Ni-TiC HARD METAL

HEMISPHERICAL TOTAL EMITTANCE OF OXIDIZED K184B Ni-TiC HARD METAL--REFERENCE INFORMATION

Reference	Investigator	Symbol	Composition and Surface Condition	Test Method	Remarks
11	Wade and Casey		Composition: 40Ni, 40TiC, 10CbC, 4Mo, 3Al, 3Cr	Hemispherical total emittance. (Total emittance measured normally and at various angles. Normal emittance equals hemispherical emittance.) Thermopile total radiation detector. Resistance-heated specimen. Comparison blackbody. Temperatures measured with thermocouples.	Measured in air. Data taken from curves.
		□	As received, then oxidized 20 minutes at 1600 F		
		O	Polished; lapped with 3-micron and 1-micron diamond paste, then oxidized 20 minutes at 1600 F		

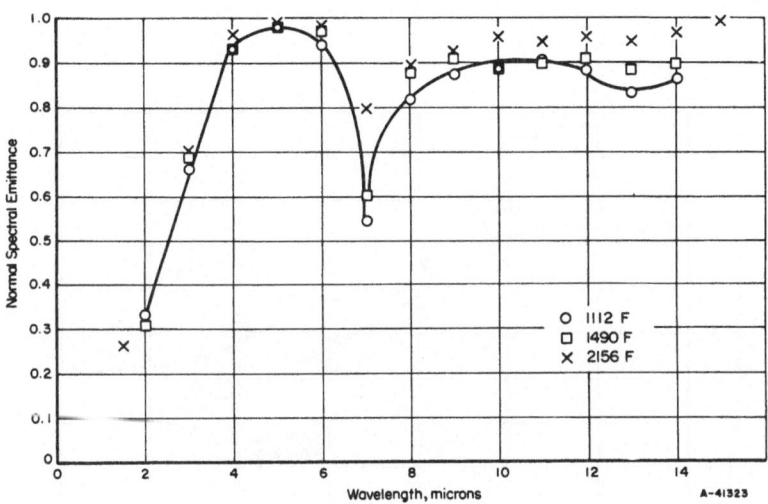

NORMAL SPECTRAL EMITTANCE OF BORON NITRIDE

NORMAL SPECTRAL EMITTANCE OF BORON NITRIDE--REFERENCE INFORMATION

Reference	Investigator	Symbol	Composition and Surface Condition	Test Method	Remarks
3	Blau, Marsh, Martin, Jasperse, and Chaffee		Boron nitride Purity and surface condition not given	Normal spectral emittance. Specimen mounted in wall of cylindrical Globar (SiC) heater. Comparison blackbody hole in heater wall. Monochromator and thermocouple detector. Temperatures measured with thermocouples.	Measured in air. Data taken from curves. (Curve drawn through 1112 F points only.)
		O	Measured at 1112 F		
		□	Measured at 1490 F		
		X	Measured at 2156 F		

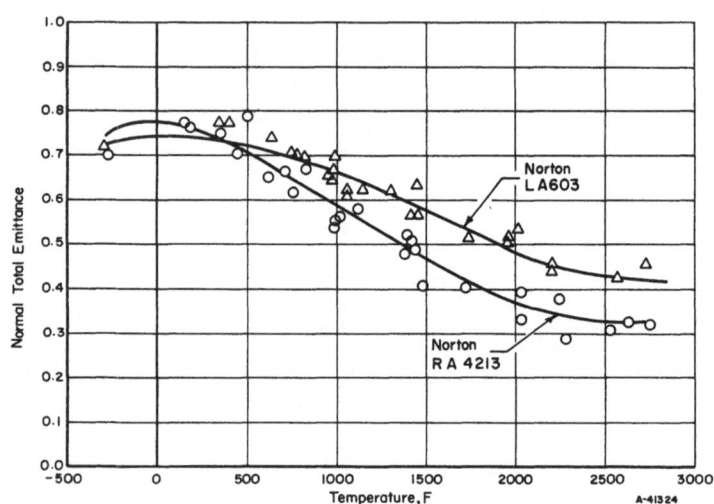

NORMAL TOTAL EMITTANCE OF ALUMINUM OXIDE

NORMAL TOTAL EMITTANCE OF ALUMINUM OXIDE--REFERENCE INFORMATION

Reference	Investigator	Symbol	Composition and Surface Condition	Test Method	Remarks
2	Olson and Morris	△	Norton LA603 Aluminum oxide	Normal total emittance. Furnace-heated specimen. Comparison blackbody.	Measured in air. Data taken from curves.
		O	Norton RA4213 Aluminum oxide	Temperatures measured with thermocouples Thermistor detector.	
			Surface condition not given		

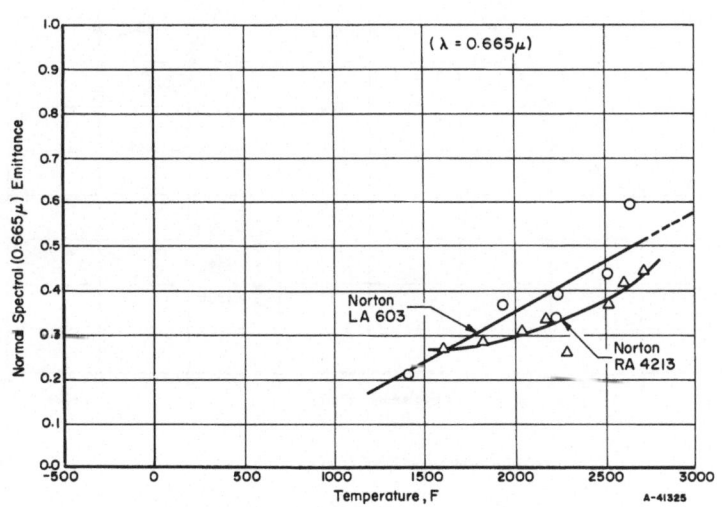

NORMAL SPECTRAL EMITTANCE OF ALUMINUM OXIDE

NORMAL SPECTRAL EMITTANCE OF ALUMINUM OXIDE--REFERENCE INFORMATION

Reference	Investigator	Symbol	Composition and Surface Condition	Test Method	Remarks
2	Olson and Morris	O	Norton LA603 Aluminum oxide	Normal spectral emittance. Furnace-heated specimen. Comparison blackbody.	Measured in air. Data taken from curves.
		△	Norton RA4213 Aluminum oxide	Commercial radiation detector and filter system for peak response at 0.665μ. Temperatures measured with thermocouples.	(λ = 0.665μ)

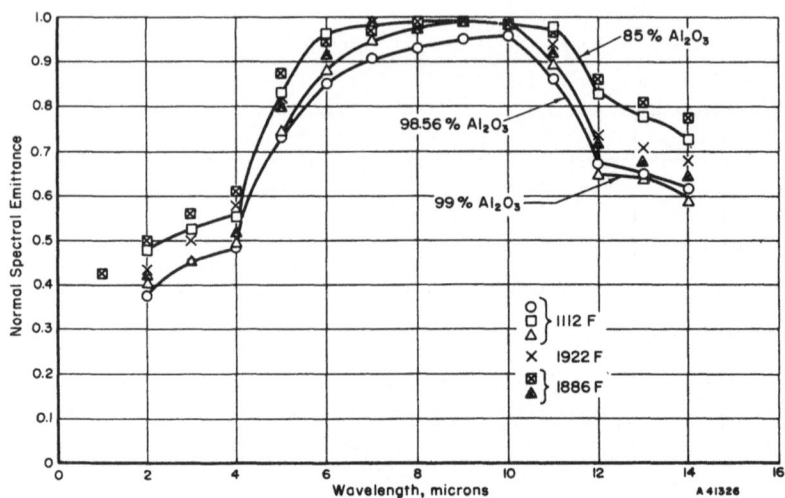

NORMAL SPECTRAL EMITTANCE OF ALUMINUM OXIDE

NORMAL SPECTRAL EMITTANCE OF ALUMINUM OXIDE--REFERENCE INFORMATION

Reference	Investigator	Symbol	Composition and Surface Condition	Test Method	Remarks
3	Blau, Marsh, Martin, Jasperse, and Chaffee		Aluminum oxide	Normal spectral emittance. Specimen mounted in wall of cylindrical Globar (SiC) heater. Comparison blackbody hole also in heater. Temperatures measured with thermocouples. Monochromator and thermocouple detector.	Measured in air. Data taken from curves. (Curves are drawn through the 1112 F points only.)
			Diamond wheel finish as supplied by manufacturer		
			TWA No. 2 (Norton A 402) 98.56% Al_2O_3		
		O	Measured at 1112 F		
		X	Measured at 1922 F		
			Coors AD85 85% Al_2O_3		
		□	Measured at 1112 F		
		⊠	Measured at 1886 F		
			Coors AD99 99% Al_2O_3		
		△	Measured at 1112 F		
		▲	Measured at 1886 F		

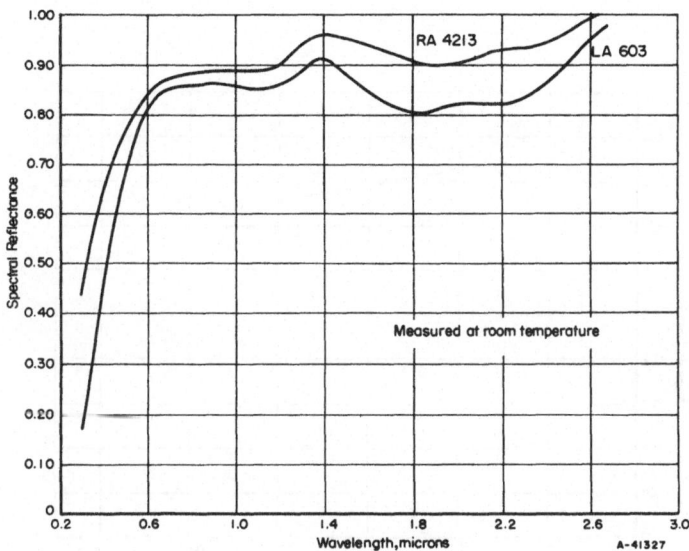

SPECTRAL REFLECTANCE OF ALUMINUM OXIDE

SPECTRAL REFLECTANCE OF ALUMINUM OXIDE--REFERENCE INFORMATION

Reference	Investigator	Symbol	Composition and Surface Condition	Test Method	Remarks
2	Olson and Morris		Aluminum oxide Norton RA4213 and LA603 Surface condition not given	Spectral reflectance. Incident radiation 9 degrees from normal to specimen surface. Integrating sphere reflectometer. Monochromator and lead sulphide detector. Normal (9 degrees) illumination diffuse reflection.	Measured in air at room temperature. Data taken from curves.

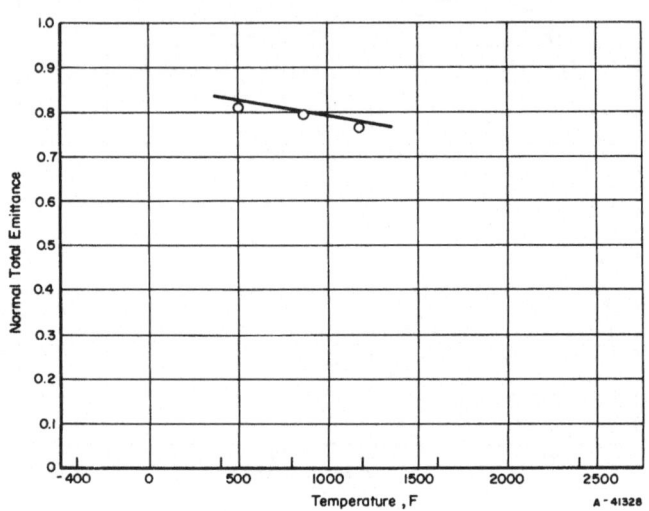

NORMAL TOTAL EMITTANCE OF BERYLLIUM OXIDE

NORMAL TOTAL EMITTANCE OF BERYLLIUM OXIDE—REFERENCE INFORMATION

Reference	Investigator	Symbol	Composition and Surface Condition	Test Method	Remarks
2	Olson and Morris	O	Beryllium oxide	Normal total emittance. Furnace-heated specimen. Comparison blackbody. Thermistor detector. Temperatures measured with thermocouples.	Measured in air. Data taken from curves.

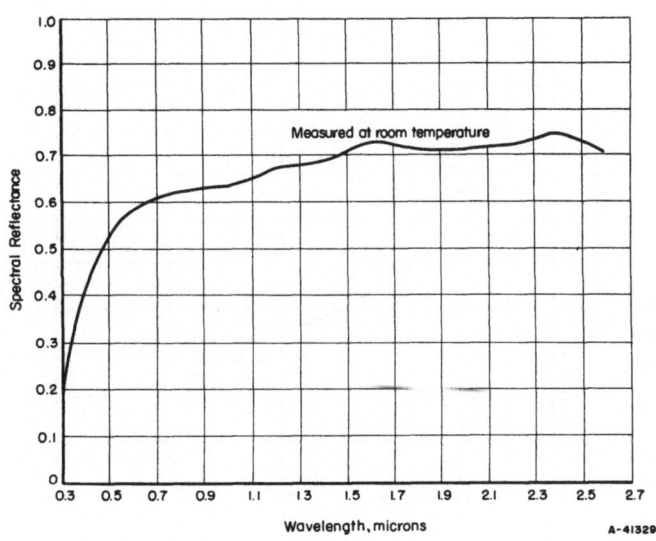

SPECTRAL REFLECTANCE OF BERYLLIUM OXIDE

SPECTRAL REFLECTANCE OF BERYLLIUM OXIDE—REFERENCE INFORMATION

Reference	Investigator	Symbol	Composition and Surface Condition	Test Method	Remarks
7	Betz, Olson, Schurin, and Morris		Beryllium oxide Purity not given As received condition	Spectral reflectance. Incident radiation 9 degrees from normal to specimen surface. Integrating sphere reflectrometer. Monochromator, and lead sulphide detector. Normal (9 degrees) illumination and diffuse reflection.	Measured in air at room temperature. Data taken from curves.

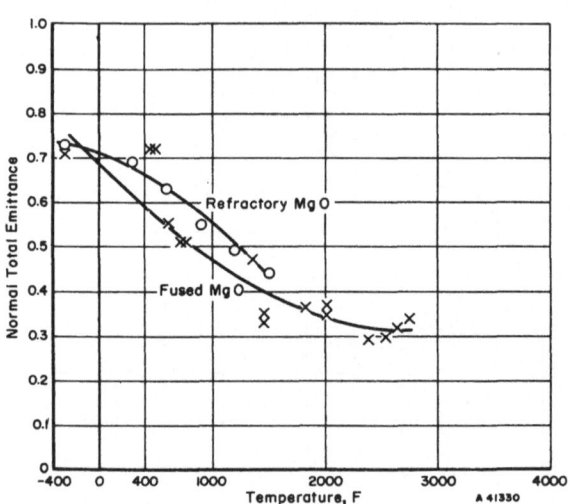

NORMAL TOTAL EMITTANCE OF MAGNESIUM OXIDE

NORMAL TOTAL EMITTANCE OF MAGNESIUM OXIDE--REFERENCE INFORMATION

Reference	Investigator	Symbol	Composition and Surface Condition	Test Method	Remarks
2	Olson and Morris	✕	Fused magnesium oxide obtained from the National Bureau of Standards. Surface condition not given	Normal total emittance. Furnace-heated specimen. Thermistor detector. Comparison blackbody. Temperatures measured with thermocouples.	Measured in air. Data taken from curve.
8	Olson and Morris	O	Refractory magnesium oxide Composition and surface condition not given	(Same as above.)	(Same as above.)

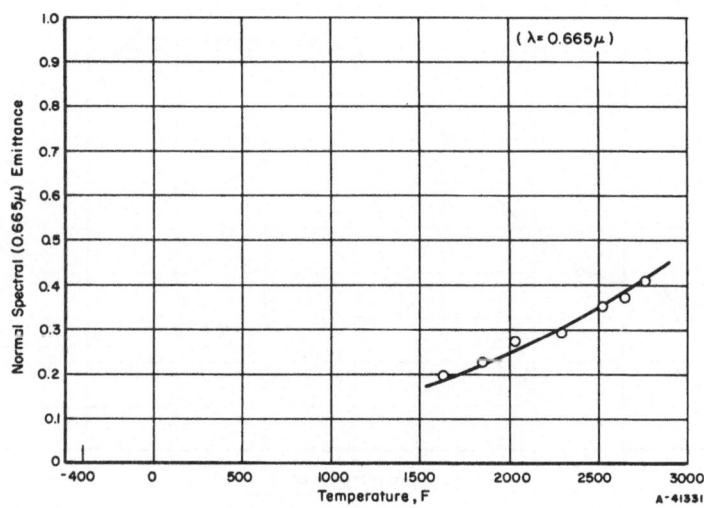

(λ = 0.665μ)

NORMAL SPECTRAL EMITTANCE OF MAGNESIUM OXIDE

NORMAL SPECTRAL EMITTANCE OF MAGNESIUM OXIDE--REFERENCE INFORMATION

Reference	Investigator	Symbol	Composition and Surface Condition	Test Method	Remarks
2	Olson and Morris	O	Fused magnesium oxide obtained from National Bureau of Standards. Surface condition not given	Normal spectral emittance. Furnace-heated specimen. Comparison blackbody. Commercial detector and filter system for peak response at 0.665μ. Temperatures measured with thermocouples.	Measured in air. Data taken from curves. (λ= 0.665μ)

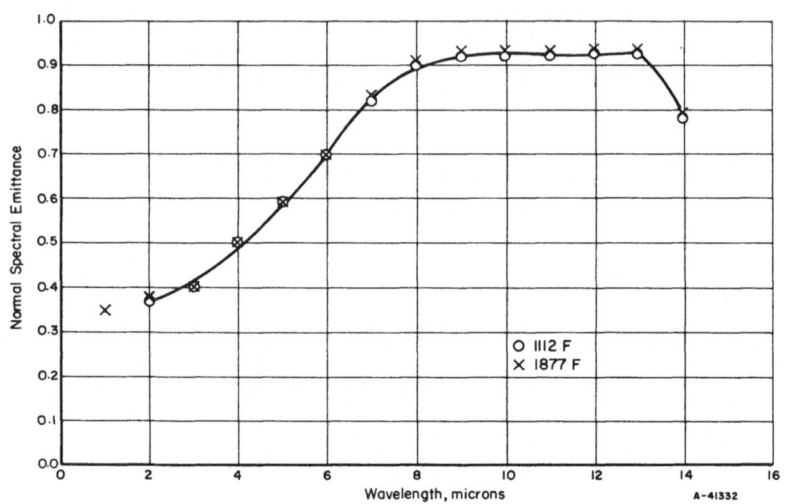

NORMAL SPECTRAL EMITTANCE OF MAGNESIUM OXIDE

NORMAL SPECTRAL EMITTANCE OF MAGNESIUM OXIDE--REFERENCE INFORMATION

Reference	Investigator	Symbol	Composition and Surface Condition	Test Method	Remarks
3	Blau, Marsh, Martin, Jasperse and Chaffee		Magnesia (MgO) Norton RM4473 Purity: 97% MgO, 1.3-1.5% CaO	Normal spectral emittance. Specimen mounted in wall of cylindrical Globar (SiC) heater. Comparison blackbody hole in heater wall. Monochromator and thermocouple detector. Temperatures measured with thermocouples.	Measured in air. Data taken from curves. (Curve drawn through 1112 F points only.)
			Surface condition not given		
		O	Measured at 1112 F		
		X	Measured at 1877 F		

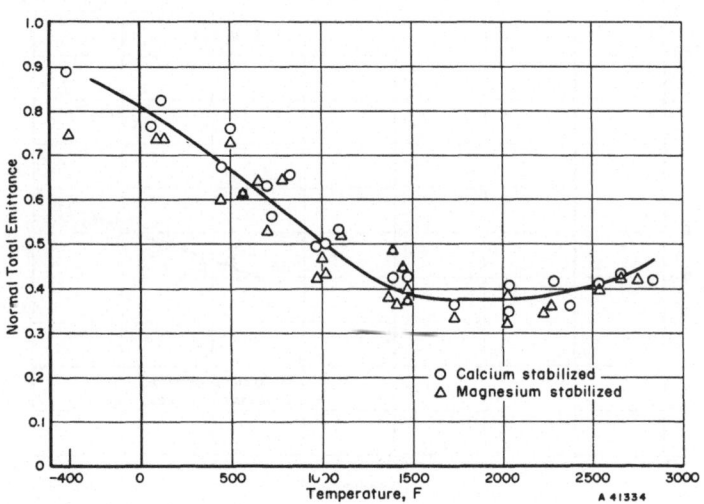

NORMAL TOTAL EMITTANCE OF ZIRCONIUM OXIDE

NORMAL TOTAL EMITTANCE OF ZIRCONIUM OXIDE--REFERENCE INFORMATION

Reference	Investigator	Symbol	Composition and Surface Condition	Test Method	Remarks
2	Olson and Morris		Zirconium oxide	Normal total emittance. Furnace-heated specimen. Comparison blackbody. Thermistor detector. Temperatures measured with thermocouples.	Measured in air. Data taken from curves.
		O	Calcium stabilized		
		△	Magnesium stabilized		

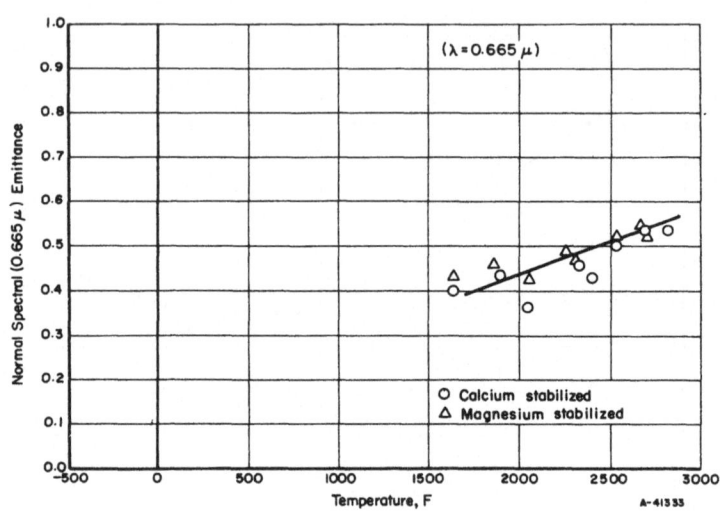

NORMAL SPECTRAL EMITTANCE OF ZIRCONIUM OXIDE

NORMAL SPECTRAL EMITTANCE OF ZIRCONIUM OXIDE--REFERENCE INFORMATION

Reference	Investigator	Symbol	Composition and Surface Condition	Test Method	Remarks
2	Olson and Morris		Zirconium oxide	Normal spectral emittance.	Measured in air.
		O	Calcium stabilized	Furnace-heated specimen.	Data taken from
		△	Magnesium stabilized	Comparison blackbody.	curves.
				Commercial detector and	
				filter system for peak	$(\lambda = 0.665 \mu)$
				response at 0.665μ.	
				Temperatures measured	
				with thermocouples.	

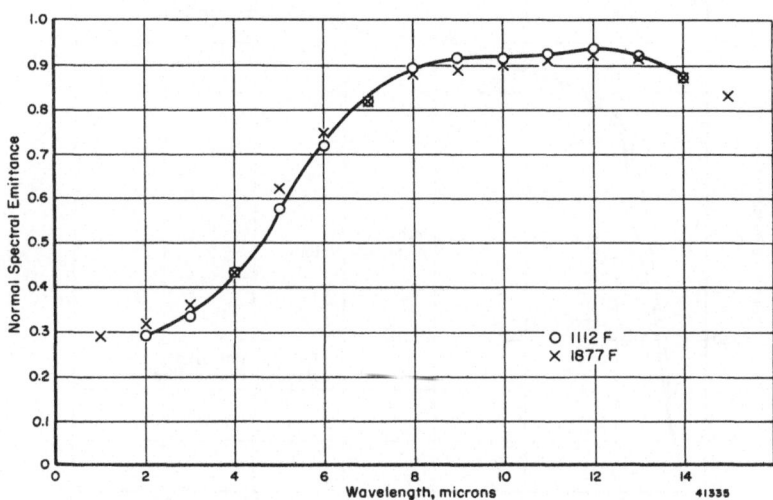

NORMAL SPECTRAL EMITTANCE OF ZIRCONIUM OXIDE

NORMAL SPECTRAL EMITTANCE OF ZIRCONIUM OXIDE--REFERENCE INFORMATION

Reference	Investigator	Symbol	Composition and Surface Condition	Test Method	Remarks
3	Blau, Marsh, Martin, Jasperse, and Chaffee		Zirconia (ZrO$_2$) Norton RZ 5601 Purity: 92% ZrO$_2$, .4.5% CaO Surface condition not given	Normal spectral emittance. Specimen mounted in wall of cylindrical Globar (SiC) heater. Comparison blackbody hole in heater wall. Monochromator and thermocouple detector. ⌐emperatures measured with thermocouples.	Measured in air. Data taken from curves. (Curves drawn through 1112 F points only.)
		O	Measured at 1112 F		
		X	Measured at 1877 F		

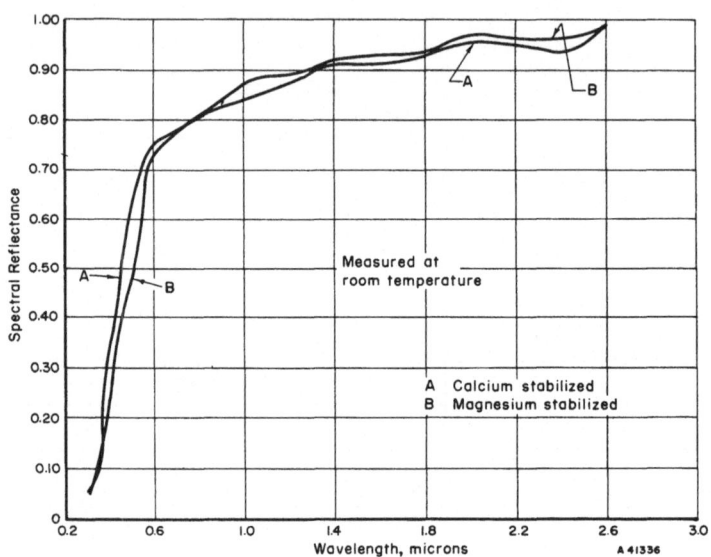

SPECTRAL REFLECTANCE OF ZIRCONIUM OXIDE

SPECTRAL REFLECTANCE OF ZIRCONIUM OXIDE--REFERENCE INFORMATION

Reference	Investigator	Symbol	Composition and Surface Condition	Test Method	Remarks
2	Olson and Morris		Zirconium oxide Calcium stabilized and magnesium stabilized Purity and surface condition not given	Spectral reflectance. Incident radiation 9 degrees from normal to specimen surface. Integrating sphere reflectometer. Monochromator and lead sulphide detector. Normal (9 degrees) illumination. Diffuse reflection.	Measured in air at room temperature. Data taken from curves.

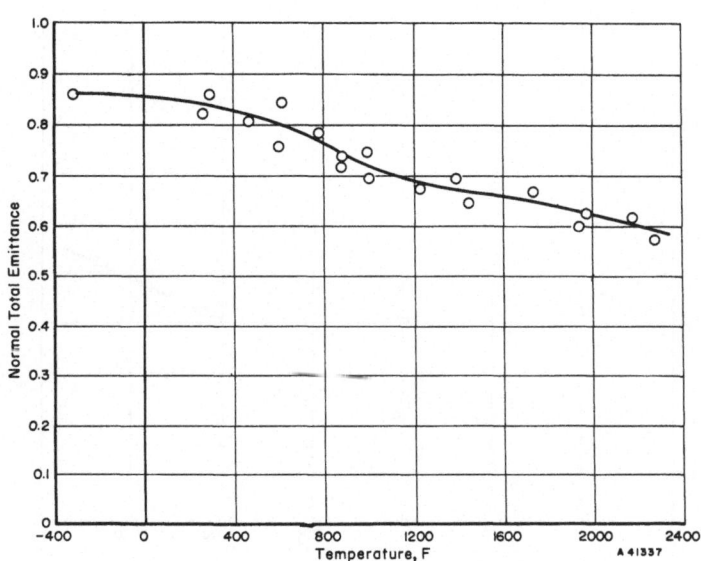

NORMAL TOTAL EMITTANCE OF PYROCERAM 9606

NORMAL TOTAL EMITTANCE OF PYROCERAM 9606--REFERENCE INFORMATION

Reference	Investigator	Symbol	Composition and Surface Condition	Test Method	Remarks
2	Olson and Morris	O	Pyroceram 9606, surface condition not given	Normal total emittance. Furnace-heated specimen. Comparison blackbody. Thermistor detector. Temperatures measured with thermocouples.	Measured in air. Data taken from curves.

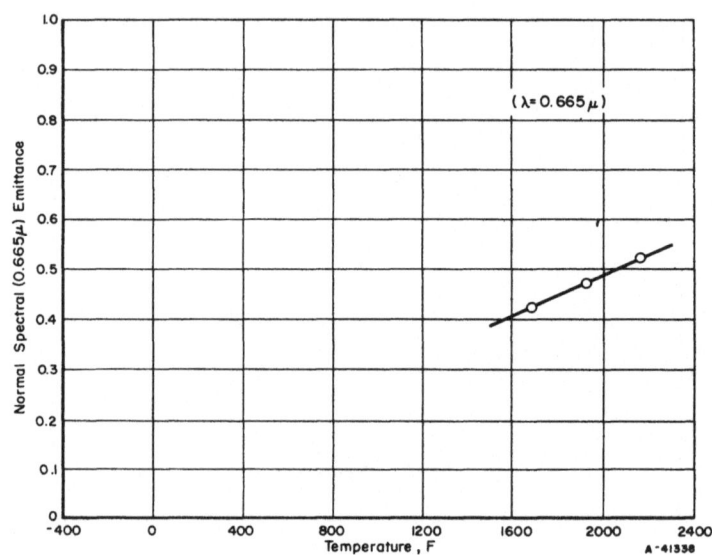

NORMAL SPECTRAL EMITTANCE OF PYROCERAM 9606

NORMAL SPECTRAL EMITTANCE OF PYROCERAM 9606--REFERENCE INFORMATION

Reference	Investigator	Symbol	Composition and Surface Condition	Test Method	Remarks
2	Olson and Morris	O	Pyroceram 9606 Surface condition not given	Normal spectral emittance. Furnace-heated specimens. Comparison blackbody. Commercial detector and filter system for peak response at 0.665μ. Temperatures measured with thermocouples.	Measured in air. Data taken from curves. $(\lambda = 0.665\mu)$

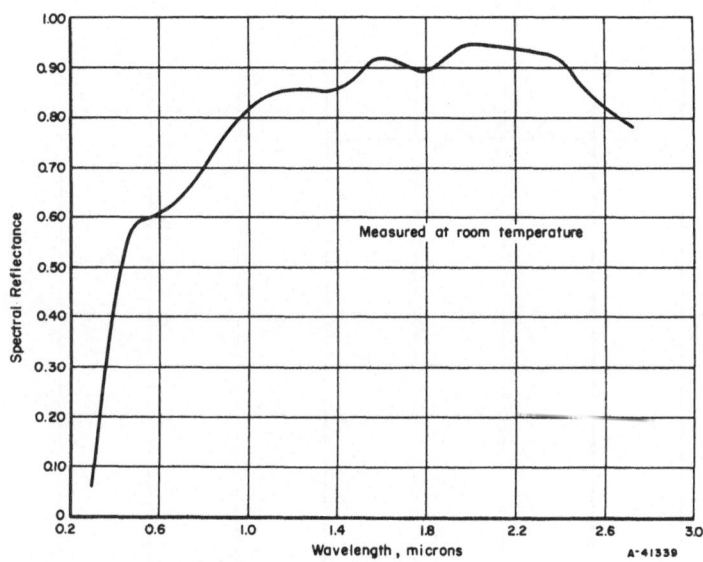

SPECTRAL REFLECTANCE OF PYROCERAM 9606

SPECTRAL REFLECTANCE OF PYROCERAM 9606--REFERENCE INFORMATION

Reference	Investigator	Symbol	Composition and Surface Condition	Test Method	Remarks
2	Olson and Morris		Pyroceram 9606 Surface condition not given	Spectral reflectance. Incident radiation 9 degrees from normal to specimen surface. Integrating sphere reflectometer. Monochromator and lead sulphide detector. Normal (9 degrees) illumination. Diffuse reflection.	Measured in air at room temperature. Data taken from curves.

464

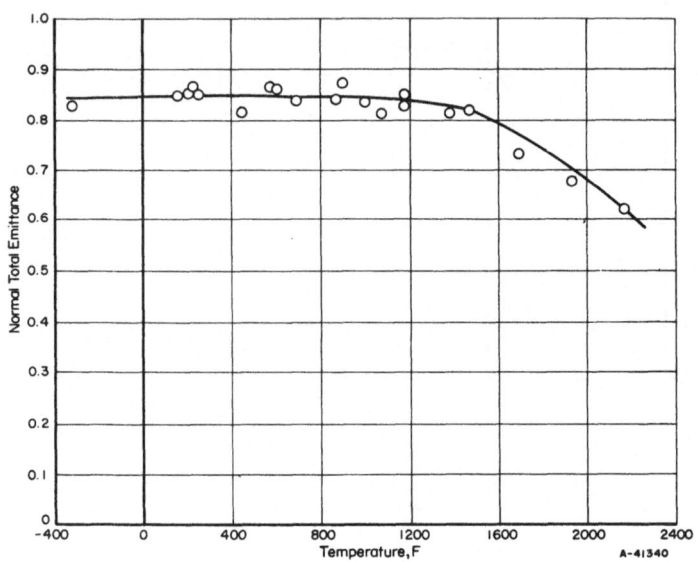

NORMAL TOTAL EMITTANCE OF PYROCERAM 9608

NORMAL TOTAL EMITTANCE OF PYROCERAM 9608--REFERENCE INFORMATION

Reference	Investigator	Symbol	Composition and Surface Condition	Test Method	Remarks
2	Olson and Morris	O	Pyroceram 9608 Surface condition not given	Normal total emittance. Furnace-heated specimen. Comparison blackbody. Thermistor detector. Temperatures measured with thermocouples.	Measured in air. Data taken from curve.

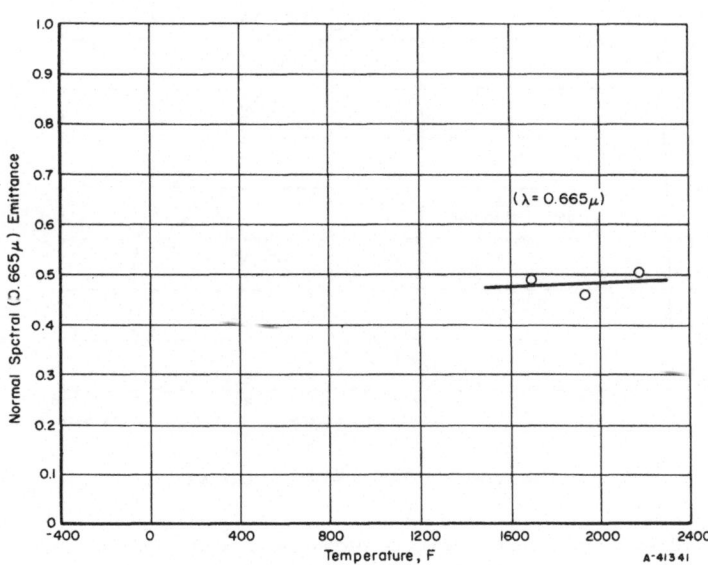

NORMAL SPECTRAL EMITTANCE OF PYROCERAM 9608

NORMAL SPECTRAL EMITTANCE OF PYROCERAM 9608--REFERENCE INFORMATION

Reference	Investigator	Symbol	Composition and Surface Condition	Test Method	Remarks
2	Olson and Morris	O	Pyroceram 9608 Surface condition not given	Normal spectral emittance. Furnace-heated specimen. Comparison blackbody. Commercial detector and filter system for peak response at 0.665μ. Temperatures measured with thermocouples.	Measured in air. Data taken from curves. (λ= 0.665μ)

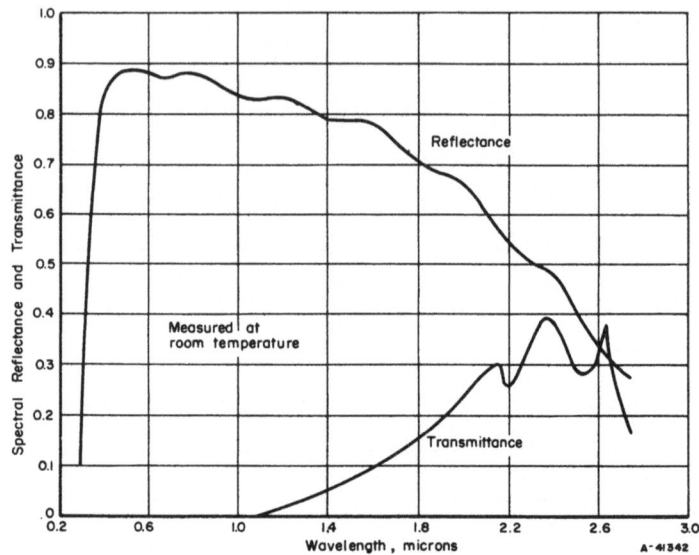

SPECTRAL REFLECTANCE AND TRANSMITTANCE OF PYROCERAM 9608

SPECTRAL REFLECTANCE AND TRANSMITTANCE OF PYROCERAM 9608--REFERENCE INFORMATION

Reference	Investigator	Symbol	Composition and Surface Condition	Test Method	Remarks
2	Olson and Morris		Pyroceram 9608 Surfaces reasonably flat and parallel	Spectral reflectance. Incident radiation 9 degrees from normal to specimen surface. Integrating sphere re- flectometer. Monochromator and lead sulphide detector. Normal (9 degrees) illumination. Diffuse reflection. Spectral Transmittance. Normal specimen position filled by $MgCO_3$ or MgO block. Specimen placed in entrance beam to sphere. Diffuse transmission.	Measured in air at room temperature. Data taken from curves.

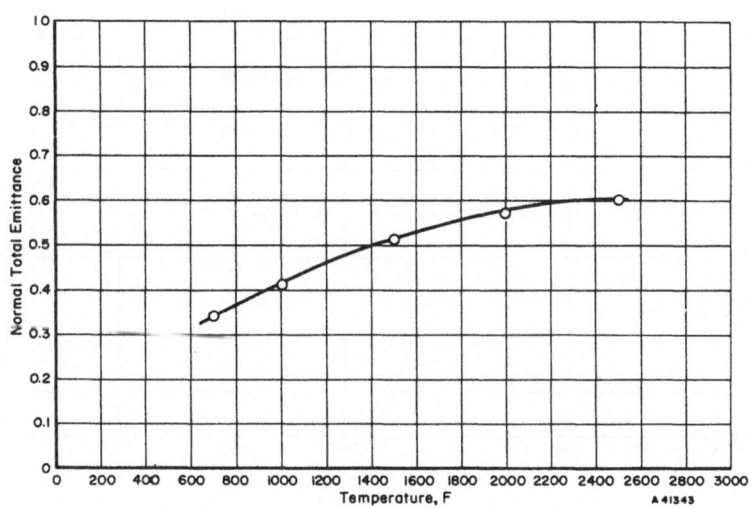

NORMAL TOTAL EMITTANCE OF MOLYBDENUM DISILICIDE

NORMAL TOTAL EMITTANCE OF MOLYBDENUM DISILICIDE--REFERENCE INFORMATION

Reference	Investigator	Symbol	Composition and Surface Condition	Test Method	Remarks
1	Anthony and Pearl	O	As received	Normal total emittance. Induction-heated specimen. Thermopile detector. Comparison blackbody. Temperatures measured with thermocouples and optical pyrometer.	Measured in continuous purge of helium gas.

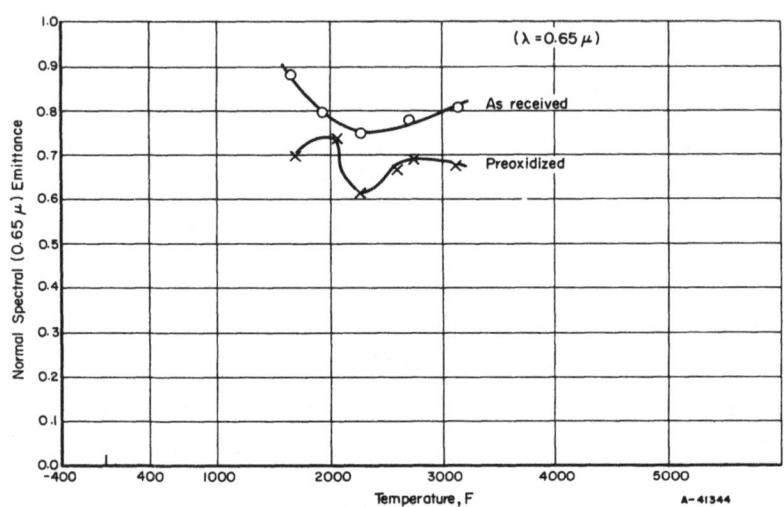

NORMAL SPECTRAL EMITTANCE OF MOLYBDENUM DISILICIDE

NORMAL SPECTRAL EMITTANCE OF MOLYBDENUM DISILICIDE--REFERENCE INFORMATION

Reference	Investigator	Symbol	Composition and Surface Condition	Test Method	Remarks
4	Blau, Chaffee, Jasperse, and Martin	O ×	Molybdenum disilicide Surface clean and smooth Preoxidized (Lower emittance for the preoxidized surface attributed to SiO_2 surface layer)	Normal spectral emittance. Induction-heated specimen. Blackbody hole drilled in specimen surface. Temperatures measured with micro-optical pyrometer.	Measured in 90% argon - 10% hydrogen atmosphere. Data taken from curves. (λ = 0.65μ)

TOTAL SOLAR ABSORPTANCES AT SEA LEVEL AND ABOVE THE ATMOSPHERE

	Finish	Above Atmosphere	Sea Level
Graphite-National GBE	(F)	0.850	0.863
Graphite-National GBE	(B)	0.869	0.877
Graphite-National GBH	(M)	0.881	0.887
Graphite-National GBH	(R)	0.885	0.891
Graphite-Speer 3474D	(M)	0.853	0.858
Graphite-Speer 3474D	(R)	0.866	0.871
Graphite-Speer 7087	(M)	0.908	0.911
Graphite-Speer 7087	(R)	0.916	0.918
Beryllium Oxide (Refractory)	(R)	0.421	0.403
Magnesium Oxide (Refractory)	(R)	0.168	0.141

TOTAL SOLAR ABSORPTANCE OF BERYLLIUM OXIDE, MAGNESIUM OXIDE AND THREE GRAPHITES--REFERENCE INFORMATION

Reference	Investigator	Symbol	Composition and Surface Condition	Test Method	Remarks
	Betz, Olson, Schurin, and Morris		Surface finishes: B* back F* front M fine milling machine cut R as received from supplier. * back and front surfaces arbitrarily assigned to graphite sample. Sides appeared different to the eye.	Solar absorptance calculated by method of truncated weighted ordinate integration using spectral reflectance vs wavelength curves and solar energy distribution curves over the limits of 0.3 to 2.4 microns. Above atmosphere values corrected for 3 per cent of energy lying outside these limits.	Calculated. Data obtained from table.

REFERENCES

(1) Anthony, F. M., and Pearl, Harry A., "Investigations of Feasibility of Utilizing Available Heat Resistant Materials for Hypersonic Leading Edge Applications", Vol III – Screening Test Results and Selection of Materials, WADC TR 59-744 (July, 1960).

(2) Olson, O. H., and Morris, J. C., "Determination of Emissivity and Reflectivity Data on Aircraft Structural Materials", Part III – Techniques for Measurement, WADC TR 56-222, ASTIA AD 239302 (April, 1960).

(3) Blau, H. H., Jr., Marsh, J. B., Martin, W. S., Jasperse, J. R., and Chaffee, E., "Infrared Spectral Emittance Properties of Solid Materials", AFCRL-TR-60-416, ASTIA AD 248276 (October, 1960).

(4) Blau, H. H., Jr., Chaffee, E., Jasperse, J. R., and Martin, W. S., "High Temperature Thermal Radiation Properties of Solid Materials", AFCRC-TN-60-165, ASTIA AD 236394 (March 31, 1960).

(5) Coffman, J. A., Coulson, K. L., and Kibler, T. M., General Electric Company, Cincinnati, Ohio, preliminary information under an Air Force contract.

(6) Riethof, T. R., "High Temperature Spectral Emissivity Studies", General Electric Company MSVD, Space Sciences Laboratory, R61SD004 (January, 1961).

(7) Betz, H. T., Olson, O. H., Schurin, B. D., and Morris, J. C., "Determination of Emissivity and Reflectivity Data on Aircraft Structural Materials", Part II: Techniques for Measurement of Total Normal Spectral Emissivity, Solar Absorptivity, and Presentation of Results, WADC TR 56-222, ASTIA AD 202493.

(8) Olson, O. H., and Morris, J. C., "Determination of Emissivity and Reflectivity Data on Aircraft Structural Materials", WADC TR 56-222, Part II, Supplement I, ASTIA 202494 (October, 1958).

(9) Jain, S. C., and Krishnan, Sir F. R. S., "The Distribution of Temperature Along a Thin Rod Electrically Heated in Vacuo", Proc. Royal Soc. London, 225, 7-19 (1954).

(10) Thorn, R. J., and Simpson, O. C., "Spectral Emissivities of Graphite and Carbon", Jour. Applied Physics, 24 (5), 633-639 (May, 1953).

(11) Wade, W. R., and Casey, F. W., Jr., "Measurements of Total Hemispherical Emissivity of Several Stably Oxidized Nickel-Titanium Carbide Cemented Hard Metals From 600°F to 1600°F", NASA Memo 5-13-59L.